Evgeny Stambulchik, Annette Calisti, Hyun-Kyung Chung and
Manuel Á. González (Eds.)

Spectral Line Shapes in Plasmas

MDPI

This book is a reprint of the special issue that appeared in the online open access journal *Atoms* (ISSN 2218-2004) in 2014 (available at: http://www.mdpi.com/journal/atoms/special_issues/SpectralLineShapes).

Guest Editors

Evgeny Stambulchik
Department of Particle Physics
and Astrophysics,
Faculty of Physics,
Weizmann Institute of Science,
Rehovot 7610001, Israel

Hyun-Kyung Chung
International Atomic Energy Agency, Atomic and
Molecular Data Unit, Nuclear Data Section, P.O.
Box 100, A-1400 Vienna, Austria

Annette Calisti
Laboratoire PIIM, UMR7345,
Aix-Marseille Université - CNRS,
Centre Saint Jérôme case 232,
13397 Marseille Cedex 20, France

Manuel Á. González
Departamento de Física Aplicada, Escuela Técnica
Superior de Ingeniería Informática,
Universidad de Valladolid, Paseo de Belén 15,
47011 Valladolid, Spain

Editorial Office
MDPI AG
Klybeckstrasse 64
4057 Basel, Switzerland

Publisher
Shu-Kun Lin

Production Editor
Martyn Rittman

1. Edition 2015

MDPI • Basel

ISBN 978-3-906980-81-2

Table of Contents

Applications

Preface

1. Foreword

Line-shape analysis is one of the most important tools for diagnostics of both laboratory and space plasmas. Its reliable implementation requires sufficiently accurate calculations, which imply the use of analytic methods and computer codes of varying complexity, and, necessarily, varying limits of applicability and accuracy. However, studies comparing different computational and analytic methods are almost non-existent. The Spectral Line Shapes in Plasma (SLSP) code comparison workshop series [1] was established to fill this gap.

Numerous computational cases considered in the two workshops organized to date (in April 2012 and August 2013 in Vienna, Austria) not only serve the purpose of code comparison, but also have applications in research of magnetic fusion, astrophysical, laser-produced plasmas, and so on. Therefore, although the first workshop was briefly reviewed elsewhere [2], and will likely be followed by a review of the second one, it was unanimously decided by the participants that a volume devoted to results of the workshops was desired. It is the main purpose of this special issue.

2. Hydrogen-Like Transitions

Many calculation cases suggested for the first two SLSP workshops are for simple atomic systems: the hydrogen atom or hydrogen-like one-electron ions. Of these, the Ly-_ transition is truly the simplest; the atomic model was further reduced by neglecting the fine structure and interactions between states with different principal quantum numbers. Interestingly, this simplest system caused the largest discrepancies between results of various models presented at the first workshop [2] due to, apparently, different treatments of the ion dynamics effect. Now, Ferri et al. [3] discuss this, extending the analysis to more complex transitions with forbidden components. The ion dynamics effect is intimately related to the microfield directionality, as studied in depth by Calisti et al. [4]. Notably, the effects of the directionality of the microfield fluctuations were first researched within the framework of the "standard theory" of the plasma line broadening almost four decades ago, but have largely been forgotten. This approach is recalled and comparisons with computer simulations are made in the paper by Demura and Stambulchik [5].

Results of computer simulations will indeed be found in a majority of studies in this volume. By many scholars, such calculations are considered ab initio and their results regarded as benchmarks—at least for hydrogen-like transitions. However, Rosato et al. [6] argue that caution should be exercised in the case of very weakly coupled plasmas; in the extreme limit of the ideal plasma model, even the largest supercomputers available today might not be able to achieve convergence.

3. Isolated Lines

Isolated lines are often contradistinguished from radiative transitions in hydrogen-like species with degenerate energy levels. Alexiou et al. [7] briefly summarize the theoretical aspects of isolated-line broadening and then delve into a detailed comparison of Stark widths and shifts of Li-like 2s–2p transitions as calculated by various approaches. One of these approaches is the semiclassical perturbation (SCP) method, the work horse behind the STARK-B database. A complete up-to-date description of SCP is presented by Sahal-Bréchot et al. [8]. Another approach, also included in the comparison [7], is based on the relativistic Dirac R-matrix method, and is described by Duan et al. [9] with a focus on the B III 2s–2p doublet.

Koubiti et al. [10] present a comparison of various line-shape computational methods applied to the case of a plasma-broadened isolated line subjected to magnetic field. Furthermore, this study covers one of the two challenges introduced at the second SLSP workshop, where participants were asked to explain previously unpublished experimental data based on best-fit spectra of their models.

4. Applications

One can hardly overestimate the significance of line-shape calculations for diagnostics of laboratory, space, and industrial plasmas. Dimitrijević and Sahal-Bréchot [11] show and discuss numerous examples of such studies, where the SCP approach [8] was used. In the article of Omar et al. [12], the authors use a few theoretical methods and computer simulations to calculate the shapes of a He I line and compare them to experimental line profiles, allowing inference of the plasma parameters. Lisitsa et al. [13] introduce a new method able to describe penetration of a neutral atomic beam into low-density inhomogeneous fusion plasmas, and provide sample calculations suitable for ITER diagnostics. Spectral line features caused by Langmuir waves and charge-exchange processes are discussed by Dalimier et al. [14], who also suggest several spectral lines for prospective studies of laser-produced plasmas.

5. Conclusions

For the first two SLSP workshops, participants submitted in total over 1,500 line-shape calculations. The studies collected in this Special Issue explore only a part of this immense work. Research is ongoing, and we expect more publications soon.

The next workshop is scheduled for March 2015 in Marseille, France [1].

Evgeny Stambulchik, Annette Calisti,

Hyun-Kyung Chung and Manuel Á. González

Guest Editors

Reprinted from *Atoms*. Cite as: Stambulchik, E.; Calisti, A.; Chung, H.-K.; González, M.A. Special Issue on Spectral Line Shapes in Plasmas. *Atoms* **2014**, *2*, 378-381.

Acknowledgements

The organizational and financial support from the International Atomic Energy Agency for conducting the SLSP workshops is highly appreciated.

Conflicts of Interest

The authors declare no conflict of interest.

References

1. Spectral Line Shapes in Plasmas workshops. Available online: http://plasmagate.weizmann. ac.il/slsp/ (accessed on 16 July 2014).
2. Stambulchik, E. Review of the 1st Spectral Line Shapes in Plasmas code comparison workshop. High Energy Density Phys. 2013, 9, 528–534.
3. Ferri, S.; Calisti, A.; Mossé, C.; Rosato, J.; Talin, B.; Alexiou, S.; Gigosos, M.A.; González, M.Á.; González-Herrero, D.; Lara, N.; Gomez, T.; Iglesias, C.A.; Lorenzen, S.; Mancini, R.C.; Stambulchik, E. Ion dynamics effect on Stark broadened line shapes: A cross comparison of various models. Atoms 2014, 2, 299–318.
4. Calisti, A.; Demura, A.; Gigosos, M.A.; González-Herrero, D.; Iglesias, C.A.; Lisitsa, V.S.; Stambulchik, E. Influence of microfield directionality on line shapes. Atoms 2014, 2, 259–276.
5. Demura, A.; Stambulchik, E. Spectral-kinetic coupling and effect of microfield rotation on Stark broadening in plasmas. Atoms 2014, 2, 334–356.
6. Rosato, J.; Capes, H.; Stamm, R. Ideal Coulomb plasma approximation in line shape models: Problematic issues. Atoms 2014, 2, 253–258.
7. Alexiou, S.; Dimitrijević, M.S.; Sahal-Brechot, S.; Stambulchik, E.; Duan, B.; González Herrero, D.; Gigosos, M.A. The Second Workshop on Lineshape Code Comparison: Isolated lines. Atoms 2014, 2, 157–177.
8. Sahal-Bréchot, S.; Dimitrijević, M.S.; Ben Nessib, N. Widths and shifts of isolated lines of neutral and ionized atoms perturbed by collisions with electrons and ions: An outline of the semiclassical perturbation (SCP) method and of the approximations used for the calculations. Atoms 2014, 2, 225–252.
9. Duan, B.; Bari, M.A.; Wu, Z.; Yan, J. Electron-impact widths and shifts of B III 2p–2s lines. Atoms 2014, 2, 207–214.
10. Koubiti, M.; Goto, M.; Ferri, S.; Hansen, S.; Stambulchik, E. Line-shape code comparison through modeling and fitting of experimental spectra of the C II 723-nm line emitted by the ablation cloud of a carbon pellet. Atoms 2014, 2, 319–333.
11. Dimitrijević, M.S.; Sahal-Bréchot, S. On the application of Stark broadening data determined with Semiclassical Perturbation approach. Atoms 2014, 2, 357–377.
12. Omar, B.; González, M.Á.; Gigosos, M.A.; Ramazanov, T.S.; Jelbuldina, M.C.; Dzhumagulova, K.N.; Zammit, M.C.; Fursa, D.V.; Bray, I. Spectral line shapes of He I line 3889 Å. Atoms 2014, 2, 277–298.

13. Lisitsa, V.S.; Kadomtsev, M.B.; Kotov, V.; Neverov, V.S.; Shurygin, V.A. Hydrogen spectral line shape formation in the SOL of fusion reactor plasmas. Atoms 2014, 2, 195–206.
14. Dalimier, E.; Oks, E.; Renner, O. Review of Langmuir-wave-caused dips and charge exchange-caused dips in spectral lines from plasmas and their applications. Atoms 2014, 2, 178–194.

Article

Ion Dynamics Effect on Stark-Broadened Line Shapes: A Cross-Comparison of Various Models

Sandrine Ferri [1,*], **Annette Calisti** [1], **Caroline Mossé** [1], **Joël Rosato** [1], **Bernard Talin** [1], **Spiros Alexiou** [2], **Marco A. Gigosos** [3], **Manuel A. González** [3], **Diego González-Herrero** [3], **Natividad Lara** [3], **Thomas Gomez** [4], **Carlos Iglesias** [5], **Sonja Lorenzen** [6], **Roberto C. Mancini** [7] **and Evgeny Stambulchik** [8]

[1] Aix-Marseille Université, CNRS, PIIM UMR7345, 13397 Marseille, France;
E-Mails: annette.calisti@univ-amu.fr (A.C.); caroline.mosse@univ-amu.fr (C.M.);
joel.rosato@univ-amu.fr (J.R.); bernard.talin@orange.fr (B.T.)

[2] TETY, University of Crete, 71409 Heraklion, TK 2208, Greece; E-Mail: moka1@otenet.gr (S.A.)

[3] Department de Óptica y Física Applicada, Universidad de Valladolid, Valladolid 47071, Spain;
E-Mails: gigosos@coyanza.opt.cie.uva.es (M.A.G.); manuelgd@termo.uva.es (M.A.G.);
diegohe@opt.uva.es (D.G.-H.); nati@opt.uva.es (N.L.)

[4] Department of Astronomy, University of Texas, Austin, TX 78731, USA;
E-Mail: gomezt@astro.as.utexas.edu (T.G.)

[5] LLNL, Livermore, CA 94550, USA; E-Mail: iglesias1@llnl.gov (C.I.)

[6] Institut für Physik, Universität Rostock, D-18051 Rostock, Germany;
E-Mail: sonja.lorenzen@uni-rostock.de (S.L.)

[7] Physics Dept., University of Nevada, Reno, NV 89557, USA; E-Mail: rcman@unr.edu (R.C.M.)

[8] Faculty of Physics, Weizmann Institute of Science, Rehovot 7610001, Israel;
E-Mail: Evgeny.Stambulchik@weizmann.ac.il (E.S.)

* Author to whom correspondence should be addressed; E-Mail: sandrine.ferri@univ-amu.fr;
Tel.: +33-49128-8623.

Received: 30 April 2014; in revised form: 10 June 2014 / Accepted: 16 June 2014 / Published: 4 July 2014

Abstract: Modeling the Stark broadening of spectral lines in plasmas is a complex problem. The problem has a long history, since it plays a crucial role in the interpretation of the observed spectral lines in laboratories and astrophysical plasmas. One difficulty is the characterization of the emitter's environment. Although several models have been proposed over the years, there have been no systematic studies of the results, until now. Here, calculations from stochastic models and numerical simulations are compared for the Lyman-α and -β lines in neutral hydrogen. Also discussed are results from the Helium-α and -β lines of Ar XVII.

Keywords: Stark broadening; line shapes; plasmas; numerical simulations; models

1. Introduction

Line shape analysis is one of the most important tools for plasma diagnostics, as it provides information on the underlying physical processes involved in the line formation. With the increasing number of applications in different areas of plasma physics, the modeling of line broadening from neutral or charged emitters has been in perpetual development and remains a keystone in plasma spectroscopy [1].

In the formation of a line shape, Stark broadening is the most computationally challenging contribution, since the main difficulty is to properly characterize the emitter environment. It involves a complex combination of atomic physics, statistical mechanics and detailed plasma physics [2]. In particular, it is well known that the quasi-static ion approximation can lead to discrepancies with experimental data near the line center. This happens whenever the electric microfields produced at the emitter by the surrounding ions fluctuate during the inverse half-width at half-maximum (HWHM) time scale. The first attempts to account for ion dynamics in theoretical models were done in the 1970s, followed by experimental proof (see the historic introduction in [3] and the references therein). Since then, several models based on stochastic or collisional approaches have been developed, together with numerical simulations ([4] and the references therein). Necessarily, their limit of applicability, accuracy and, thus, results differ from one another, and up to now, no systematic comparison have existed [5].

The purpose here is to present cross-comparisons of different models that account for the ion dynamics effect. The line shape formalism is briefly recalled in Section 2, which serves to introduce notation. The specifics of the various models and numerical simulations are also presented in this section. We review the simulations Euler–Rodrigues (ER)-simulation [6], HSTRK [7], HSTRK_frequency separation technique (FST) [8], SimU [9,10], Xenomorph [11] and the models QuantST.MMM (MMM—model microfield method) [12], quasicontiguous (QC)-frequency fluctuation model (FFM) [13], multi-electron line-shape (MELS) [14], multi-electron radiator line-shape (MERL) [15,16], PPP [17], ST-PST [18] and UTPP [19] that have been used for the present purpose. The ion dynamics effect on the hydrogen Lyman-α and -β lines is discussed in Section 3.1, demonstrating the difficulty of such modeling even for these well-known lines. In Section 3.2, results on helium-α and -β lines of Ar XVII produced by the two stochastic models (Boerker–Iglesias–Dufty (BID) [20] and FFM [4,21]) are discussed with the help of the numerical

3

simulation (SimU). The reliability of such calculations is of interest in the diagnostics of inertial confinement fusion core plasma conditions. Conclusions are given in Section 4.

2. Theory, Models and Simulations

We recall that the line shape is given by:

$$I(\omega) = \frac{1}{\pi} Re \int_0^\infty dt \, e^{i\omega t} C(t) \tag{1}$$

where $C(t)$ is the autocorrelation function of the radiator dipole operator $\mathbf{d}$, which can be expressed in Liouville space as:

$$C(t) = \ll \mathbf{d}^\dagger | \mathbf{U}(t) | \mathbf{d}\rho_0 \gg \tag{2}$$

where the double bra and ket vectors are defined as usual in Liouville space. Here, ρ_0 is the density operator for the emitter only at the thermodynamical equilibrium and $\mathbf{U}(t) = \{U_l(t)\}_{l \in F}$ is the bath averaged evolution operator of the emitter. l belongs to a measurable functional space, $\{F\}$, which provides a statistical method for the calculation of average quantities. The main problem is to determine $\mathbf{U}(t)$. One has thus:

- to find the time evolution of $U_l(t)$ for a given microfield configuration, which means solving the following equation:

$$\frac{dU_l(t)}{dt} = -i[L_0 - \mathbf{d} \cdot \mathbf{F}_l(t)] \, U_l(t), \quad U_l(0) = 1 \tag{3}$$

where L_0 represents the Liouvillian of the unperturbed radiator and $\mathbf{d} \cdot \mathbf{F}_l(t)$ represents the Stark effect that connects the dipole operator $\mathbf{d}$ to the microfield created by surrounding charged particles $\mathbf{F}_l$ (including ions and electrons),

- and to average it over a statistical ensemble of the microfields $\{\ \}_{l \in F}$.

In its general form, the problem cannot be treated analytically. Nevertheless, $\mathbf{U}(t)$ can be obtained by numerical simulation integrating Equation (3) on simulated sampling of microfield histories. Usually, such a calculation is split into two independent steps [3]. First, the plasma particle trajectories are obtained by a numerical solution of Newton's equations of motion or an alternative method. Knowing the trajectory of each particle, the electric fields at the emitter are evaluated and stored to be used in the second step. Then, the line shape simulation follows: a step-by-step integration of Equation (3) is performed using these field histories. The evolution operator of the emitter is calculated, and the whole procedure is repeated several times in order to average over a representative sample set of independent perturbing field histories $\{f_1, f_2...f_N\}$. As a result, $C(t)$ is given by:

$$C(t) = \frac{1}{N} \sum_{i=1}^N C_i(t) \tag{4}$$

and the line shape is obtained by a Fourier transform of $C(t)$. Although all line shape simulations are based on the same scheme, we will see in the next section that they can differ slightly depending on the details of the models.

Alternatively, efficient analytical models based on fundamental assumptions and approximations have been developed [1]. In the standard theory (ST), the line shape calculation is based on the separation between the ions and the electrons due to the radically different dynamical properties of the microfields they create. Indeed, the typical fluctuation rate of the electric field created by perturber species p with a velocity relative to the center of mass v_p and a density n_p is defined by:

$$\nu_p = v_p/d_p \tag{5}$$

where $d_p = (3/4\pi n_p)^{1/3}$ is a typical interparticle distance. Assuming equal temperature for ions and electrons and plasma neutrality, one has [3]:

$$\frac{\nu_e}{\nu_i} \sim \left(\frac{\mu_i}{\mu_e}\right)^{1/2} Z_i^{1/3}. \tag{6}$$

Thus, the perturbation due to the electrons (with reduced mass μ_e) is nearly two orders of magnitude faster than that of ions (with reduced mass μ_i and charge Z_i). This allows for treating the electrons and the ions in a different way. The fast electrons are assumed to perturb the emitter by means of collisions, treated in the impact approximation, and the slow ions are assumed to be quasi-static. This results in a quantum-emitter system perturbation operator $l = -\mathbf{d} \cdot \mathbf{F}_{i,l} + i\phi_e$, containing a non-Hermitian homogeneous electron-impact broadening contribution ϕ_e and the ion microfield interaction $-\mathbf{d} \cdot \mathbf{F}_{i,l}$, which has to be numerically averaged with a static-field probability distribution $Q(\mathbf{F}_i)$, or because of isotropy, with $dW(F_i) = 4\pi F_i^2 Q(\mathbf{F}_i)dF_i$. The later can be calculated numerically in the ideal gas limit for perturbing ions [22] or using more sophisticated models that account for ion correlations [23]. Using the set of above assumptions, the quasi-static line shape is written as:

$$I_s(\omega) = -\frac{1}{\pi} Im \ll \mathbf{d}^\dagger | \int d\mathbf{F}_i \, Q(\mathbf{F}_i) \, G_s(\omega, \mathbf{F}_i) \, |\mathbf{d}\rho_0 \gg \tag{7}$$

in which the resolvent operator is given by:

$$G_s(\omega, \mathbf{F}_i) = (\omega - L_0 + \mathbf{d} \cdot \mathbf{F}_i - i\phi_e)^{-1}. \tag{8}$$

Although the electrons are often well described within the impact approximation, a quasi-static treatment of the ions can lead to large errors for plasma conditions, such as the ion microfields fluctuate during the inverse HWHM time scale. In the next section, we briefly review the simulations and the models that have been developed to account for the ion dynamics effect and that have been used for the present cross-comparisons.

2.1. The Numerical Simulations

The results from four numerical simulation codes based on different models have been submitted. They differ either in the way they model the motions of the plasma particles or in the procedure for the integration of the Schrödinger equation.

In the **ER-simulation**, the simulated plasma is an electrically neutral ensemble of statistically independent charged particles made of N_i ions and N_e electrons moving along straight line trajectories within a spherical volume. An emitter is assumed to be placed at the center of such a box. The temporal evolution of the whole system is measured along a discrete time axis from zero to a definite number of times of a fixed increment. Every temporal state is given by the set of values of the positions and velocities of the particles in the system. At every time step, the electric field produced by ions and electrons is calculated using Coulomb's law or a Debye-screened field. This electric field is an input to the Schrödinger equation that computes the emitter time evolution operator. For hydrogen and when the no-quenching approximation is considered, the atom state is described with the Euler–Rodrigues parameters [24].

The **HSTRK** and **HSTRK_FST** codes also use the Gigosos–Cardeñoso approach [25]. Both codes rely on the Hegerfeld–Kesting–Seidel method of collision-time statistics [26] and compute $C(t)$. Depending on the appropriate option, HSTRK can do an electron only, ion only or joint simulation, but one can also do combinations, e.g., electron simulation and quasi-static ions or impact electrons and ion simulation. For the Fourier transform, if a long-time exponential behavior is detected for times $t > \tau$, then the contribution to the Fourier transform of the (τ, ∞) region is computed analytically using the detected exponential decay and added to $\int_0^\tau dt C(t) e^{\iota \omega t}$. τ is determined via start-up runs, e.g., a run with a small number of configurations is done to obtain a rough idea of the HWHM and τ is adjusted to cover at least a number of inverse HWHMs. The integral is done by Filon's rule [27].

HSTRK_FST implements the frequency separation technique, which first identifies the "impact" phase space of ion perturbers (e.g., impact parameters and velocities), which produce a width much less (in these runs, "much less" was 10-times less) than the field fluctuation frequency. This meant:

$$HWHM(\Omega) = 0.1\,\Omega \qquad (9)$$

where the HWHM is computed by including all ion perturbers with impact parameter ρ and velocity $v > \Omega\rho$. Hence, the calculation is essentially the same, except that only slow ions $v < \Omega\rho$ are included in the simulation. The $C(t)$ obtained from the simulation of these slow ions is then multiplied by $e^{-HWHM(\Omega)t}$, and the Fourier transform is taken as in HSTRK. The use of a pure exponential form for the rapidly fluctuating (impact) part is a consequence of using the complete collision assumption for solving the impact part [28,29] and results in a $C(t)$ that is not correct for very short times. This is manifested in the (far) wing behavior of the HSTRK_FST profiles and can be remedied by using the incomplete collision formulas of the above-cited analytical solutions.

SimU is a combination of two codes: a molecular dynamics (MD) simulation of variable complexity and a solver for the evolution of an atomic system with the MD field history used as a (time-dependent) perturbation. A technical difference from other numerical simulation methods is the way the spectrum is calculated. Instead of employing the dipole autocorrelation function via Equation (1), SimU calculates the Fourier transform of the dipole matrix:

$$\vec{d}(\omega) = \int_0^\infty dt\, e^{-i\omega t}\vec{d}(t) \tag{10}$$

and then uses it directly instead of $C(t)$:

$$I^\lambda(\omega) \propto \frac{1}{2\pi}\sum_i \rho_i \sum_f \omega_{fi}^4 |\vec{e}_\lambda \cdot \langle \vec{d}_{fi}(\omega)\rangle|^2 \tag{11}$$

where $\vec{e}_\lambda$ is the light polarization direction and each initial state i is assigned a population factor ρ_i. Similarly to other methods, this procedure is repeated many times and averaged (*cf.* Equation (4)).

The recently developed code, **Xenomorph**, is based on the models of Gigosos and González [30], where a straight line assumption is made. A general Schrödinger solver described in [31] is used to obtain the eigenvalues $E_n(t)$ and eigenvectors $|n(t)\rangle$ at every time step of the simulation. The emitter time evolution operator is then evaluated:

$$U_l(t + \Delta t) = \left\{ \sum_n e^{-iE_n(t)\Delta t/\hbar}n(t)\rangle\langle n(t)| \right\} U_l(t) \tag{12}$$

and is used to obtain the dipole matrix. The Fourier transform of the latter is computed to obtain the line shape function, as is done in SimU (*cf.* Equations (10) and (11)).

2.2. The Models

The main difficulty in introducing the ion dynamics in the Stark line shape calculations is to develop a model that provides a sufficiently accurate solution of the evolution Equation (2) assuming an idealized stochastic process that conserves the statistical properties of the "real" interaction between the microfields and the radiating atom.

A successful model developed for neutral emitters—the model microfield method (MMM), due to Brissaud and Frisch [32,33]—involves stochastic fields that are constant in a given time interval and suddenly jump from one value to the next one at random times. The amplitudes of the field sequences are determined in order to be consistent with the static properties of the microfield, *i.e.*, the static-field probability distribution $Q(\mathbf{F})$. The jumping frequency $\nu(\mathbf{F})$ has to be chosen properly in order to reproduce the dynamics properties of the microfields represented by their autocorrelation function $< \mathbf{F}(t) \cdot \mathbf{F}(0) >$. In **QuantSt.MMM**, MMM (for ions) is combined with a quantum-statistical approach to calculate pressure broadening due to plasma electrons. The perturbation by electrons is considered to second order in the potential [34,35].

MELS and **MERL** are based upon the **BID** model. The latter derives from the MMM, but its formulation is based on statistical mechanics [36] and provides a unified description of radiative and transport properties for charged emitters [20]. The stochastic line shape is written as:

$$I_d(\omega) = -\frac{1}{\pi}Im \ll \mathbf{d}^\dagger | \frac{\int d\mathbf{F}Q(\mathbf{F}_i)G_{BID}(\omega,\mathbf{F}_i)}{1 + i\nu(\omega)\int d\mathbf{F}Q(\mathbf{F}_i)G_{BID}(\omega,\mathbf{F}_i)} |\mathbf{d}\rho_0 \gg \tag{13}$$

in which the resolvent is given by:

$$G_{BID}(\omega, \mathbf{F}_i) = (\omega - L_0 + \mathbf{d} \cdot \mathbf{F}_i - i\nu(\omega))^{-1} \tag{14}$$

The jumping frequency $\nu(\omega)$ is chosen as:

$$\nu(\omega) = \frac{\nu_0}{1 + i\omega\tau}. \tag{15}$$

where the two parameters ν_0 and τ are defined in this model by the low- and high-frequency limits of the momentum autocorrelation function. Here, τ is assumed to be null.

Another approach is the frequency fluctuation model (**FFM**), on which the **PPP** code and, recently, the **QC-FFM** code rely. The latter is a hybrid model using the quasi-contiguous approximation [37] for H-like transitions and the FFM for modeling the microfield dynamics effect. The FFM relies on a different idealization of the stochastic process than MMM and BID. Here, the quantum system perturbed by a time-dependent microfield behaves like a set of field-dressed two-level transitions (SDT) subject to a collision-type mixing process. More precisely, the fluctuation mechanism of these SDT obeys a stationary Markov process defined by the instantaneous probability of states $p_j = a_j / \sum_k a_k$ (a_j being the intensity of the SDT, j) and the transition rates between these states $\mathbf{W}_{k,j} = -\Gamma_j \delta_{k,j} + W_{k,j}$, where $\Gamma_{k,j} = \nu \delta_{i,j}$ and $W_{k,j} = \nu p_j$.

The typical fluctuation rate ν_{FFM} of the electric field, given by Equation (5), is used. Working in the Liouville space of the dressed two-level radiators, the line shape is written as [38]:

$$I_d(\omega) = \frac{1}{\pi} Re \sum_{j,k} i \ll D_k | G_{FFM}(\omega) | D_j p_j \gg \tag{16}$$

with the resolvent:

$$G_{FFM}(\omega) = (\omega - \mathbf{L}_\omega + i\mathbf{W})^{-1} \tag{17}$$

where $\mathbf{L}_\omega$ is the Liouville operator involving the transition energies of the SDT (ω_i) and D_i are the matrix elements of the dipole moment for the SDT. Due to the particular form of the matrix of transition rates $\mathbf{W}$, the dynamic line shape is written as [4]:

$$I_d(\omega) = \frac{\sum_k a_k}{\pi} Re \frac{\sum_k \frac{p_k}{\nu + i(\omega - \omega_k)}}{1 - \nu \sum_k \frac{p_k}{\nu + i(\omega - \omega_k)}} \tag{18}$$

Despite the fact that the two stochastic models lead to different functional forms, it follows that both BID and FFM recover the static limit for $\nu_{BID} = 0$ in Equation (14) and for $\nu_{FFM} = 0$ in Equation (18). In the opposite limit, both models recover the fast fluctuation limit ($\nu \to \infty$) that should approximate the "no ions" profile. However, BID recovers the impact limit in the line center whenever ν is large, while the FFM does not (see [39] for a more detailed discussion). We note that QC-FFM uses the FFM approximation for ions and electrons alike. For the latter, correctly approaching the impact approximation in the fast fluctuation limit becomes especially important. To this end, a modification to the effective fluctuation rate was introduced:

$$\tilde{\nu} = \nu + \frac{\nu^2}{\nu_0} \tag{19}$$

where ν_0 is an empirically obtained constant (for details, see [13]).

Two other models based on the collisional approach have been used, too. The **ST-PST** model is based on the standard theory with a number of options. Specifically, apart from the pure ST results, ST-PST can (and by default does) also compute the results of ST with penetrating collisions correctly accounted for analytically [18]. In addition, an **FST-FFM** calculation is also done [8]: first, an Ω is determined, exactly as described above for HSTRK_FST. Next, the FFM is applied to the field that excludes the fast, impact part. Last, the two profiles are convolved. As a result, the impact limit is correctly built in and recovered, hence extending the FFM validity without sacrificing its speed. Note, however, that with the current FST implementation, which uses the completed collision assumption for the impact phase space, the far wings are not accurate, as already discussed.

The **UTPP** code is devoted to the calculation of hydrogen line shapes in regimes where the impact approximation for ions is reasonably accurate. Such a regime is attained for lines with a low principal quantum number in magnetic fusion experiments in the absence of Doppler broadening (Doppler-free line shape models were required for radiation transport simulations, e.g., [40]). In UTPP, a line shape is calculated using the following formula:

$$I(\omega) = \frac{1}{\pi} \text{Re} \ll \mathbf{d}^\dagger | \frac{1}{s + iL_0 + K(s)} | \mathbf{d}\rho_0 \gg \tag{20}$$

where $s = -i\omega$ and $K(s)$ is a collision operator calculated in a framework similar to that used in the Voslamber unified theory (Bogoliubov-Born-Green-Kirkwood-Yvon (BBGKY) hierarchy), but here adapted to ions [19]. The main advance with respect to the unified theory is that the collision operator accounts for the finite lifetime of the atom during each collision; this lifetime yields an effective range for the action of the microfield of the order of $v/\bar{\gamma}$, where $\bar{\gamma}$ is a typical matrix element of the collision operator (see the discussion in [41]). This model (and its adaptation to electrons) does not lead to a divergent collision operator if the Debye length is assumed infinite, which is in contrast to standard hydrogen models (see [42]); this makes it suitable for the presented cases, provided the perturbing species under consideration is strongly dynamic.

3. Comparisons and Discussion

To test the accuracy of the different numerical codes based either on stochastic and collisional models or numerical simulations, calculations for standardized case problems were carried out and analyzed [5]. A preselected set of transitions on a grid of electron densities (n_e) and temperatures ($T = T_e = T_i$) have been proposed, and for each case, the atomic and plasma models have been specified. In this way, various contributions that can affect the Stark broadened line shape, such as the influence of particle correlations on electric microfields, the effects of external fields, the high-n merging with continuum or the satellite broadening, have been investigated. For the present purpose, we will only focus on cases where the ion dynamics effect was studied.

3.1. Hydrogen Lyman-α and Lyman-β Lines

The following examples consider the hydrogen Lyman-α and Lyman-β lines in an ideal plasma consisting of protons for electron densities $n_e = 10^{17} - 10^{19}$ cm^{-3} and temperatures $T = 1 - 100$ eV. These cases are not necessarily practical, but permit basic comparisons to assess the influence of ion dynamics on the line profiles. Here, only pure ionic linear Stark effect is considered ($\Delta n \neq 0$ interactions are ignored) and the fine structure is not taken into account. The concept of ideal plasma means that unscreened particles moving along straight path trajectories are considered in the numerical simulations, and the Holtsmark static-field distribution function [22] is used in the models.

An overall comparison of the results is presented in Figure 1. For each subcase (determined by a combination of (n_e, T)) and for each code, ratios between the full-width at half-maximum (FWHM) and an average of FWHM of all submitted results have been evaluated [5]:

$$R_i = \frac{\text{FWHM}}{<\text{FWHM}>} \tag{21}$$

The graph is divided in two regions: the left side corresponds to results for the Lyman-α line and the right side to the Lyman-β line. Each region is divided into three sub-regions that correspond to the three densities chosen. Finally, in each sub-region, each set of results corresponds to the temperatures, $T = 1, 10, 100$ eV, respectively. For the Lyman-α case, the results present a large dispersion, deviating from the average by more than a factor of five in each direction. In contrast, the scatter for the Lyman-β shows a rather good agreement between the codes. In fact, these two lines present a completely different behavior concerning the ion dynamics effect.

Figure 1. Overall comparison of the workshop results of the ion dynamics effect on Lyman-α and -β hydrogen lines. For each subcase, *i.e.*, different pairs of (n_e, T), the scatter of ratios between the different results and an average value is plotted. The different symbols correspond to: (black dot) SimU; (red square) UTPP; (blue triangle) PPP; (blue asterisk) Xenomorph; (cyan open triangle) HSTRK; (cyan triangle) HSTRK_FST; (red diamond) ER-simulation; (green circle) QuantST.MMM; (black cross) QC-FFM.

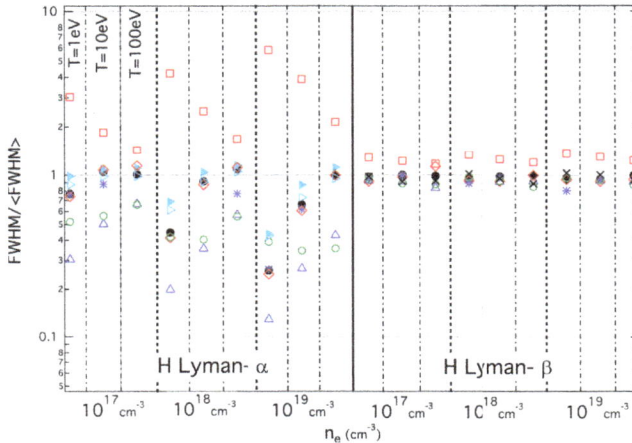

3.1.1. The Lyman-α Line

The static Stark effect of the Lyman-α line (as all the $\Delta n = n - n' = 1$ lines, where n and n' are the principal quantum number of the upper and lower states, respectively) features a strong unshifted component that is highly sensitive to the ion dynamics effect. Thus, even though the Lyman-α line is the simplest case from the atomic structure point of view, it presents a non-trivial Stark-broadening behavior.

In Figure 2, only results from the numerical simulations are plotted for the sake of clarity. One sees that in the range of 1 to 100 eV, the simulations either predict that the width increases when the plasma temperature increases (for the fixed density $n_e = 10^{19}$ cm^{-3}, they present a temperature dependency as $\sim T^{1/3}$) or predict that the width is mostly insensitive to the temperature's rise (for the fixed density $n_e = 10^{17}$ cm^{-3}). Concerning the dependence on the plasma density, the width, which is mainly due to the width of the central component for $T = 1$ eV, increases as $n_e^{1/3}$. For $T = 100$ eV, the results show a $n_e^{2/3}$ dependence, corresponding to the quasi-static behavior of the lateral components [2]. We mention, however, that the cutoff of the Coulomb interaction at a finite box size may not accurately reproduce an ideal plasma [42].

Figure 2. Lyman-α ion FWHMs as a function of (**a**) of T at fixed densities and (**b**) of n_e at fixed temperatures. The ideal, one-component plasma consisting of protons is assumed. Only results from numerical simulations are presented: (red circle) ER-simulation; (blue square) HSTRK; (black dot) SimU; (green asterisk) Xenomorph.

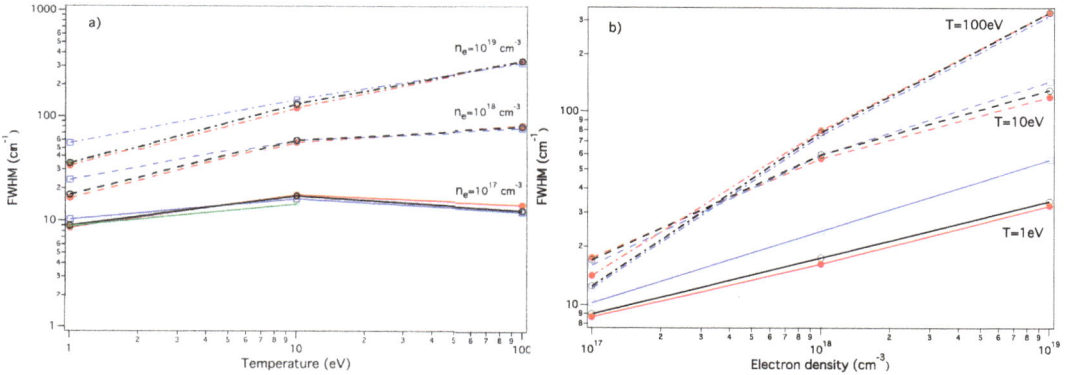

In general, the highest temperature results in the best agreement among simulation codes, for all densities. For lower temperature differences are more discernible, with the most discernible being the appearance of shoulders in ER-simulation and SimU for the highest density and lowest temperature and the lack of such shoulders in HSTRK. This is a general trend at the lowest temperature of 1 eV for all densities, with HSTRK producing significantly larger widths than both ER-simulation and SimU.

The dispersion of the results of the various models demonstrates the difficulties in accurately treating the ion dynamics effect (see Figure 3). In every studied case, the PPP displays a weaker ion dynamics effect on this line, probably due to an incomplete description of this effect on the central component. The FFM mixes the unshifted components with the Stark-shifted components with a unique fluctuation rate. Yet, the unshifted components are not sensitive to the microfield intensity, but only to its rotation, whereas the Stark-shifted components are sensitive to the microfield vibration [43]. A more detailed discussion on the influence of the microfield directionality in the line shape is presented in a separate study [44].

Concerning the description of the ion dynamics effect in terms of microfields mixing, the QuantST.MMM results compare less favorably to the simulations, especially in the far wings.

As already discussed, the far wings of HSTRK_FST are not reliable in this version, due to the complete collision assumption used in the computation of the impact part. This is an artifact of this assumption rather than an inherent limitation of the method.

The UTPP code yields a line width systematically larger than the results obtained from other codes or models and, in particular, the results from numerical simulations. If the latter give reference profiles, this result is expected in general, because the plasma conditions are such that static effects with simultaneous strong collisions are important. However, the application of the UTPP to the electron broadening (not presented here) also indicates a significant discrepancy, with an overestimate of the numerical simulation results by a factor of two. It has been suggested that this discrepancy stems from the fact that the simulations that use a box actually miss a significant contribution to the line broadening, due to the far perturbers, namely, those inside the $v/\bar{\gamma}$ sphere, but outside the simulation box. It is quite difficult to test this argument by enlarging the simulation box up to $v/\bar{\gamma}$, because this would imply a very large number of particles (up to several billions). An adaptation of UTPP able to account for a plasma of finite size has been performed and has led to a line shape in good agreement with the simulations [42]. This could suggest that an artificial setting of an infinite Debye length in the numerical simulations able to work with an infinite Debye length requires a careful interpretation of the results.

Figure 3. Lyman-α line shape in ideal ionic one component plasma (OCP) calculated for (**a**) the more dynamical regime ($n_e = 10^{17}$ cm^{-3} and $T = 100$ eV) and (**b**) the more static regime ($n_e = 10^{19}$ cm^{-3} and $T = 1$ eV): SimU (black dash); ER-simulation (red dash); HSTRK_FST (blue dot-dash); PPP (solid cyan); QuantST.MMM (solid purple); UTPP (solid green).

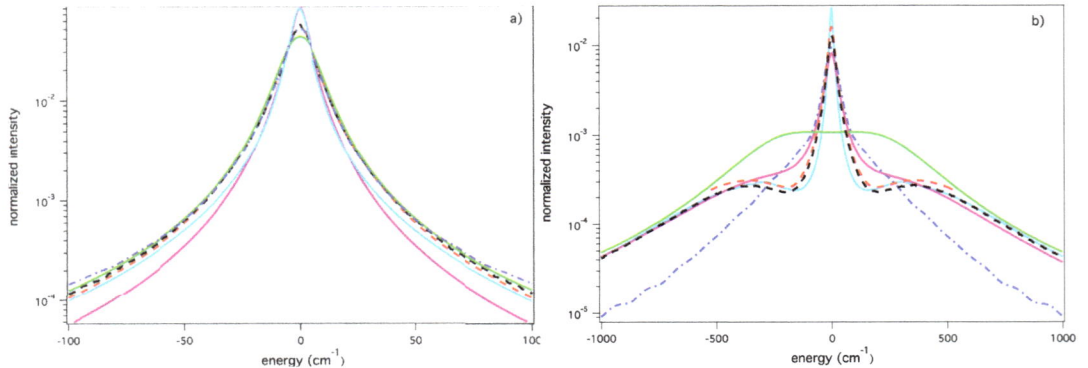

3.1.2. The Lyman-β Line

The static profile of the Lyman-β line (as all of the $\Delta n = 2$ lines) normally shows a dip at the line center. One sees in Figure 4 that, due to the ion dynamics effect, the simulations fill this dip, and the width increases with increasing temperature. This trend is seen for plasma conditions that correspond to typical microfield fluctuation rate values (see Equation (5)) smaller than the splitting of the two Stark components measured in the static case. Here, for $n_e = 10^{17}$ cm^{-3}, the Stark splitting of the static line shape is equal to 5.9×10^{-3} eV, and the typical fluctuation rate is equal to $\hbar\nu = 6.8 \times 10^{-4}$ eV and $\hbar\nu = 2.5 \times 10^{-3}$ eV for $T = 1$ eV and $T = 10$ eV, respectively. For $T = 100$ eV, $\hbar\nu = 2.2 \times 10^{-2}$ eV, *i.e.*, three-times greater that the Stark splitting in the static case. The two components merge, leading to a line shape that is narrower than the one calculated for $T = 10$ eV, as is seen in Figure 4a. Note that for an infinite fluctuation rate, the line shape becomes the Dirac δ-function.

The agreement between the Lyman-β FWHM results of different codes is much better than that for Lyman-α, as is shown in Figure 1. Nevertheless, the concept of FWHM is not really adequate for such a line with a dip in the center. A better way to discuss the ion dynamics effect on a Lyman-β line would be the measure of the relative dip given by:

$$D_{dip} = \frac{I_{max} - I_{\omega_0}}{I_{max}} \tag{22}$$

where I_{max} and I_{ω_0} are the maximum intensity and the intensity at the center of the line, respectively. Table 1 shows the relative dip from the different codes for $n_e = 10^{17}$ cm^{-3}, while the line shapes are shown in Figure 5.

Obviously, the QC approximation, and, hence, the QC-FFM method, is inherently unable to reproduce the central structure (a peak or a dip) of a low-n spectral line. However, the wings of such lines, as well as entire profiles of higher-n transitions, show a very good agreement with numerical simulations [13].

Figure 4. The ion dynamics effect on the Lyman-β line for different values of T obtained by SimU: (solid red $T = 1$ eV; (green dash) $T = 10$ eV and (blue dot-dash) $T = 100$ eV at a fixed (**a**) $n_e = 10^{17}$ cm^{-3} and (**b**) $n_e = 10^{19}$ cm^{-3}. The ideal one-component plasma consisting of protons is assumed.

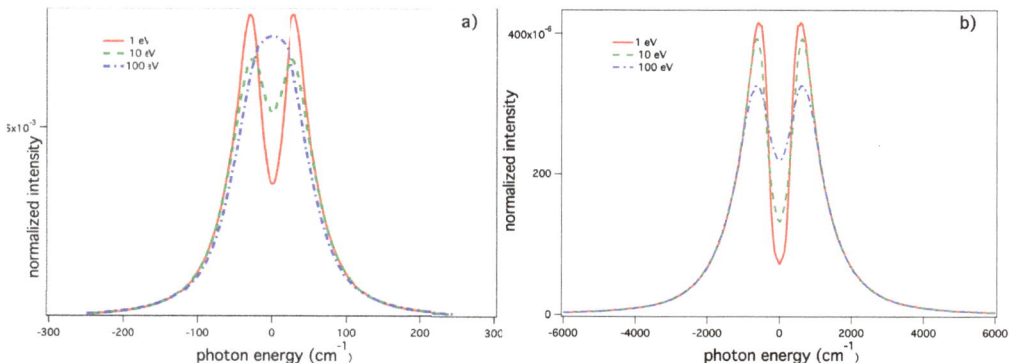

Figure 5. Lyman-β line for $n_e = 10^{17}$ cm^{-3} and $T = 10$ eV: SimU (black dash); ER-simulation (red dash); Xenomorph (blue dot-dash); PPP (solid cyan); QC-FFM (solid orange); QuantST. MMM (solid purple); UTPP (solid green).

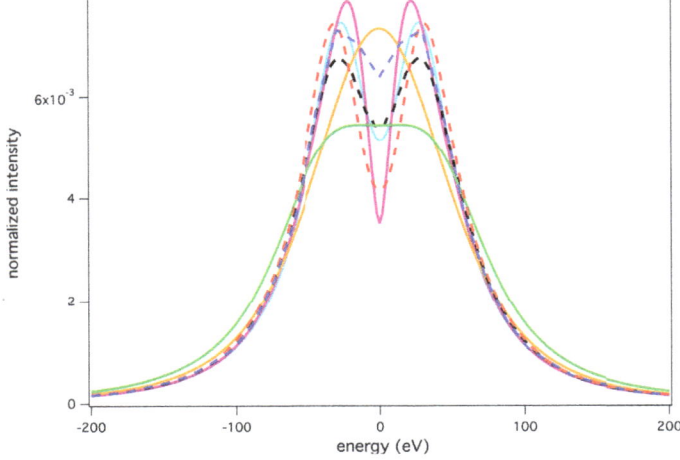

Table 1. The relative dip (%) measured on the Lyman-β line from the different codes for $n_e = 10^{17}$ cm^{-3}.

T (eV) =	1	10	100
ER-simulation	75	44	10
SimU	56	19	0
Xenomorph	56	14	/
PPP	70	31	0
QuantSt.MMM	71	55	32
UTPP	0.6	0.6	0

3.2. Argon He-α and He-β Lines

The argon H- and He-like lines are observed in inertial confinement fusion implosion core plasmas when a tracer amount of argon is added to the deuterium gas fill to diagnose the plasma conditions [45]. Such a diagnostic relies on the temperature sensitivity of the satellite line;s relative intensity to the resonance one and the density dependence in the Stark broadening of both satellite and resonance line profiles [46]. Moreover, they are sensitive to the ion dynamics effects and present a challenge for theoretical models [47]. We only focus here on the He-α and He-β lines. A specific study of the effect of satellite line shapes on the He-β line can be found elsewhere [48].

Two electron densities, $n_e = 5 \times 10^{23}$ cm^{-3} and $n_e = 2 \times 10^{24}$ cm^{-3}, and a plasma temperature of $T = 1$ keV were selected for this comparison. Plasma ions are deuterons with 0.1% Argon XVII.

The MELS and MERL (BID) and the PPP (FFM) models submitted results, and the numerical simulation, SimU, was recently extended to describe such lines. Here, the simulation accounts for all interactions; no artificial cutoff arises as for the ideal case conditions. We consider it as a reference.

Figure 6 displays the He-α profiles calculated with the PPP code within the quasi-static approximation. For clarity, results from MELS are not plotted here, but the agreement between the two codes is very good. The small differences observed between both codes are explained by the difference in the electron broadening treatment (the impact approximation is used in PPP, while a frequency dependent collision operator is used in MELS). The quasi-static profile is the superimposition of a strong intensity component, which corresponds to the $1s2p\ ^1P_1 - 1s^2\ ^1S_0$ resonance transition, and a weak intensity component, which corresponds to the $1s2p\ ^3P_1 - 1s^2\ ^1S_0$ intercombination transition. The pure electron broadened profiles are plotted for each component for a better understanding of the ionic Stark effect on the line shapes. Both components display a pronounced quadratic Stark effect in their "blue" wing, and forbidden lines appear on top of their "red" wing.

Figure 6. The He-α line calculated within the quasi-static approximation for $T = 1\,\mathrm{keV}$ and $n_e = 2 \times 10^{24}\ \mathrm{cm}^{-3}$. (Black line) the entire profile; (blue line) resonant line profile; (red line) intercombination line profile. The pure electron-broadened profiles are plotted in dashed lines for each component.

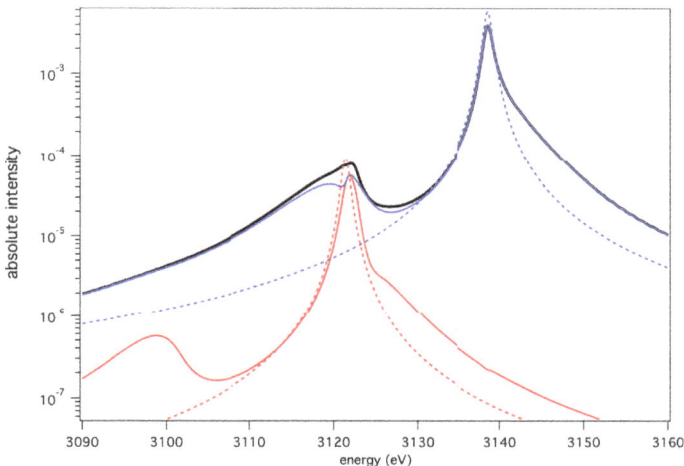

Concerning the ion dynamics effect, BID and FFM show a different behavior. BID profiles present a more pronounced deviation relative to the static calculation than the FFM. Figures 7 and 8 illustrate this for the two densities. These discrepancies cannot be explained by the use of a different fluctuation rate in both models. A specific study using the same fluctuation rate for both models shows that the BID and FFM are in good agreement for varying values of this parameter for the resonance line, but not for the intercombination line [39]. Figure 9 shows this difference using both models with the same fluctuation rate. The difference seen on the forbidden component of the intercombination line might be due to a numerical inaccuracy, because of the very weak value of its intensity.

Moreover, numerical simulation results from the SimU code do not discriminate between the stochastic models. For example, in Figure 9, both models agree with the simulation on the allowed transitions, but not on the forbidden transitions. This might be due to a different dynamics between strong microfields, which are emphasized by the quadratic Stark effect of the allowed transitions, and weak microfields, which are the cause of the linear Stark effect of the forbidden transitions.

Figure 7. The He-α line for $T = 1$ keV and $n_e = 5 \times 10^{23}$ cm^{-3}: static ions MELS (grey dash); ion dynamics BID (solid red); FFM (solid blue) and SimU (black dot).

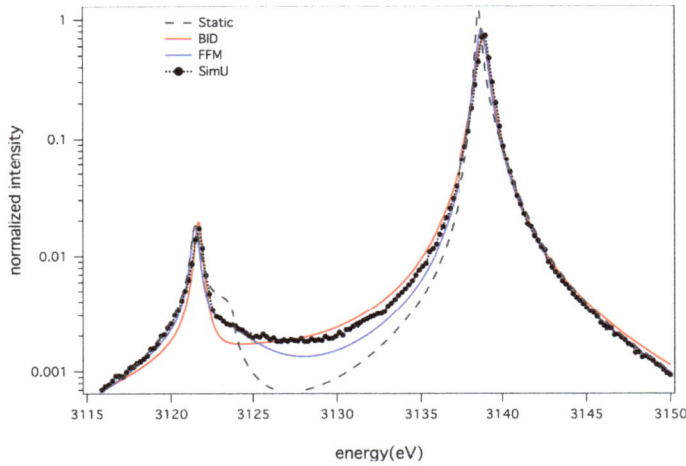

Figure 8. The He-α line for $T = 1$ keV and $n_e = 2 \times 10^{24}$ cm^{-3}: static ions (grey dash); ion dynamics BID (solid red); FFM (solid blue); and SimU (black dot).

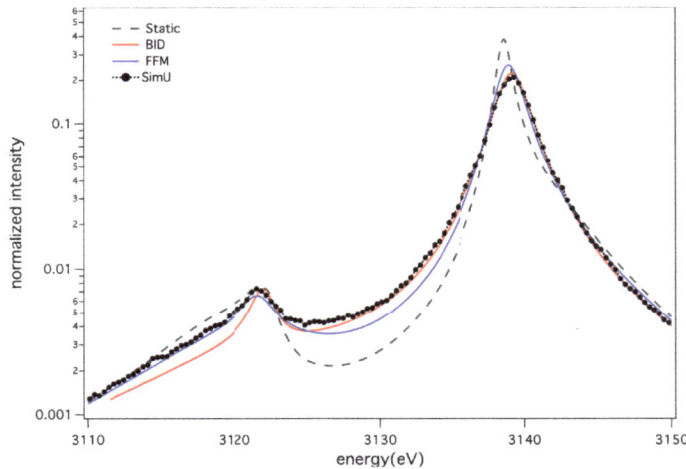

Figure 9. The He-α line for $T = 1$ keV and $n_e = 2 \times 10^{24}$ cm^{-3}: (**a**) resonance line and (**b**) intercombination line. Static ions MELS (red dash) and PPP (blue dash); ion dynamics BID (solid red); FFM (solid blue); and SimU (black dot).

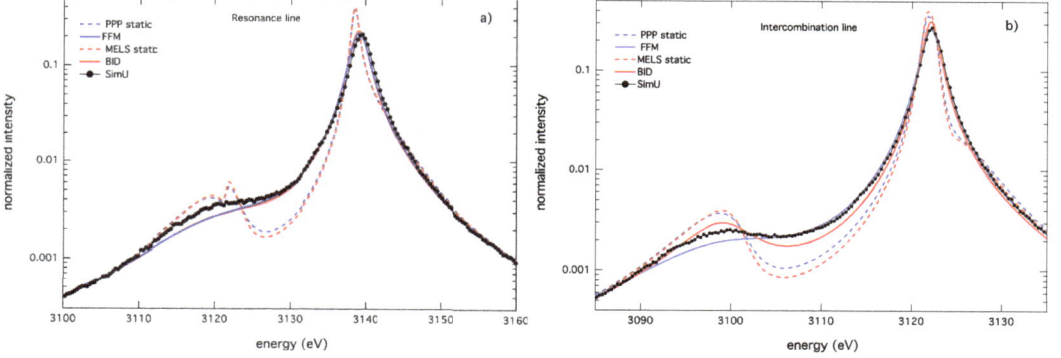

In order to explain these differences, a specific study on the pure ion-broadened profiles was carried out. As both resonance and intercombination lines present similar atomic systems, we will only focus the discussion of the resonance line. Figure 10 shows FFM profiles for different fluctuation rates and the SimU profile. It seems that different values of ν are needed to reproduce different portions of the simulated profile. A lower fluctuation rate has to be used to fit the forbidden component, whereas a higher ν is needed to reproduce the allowed component. This can be interpreted as weak and strong microfields not producing the same dynamics effect on the line shape.

Figure 10. The He-α line, the strong component for $T = 1$ keV and $n_e = 5 \times 10^{23}$ cm^{-3}: SimU (black circles); FFM with $\nu = 3$ eV (solid blue); $\nu = 5.62$ eV (solid red); and $\nu = 8$ eV (solid black).

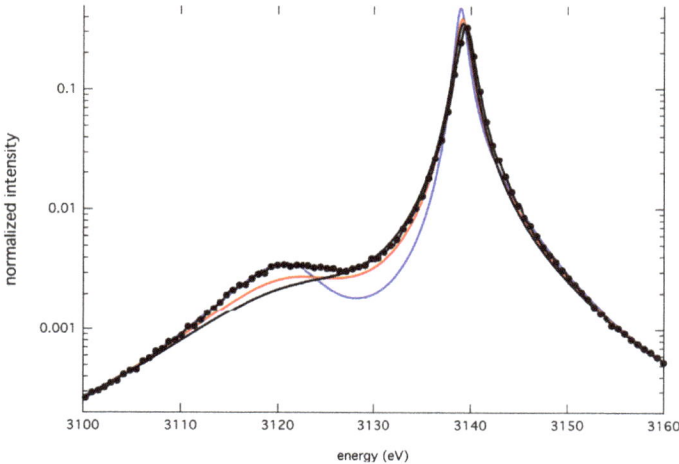

Finally, the He-β line is presented in Figure 11. At the chosen plasma conditions, as the Stark splitting of the He-β quasi-static line shape is greater than the fluctuation rate and the electron width is larger, the ion dynamics effect is less pronounced than on the He-α. Figure 11 shows SimU, BID and FFM in rather good agreement relative to the discrepancies of their quasi-static profiles. The measure of the dynamics-to-static relative depth is defined by:

$$D_{d-s} = \frac{I_{dyn}(\omega_0) - I_{stat}(\omega_0)}{I_{dyn}(\omega_0)} \qquad (23)$$

There is a fairly good agreement between the BID and the FFM (see Table 2).

Figure 11. The He-β line for $T = 1$ keV and (**a**) $n_e = 5 \times 10^{23}$ cm^{-3}; (**b**) $n_e = 2 \times 10^{24}$ cm^{-3}. Static ions: MERL (red dot), PPP (blue dot); SimU (black dot); BID (solid red); FFM (solid blue).

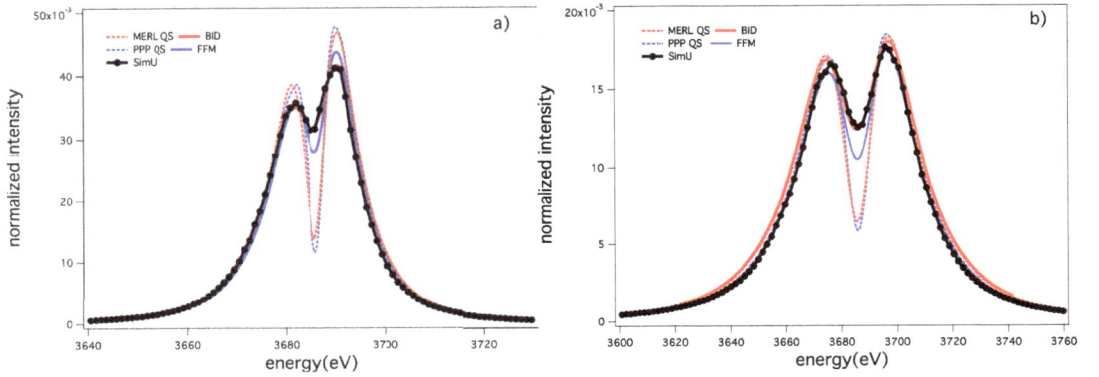

Table 2. Dynamics-to-static relative dip (%) measured on the argon He-β line for $T = 1$ keV.

Models	BID	FFM
$N_e = 5 \times 10^{23}$ cm^{-3}	58	57
$N_e = 1 \times 10^{24}$ cm^{-3}	50	51
$N_e = 2 \times 10^{24}$ cm^{-3}	47	48

4. Conclusions

Line shape calculations from different numerical codes that account for the ion dynamics effect were presented. To test the accuracy of the different codes, standardized case problems have been chosen and a systematic cross-comparison has been done. Results from four numerical simulations based on different algorithms and seven models relying on either stochastic or collisional processes, have been then submitted.

Surprisingly, the results obtained on the hydrogen Lyman-α line in an ideal OCP plasma consisting of protons presents a large dispersion. While the numerical simulations show a relatively good agreement between each other, the FFM and MMM models systematically display a weaker width than the averaged results. This can be explained by an incomplete description of the ion dynamics effect on the central component of this line. The detailed study on the influence of the microfields directionality in the line shape presented in this volume, [44] or other methods discussed here can help improve the modeling of lines with unshifted components. The overestimate of the UTPP code based on a collisional approach is explained by an incomplete description of ion static effects. The results obtained on the H Lyman-β line present a better agreement between all codes.

Concerning the ion dynamics effect on the argon He-α and -β lines, BID and FFM show a different behavior that has been attributed, up to now, to numerical inaccuracies, due to the very weak value of the line intensities. The recently developed numerical simulation code, SimU, could not help discriminate between the two models, but highlighted another problem: it seems that different values of fluctuation rates have to be used to reproduce different portions of the simulated profile. As both the linear and quadratic Stark effect, which are linked to the weak and strong values of microfields, respectively, are involved in producing the shape of this line, one can wonder if a frequency- (or field-) dependent fluctuation rate is needed to give a better description of ion dynamics on this line.

Acknowledgments

The authors would like to acknowledge the International Atomic Energy Agency B.J. Braams and H.-K. Chung for the organizational and financial support of this workshop.

Author Contributions

The present work is based on codes developed by all authors, who also participated in all aspects of this work.

Conflicts of Interest

The authors declare no conflicts of interest.

References

1. Griem, H.R. *Principles of Plasma Spectroscopy*; Cambridge University Press: Cambridge, UK, 1997.
2. Griem, H.R. *Spectral Line Broadening by Plasmas*; Academic Press: New York, NY, USA, 1974; ISBN:0-12-302850-7.
3. Stambulchik, E.; Maron, Y. Plasma line broadening and computer simulations: A mini-review. *High Energy Density Phys.* **2010**, *6*, 9–14.
4. Calisti, A.; Mossé, C.; Ferri, S.; Talin, B.; Rosmej, F.; Bureyeva, L.A.; Lisitsa, V.S. Dynamic Stark broadening as the Dicke narrowing effect. *Phys. Rev. E Stat. Nonlin. Soft Matter Phys.* **2010**, *81*, 016406, doi:10.1103/PhysRevE.81.016406.
5. Stambulchik, E. Review of the 1st Spectral Line Shapes in Plasmas code comparison workshop. *High Energy Density Phys.* **2013**, *9*, 528–534.

6. Gigosos, M.A.; Gonzalez, M.A. Comment on "A study of ion-dynamics and correlation effects for spectral line broadening in plasmas: K-shell lines". *J. Quant. Spectrosc. Radiat. Transf.* **2007**, *105*, 533–535.

7. Alexiou, S. Stark broadening of hydrogen lines in dense plasmas: Analysis of recent experiments. *Phys. Rev. E Stat. Nonlin. Soft Matter Phys.* **2005**, *71*, 066403, doi:10.1103/PhysRevE.71.066403.

8. Alexiou, S. Implementation of the Frequency Separation Technique in general lineshape codes. *High Energy Density Phys.* **2013**, *9*, 375–384.

9. Stambulchik, E.; Maron, Y. A study of ion-dynamics and correlation effects for spectral line broadening in plasma: K-shell lines. *J. Quant. Spectrosc. Radiat. Transf.* **2006**, *99*, 730–749.

10. Stambulchik, E.; Alexiou, S.; Griem, H.R.; Kepple, P.C. Stark broadening of high principal quantum number hydrogen Balmer lines in low-density laboratory plasmas. *Phys. Rev. E Stat. Nonlin. Soft Matter Phys.* **2007**, *75*, 016401, doi:10.1103/PhysRevE.75.016401.

11. Gomez, T.A.; Mancini R.C.; Montgomery M.H.; Winget D.E. White dwarf line shape theory including ion dynamics and asymmetries. **2014**, in preparation.

12. Lorenzen, S. Comparative Study on Ion-Dynamics for Broadening of Lyman Lines in Dense hydrogen Plasmas. *Contrib. Plasma Phys.* **2013**, *53*, 368–374.

13. Stambulchik, E.; Maron, Y. Quasicontiguous frequency-fluctuation model for calculation of hydrogen and hydrogen-like Stark-broadened line shapes in plasmas. *Phys. Rev. E Stat. Nonlin. Soft Matter Phys.* **2013**, *87*, 053108, doi:10.1103/PhysRevE.87.053108.

14. Iglesias, C.A.; Vijay, S. Robust algorithm for computing quasi-static Stark broadening of spectral lines. *High Energy Density Phys.* **2010**, *6*, 399–405.

15. Woltz, L.A.; Hooper, C.F., Jr. Calculation of spectral line profiles of multielectron emitters in plasmas. *Phys. Rev. A* **1988**, *38*, 4766–4771.

16. Mancini, R.C.; Kilcrease, D.P.; Woltz, L.A.; Hooper, C.F. Calculational Aspects of the Stark Line Broadening of Multielectron Ions in Plasmas. *Comput. Phys. Commun.* **1991**, *63*, 314–322.

17. Calisti, A.; Khelfaoui, F.; Stamm, R.; Talin, B.; Lee, R.W. Model for the line shapes of complex ions in hot and dense plasmas. *Phys. Rev. A* **1990**, *42*, 5433–5440.

18. Alexiou, S.; Poquérusse, A. Standard line broadening impact theory for hydrogen including penetrating collisions. *Phys. Rev. E Stat. Nonlin. Soft Matter Phys.* **2005**, *72*, 046404, doi:10.1103/PhysRevE.72.046404.

19. Rosato, J.; Capes, H.; Stamm, R. Influence of correlated collisions on Stark-broadened lines in plasmas. *Phys. Rev. E Stat. Nonlin. Soft Matter Phys.* **2012**, *86*, 046407, doi:10.1103/PhysRevE.86.046407.

20. Boercker, D.B.; Iglesias, C.A.; Dufty, J.W. Radiative and transport properties of ions in strongly coupled plasmas. *Phys. Rev. A* **1987**, *36*, 2254–2264.

21. Talin, B.; Calisti, A.; Godbert, L.; Stamm, R.; Lee, R.W.; Klein, L. Frequency-fluctuation model for line shape calculations in plasma spectroscopy. *Phys. Rev. A* **1995**, *51*, 1918–1928.

22. Holtsmark, J. Über die Verbreiterung von Spektrallinien. *Ann. Phys.* **1919**, *58*, 577–630.

23. Iglesias, C.A.; Boercker, D.B.; Iglesias, C.A. Electric field distributions in strongly coupled plasmas. *Phys. Rev. A* **1985**, *31*, 1681–1686.

24. Gigosos, M.A.; Gonzalez, M.A.; Cardeñoso, V. Computer simulated Balmer-alpha, -beta and -gamma Stark line profiles for non-equilibrium plasmas diagnostics. *Spectrochim. Acta Part B* **2003**, *58*, 1489–1504.

25. Gigosos, M.A.; Cardeñoso, V. New plasma diagnostics table of hydrogen Stark broadening including ion dynamics. *J. Phys. B-At. Mol. Opt. Phys.* **1996**, *29*, 4795–4838.

26. Hegerfeldt, G.; Kesting, V. Collision-time simulation technique for pressure-broadened spectral lines with applications to Ly-α. *Phys. Rev. A* **1988**, *37*, 1488–1496.

27. Filon, L.N.G. On a quadrature formula for trigonometric integrals. *Proc. R. Soc. Edinb.* **1928**, *49*, 38–47.

28. Pfennig, H. On the Time-Evolution Operator in the Semiclassical Theory of Stark-Broadening of hydrogen Lines. *J. Quant. Spectrosc. Radiat. Transf.* **1972**, *12*, 821–837.

29. Lisitsa, V.S.; Sholin, G.V. Exact Solution of the Problem of the Broadening of the hydrogen Spectral Lines in the One-Electron Theory. *Sov. J. Exp. Theor. Phys.* **1972**, *34*, 484–489.

30. Gigosos, M.A.; González, M.A. Calculations of the polarization spectrum by two-photon absorption in the hydrogen Lyman-α line. *Phys. Rev. E Stat. Nonlin. Soft Matter Phys.* **1998**, *58*, 4950–4959, doi:10.1103/PhysRevE.58.4950.

31. Djurović, S.; Ćirišan, M.; Demura, A.V.; Demchenko, G.V.; Nikolić, D.; Gigosos, M.A.; Gonzalez, M.Á. Measurements of H_β Stark central asymmetry and its analysis through standard theory and computer simulations. *Phys. Rev. E Stat. Nonlin. Soft Matter Phys.* **2009**, *79*, 046402, doi:10.1103/PhysRevE.79.046402.

32. Frisch, U.; Brissaud, A. Theory of Stark broadening—I soluble scalar model as a test. *J. Quant. Spectrosc. Radiat. Transf.* **1971**, *11*, 1753–1766.

33. Frisch, U.; Brissaud, A. Theory of Stark broadening—II exact line profile with model microfield. *J. Quant. Spectrosc. Radiat. Transf.* **1971**, *11*, 1767–1783.

34. Günter, S.; Hitzschke, L.; Röpke, G. Hydrogen spectral lines with the inclusion of dense-plasma effects. *Phys. Rev. A* **1991**, *44*, 6834–6844.

35. Lorenzen, S.; Omar, B.; Zammit, M.C.; Fursa, D.V.; Bray, I. Plasma pressure broadening for few-electron emitters including strong electron collisions within a quantum-statistical theory. *Phys. Rev. E Stat. Nonlin. Soft Matter Phys.* **2014**, *89*, 023106.

36. Dufty, J.W. *Spectral Lines Shapes*; Wende, B., Ed.; De Gruyter: New York, NY, USA, 1981.

37. Stambulchik, E.; Maron, Y. Stark effect of high-n hydrogen-like transitions: Quasi-contiguous approximation. *J. Phys. B-At. Mol. Opt. Phys.* **2008**, *41*, 095703, doi:10.1088/0953-4075/41/9/095703.

38. Mossé, C.; Calisti, A.; Stamm, R.; Talin, B.; Lee, R.W.; Klein, L. Redistribution of resonance radiation in hot and dense plasmas *Phys. Rev. A* **1999**, *60*, 1005–1014.

39. Iglesias, C. Efficient algorithms for stochastic Stark-profile calculations. *High Energy Density Phys.* **2013**, *9*, 209–221.

40. Rosato, J.; Reiter, D.; Kotov, V.; Marandet, Y.; Capes, H.; Godbert-Mouret, L.; Koubiti, M.; Stamm, R. Progress on radiative transfer modelling in optically thick divertor plasmas. *Contrib. Plasma Phys.* **2010**, *50*, 398–403.

41. Rosato, J.; Capes, H.; Stamm, R. Divergence of the Stark collision operator at large impact parameters in plasma spectroscopy models. *Phys. Rev. E Stat. Nonlin. Soft Matter Phys.* **2013**, *88*, 035101:1–035101:3.

42. Rosato, J.; Capes, H.; Stamm, R. Ideal Coulomb plasma approximation in line shape models: Problematic issues. *Atoms* **2014**, *2*, 277–298.

43. Lisitsa, V. Private communication.

44. Calisti, A.; Demura, A.V.; Gigosos, M.A.; GonzÃąlez-Herrero, D.; Iglesias, C.A.; Lisitsa, V.S.; Stambulchik, E. Influence of Microfield Directionality on Line Shapes. *Atoms* **2014**, *2*, 259–276.

45. Chung, H.-K.; Lee, R.W. Application of NLTE population kinetics. *High Energy Density Phys.* **2009**, *5*, 1–14.

46. Woolsey, N.C.; Hammel, B.A.; Keane, C.J.; Asfaw, A.; Back, C.A.; Moreno, J.C.; Nash, J.K.; Calisti, A.; Mossé, C.; Stamm, R.; *et al.* Evolution of electron temperature and electron density in indirectly driven spherical implosions. *Phys. Rev. E Stat. Nonlin. Soft Matter Phys.* **1996**, *56*, 2314, doi:10.1103/PhysRevE.56.2314.

47. Haynes, D.A.; Garber, D.T.; Hooper, C.F.; Mancini, R.C.; Lee, Y.T.; Bradley, D.K.; Delettrez, J.; Epstein, R.; Jaanimagi, P.A. Effects of ion dynamics and opacity on Stark-broadened argon line profiles. *Phys. Rev. E Stat. Nonlin. Soft Matter Phys.* **1996**, *53*, 1042–1050.

48. Mancini, R.C.; Iglesias, C.A.; Ferri, A.; Calisti, S.; Florido, R. The effects of improved satellite line shapes on the argon Heβ spectral feature. *High Energy Density Phys.* **2013**, *9*, 731–736.

Reprinted from *Atoms*. Cite as: Calisti, A.; Demura, A.V.; Gigosos, M.A.; González-Herrero, D.; Iglesias, C.A.; Lisitsa, V.S.; Stambulchik, E. Influence of Microfield Directionality on Line Shapes. *Atoms* **2014**, *2*, 259-276.

Article

Influence of Microfield Directionality on Line Shapes

Annette Calisti [1,*], **Alexander V. Demura** [2], **Marco A. Gigosos** [3], **Diego González-Herrero** [3], **Carlos A. Iglesias** [4], **Valery S. Lisitsa** [2] **and Evgeny Stambulchik** [5]

[1] Aix Marseille Université, CNRS, PIIM UMR 7345, 13397 Marseille, France

[2] National Research Centre "Kurchatov Institute", Moscow 123182, Russia;
E-Mails: demura45@gmail.com (A.D.); vlisitsa@yandex.ru (V.S.L.)

[3] Departamento de Óptica, Universidad de Valladolid, Valladolid 47071, Spain;
E-Mails: gigosos@coyanza.opt.cie.uva.es (M.A.G.); diegohe@opt.uva.es (D.G.-H.)

[4] Lawrence Livermore National Laboratories, P.O. Box 808, Livermore, CA 94550, USA;
E-Mail: iglesias1@llnl.gov (C.A.I.)

[5] Faculty of Physics, Weizmann Institute of Science, Rehovot 7610001, Israel;
E-Mail: Evgeny.Stambulchik@weizmann.ac.il (E.S.)

* Author to whom correspondence should be addressed; E-Mail: annette.calisti@univ-amu.fr;
Tel.: +33-4-912-827-19.

Received: 15 April 2014; in revised form: 2 June 2014 / Accepted: 5 June 2014 / Published: 19 June 2014

Abstract: In the framework of the Spectral Line Shapes in Plasmas Code Comparison Workshop (SLSP), large discrepancies appeared between the different approaches to account for ion motion effects in spectral line shape calculations. For a better understanding of these effects, in the second edition of the SLSP in August, 2013, two cases were dedicated to the study of the ionic field directionality on line shapes. In this paper, the effects of the direction and magnitude fluctuations are separately analyzed. The effects of two variants of electric field models, (i) a pure rotating field with constant magnitude and (ii) a time-dependent magnitude field in a given direction, together with the effects of the time-dependent ionic field on shapes of the He II Lyman-α and -β lines for different densities and temperatures, are discussed.

Keywords: spectral line shapes in plasmas; ion dynamics effects; ion microfield fluctuations

1. Introduction

The effect of ionic field fluctuations on spectral line shapes of hydrogen and hydrogen-like emitters has been studied for a long time by different groups. In the 1970s, the observed deviations between experiments and theories were attributed to ion motion. At the same time, the first attempts to include ion motion effects in theories appeared [1–5], and the experimental proof on hydrogen had been obtained nearly concomitantly by Wiese and co-workers [6,7]. The first N-body molecular dynamics simulations appeared in the the the late 1970s [8].

Nowadays, large differences still appear between the various approaches to take into account the effect of ion dynamics in the line shape calculations. For a better understanding of the origin of these differences highlighted during the SLSP conference in 2012 [9], a study of the ionic field directionality on line shapes has been proposed at the second edition of the SLSP workshop in 2013 [10].

In this paper, we report the study of the effects of the direction and magnitude fluctuations of the ionic field analyzed separately. To this end, "rotational" and "vibrational" microfields have been defined as:

$$\vec{F}_{rot}(t) = F_0 \frac{\vec{F}(t)}{|\vec{F}(t)|} \tag{1}$$

and:

$$\vec{F}_{vib}(t) = \vec{n}_z |\vec{F}(t)| \tag{2}$$

with F_0 the Holtsmark field ($F_0 = 2.603 Z e n^{2/3}$ with Z the ionic charge number, e the elementary charge in statcoulomb, n the ion number density in cm^{-3}) and $\vec{F}(t)$ the field created at the emitter by the surrounding ion charges interacting through an electron-Debye-screened Coulomb potential. The considered plasmas consist of an impurity (or a small percentage) of hydrogenic helium in a bulk of protons. All of the charged particles interact together through an effective Yukawa potential to account for the influence of electrons on the ionic structure.

The effects of these two variants of electric field models together with the effects of $\vec{F}(t)$ on spectral line shapes are discussed for the He II Lyman-α and -β lines for different densities and temperatures. The spectral profiles of these two lines present very different behaviors when the fluctuating electric fields are taken into account due to the existence or absence of an unshifted central component. They are good candidates for studies comparing the different approaches, such as simulation modeling [11–14] or kinetics models [15–19], developed to account for ionic field fluctuations.

2. Spectral Line Shape Calculations

The spectral line shape is related to the Fourier transform of the radiator dipole operator correlation function, $C_{dd}(t) = < \vec{d}(t) \cdot \vec{d}(0) >$, by:

$$I(\omega) = \Re e \frac{1}{\pi} \int_0^\infty dt e^{i\omega t} < \vec{d}(t) \cdot \vec{d}(0) > \tag{3}$$

The correlation function, $C_{dd}(t)$, of the radiator dipole operator, $\vec{d}$, can be written in Liouville space as [20,21]:

$$C_{dd}(t) = \ll \vec{d^\dagger} | \{U_l(t)\}_{\text{bath}} | \vec{d}\rho_0 \gg \tag{4}$$

where the double bra and ket vectors are defined as usual in Liouville space, ρ_0 is the equilibrium density matrix and $\{U_l(t)\}_{\text{bath}}$ is the bath-averaged evolution operator of the emitter. $U_l(t)$ is a solution of the following stochastic Liouville equation (SLE):

$$\frac{dU_l(t)}{dt} = -iL_l.U_l(t) \tag{5}$$

with the condition $U(0) = I$, the identity operator. L_l designates the Liouvillian of the radiator in the bath. We have $L_l = L_0 + l(t)$, where L_0 is the Liouvillian of the free radiator and $l(t)$ a random perturbation of the thermal bath (the plasma). The most difficult part of a line-broadening problem is to identify correctly the environment of the emitter, $l(t)$. In particular, accounting for the fluctuations of electric fields produced at emitters, by moving electrons and ions, has been of constant interest for both the experimental and theoretical points of view since the 1960s [22]. In the standard theory [23], due to their great difference of mass, ions and electrons are treated in different ways, leading to:

$$L_l(t) = L_0 - \vec{d} \cdot \vec{F}_l(t) - i\Phi \tag{6}$$

where $\vec{F}_l(t)$ is the electric field produced by surrounding ions in a given configuration l and Φ is the electronic collisional operator.

In this paper, in order to enhance the ionic field fluctuation effects on spectral line shapes, the electric fields produced at emitters by moving electrons is neglected. Thus, $L_l(t)$ is reduced to: $L_l(t) = L_0 - \vec{d} \cdot \vec{F}_l(t)$. Additionally, calculations of the spectral profile have been performed assuming three variants for the fluctuating electric field, $\vec{F}_l(t)$:

- The time-dependent field created by the protons interacting through a Debye-screened Coulomb potential;
- A pure rotating field following Equation (1);
- A pure vibrating field following Equation (2).

2.1. The Different Codes and Approaches

A straightforward way to take into account the fluctuations of the electric field at the emitter is the numerical simulation, which solves the Schrödinger equation describing the time evolution of the emitter wave functions in the time-dependent field of surrounding charges produced by molecular dynamics (MD) or an alternative technique and then average over a statistically representative number of configurations to obtain the final result. In the following, the MD simulation results will be considered as benchmarks.

Four types of simulation codes are involved in the present study:

- The SimU code [11]: The perturbing fields are simulated by the particle field generator, where the motion of a finite number of plasma particles (electrons and ions) is calculated assuming that classical trajectories are valid. Then, using this field as a perturbation, the radiator dipole oscillating function is calculated by the Schrödinger solver. Finally, using the fast Fourier transformation (FFT) method, the power spectrum of the radiator dipole oscillating function is evaluated, giving the spectral line shape. The results of repeated runs of this procedure are then averaged to obtain a smooth spectrum. Although, in principle, the particle field generator may account for interactions between all particles, for the cases presented in this study, perturbing protons were modeled as reduced-mass Debye quasiparticles interacting only with the stationary radiator via the Debye potential.

- The BinGo code [12]: This code uses standard classical MD simulation to compute the perturbing fields. In this work, the plasma model consists of classical point ions interacting together through a Coulombic potential screened by electrons and localized in a cubic box with periodic boundary conditions. Newton's equations of particle motion are integrated by using a velocity-Verlet algorithm using a time-step consistent with energy conservation. The simulated time-depending field histories are used in a step-by-step integration of the Schrödinger equation to obtain $U_l(t)$ and, thus, $C_{dd,l}(t)$ in the Liouville space. An average over a set of histories is necessary to evaluate $C_{dd}(t)$. Again, the spectral line shape is obtained using FFT.

- The Euler–Rodrigues (ER)-simulation code [13]: The plasma model for the simulation of time-dependent field histories consists of an emitter at rest in the center of a spherical volume and set in a bath of statistically independent charged quasi-particles moving along straight line trajectories. A reinjection technique ensures statistical homogeneity and stability. The simulated electric field histories are used in a solver for the evolution of the atomic system. For hydrogen, if the SO (4) symmetry is valid, Euler–Rodrigues (ER) parameters are used; otherwise the diagonalization process is done using Jacobi's method.

- The DM-simulation code [14]: This code uses the same solver as the ER-simulation code, but the time-dependent field histories are simulated using the MD simulation technique in order to account for the particle interactions.

Even though these numerical simulation techniques have, successfully, been used as model laboratory experiments to compare with line shapes resulting from other method calculations or experiments, they can be impractical when the relevant atomic structure becomes too complex. Several approaches and models have been developed to overcome this difficulty [22] and implemented in numerical codes. Three different models (or codes) have contributed to the present study:

- The multi-electron line shape (MELS) code [24]: The "standard" theory (quasi-static ions and impact electrons) and the Boerker-Iglesias-Dufty (BID) model [15] to account for ion dynamics effects. The microfield distribution is from the adjustable-parameter exponential approximation (APEX) model [25,26].

- The PPP code [27]: The Stark broadening is taken into account in the framework of the standard theory by using the static ion approximation and an impact approximation for the electrons or including the effects of ionic perturber dynamics by using the fluctuation frequency model [16,17]. The microfield distribution functions required are calculated using the APEX model or an external MD simulation code.

- Quasicontiguous (QC)-frequency fluctuation model (FFM) [18,19]: The static line profile for a fixed field is represented by a rectangular shape, which is then integrated over the microfield distribution with the dynamic properties of microfields accounted for via FFM.

2.2. Plasma Characteristics

We consider a plasma containing an impurity (or a small percentage) of He II in protons at two temperatures, ($T = 1$ and $T = 10$ eV) and two ion number densities ($n = 10^{18}$ and $n = 10^{19}$ cm^{-3}).

The plasma conditions are summarized in Table 1 together with some dimensionless parameters useful for characterizing the plasma, such as Γ, the coupling plasma parameter, α, the ratio of the interparticle distance to the screening length and typical frequency values, such as the plasma ion frequency, ω_{pi}, and an estimate of the characteristic frequency of the perturbing field, ν_{dyn}. With the interparticle distance and the screening length given by $r_0 = (3/4\pi n)^{1/3}$ and $\lambda_D = \sqrt{\dfrac{k_B T}{4\pi ne^2}}$, respectively, we have $\Gamma = e^2/(r_0 kT)$, $\alpha = \dfrac{r_0}{\lambda_D}$, $\omega_{\mathrm{pi}} = \sqrt{\dfrac{4\pi ne^2}{m}}$, m being the proton mass and $\nu_{dyn} = v/r_0$, where v is the thermal velocity in the radiator's reference frame.

Table 1. Plasma characteristics.

N_e (cm^{-3})	T(eV)	Γ	α	ω_{pi}(rad/s)	ν_{dyn}(rad/s)
10^{18}	1	0.23	0.83	1.32×10^{12}	1.77×10^{12}
10^{18}	10	0.02	0.26	–	5.57×10^{12}
10^{19}	1	0.50	1.22	4.16×10^{12}	3.80×10^{12}
10^{19}	10	0.05	0.39	–	1.20×10^{13}

Figure 1a illustrates the temporal variation of the ionic field at the emitter for $n = 10^{18}$ cm^{-3} and $T = 1$ eV. In the same figure, the graph in the subwindow shows the field autocorrelation function, $C_{FF}(t)$ whose decay gives an estimate of the time scale of the field fluctuations. Figure 1b shows the corresponding rotational z-component (full line) and the vibrational fields (dash-dot).

The characteristic time scale of the field fluctuations has to be compared to the radiator time of interest, which is usually determined by the inverse of the HWHM (half width at half maximum) due to all broadening mechanisms of the considered line profile. This time depends on the plasma conditions and on the atomic data of the considered line and can be considered as the physical time of interest characterizing the response time of a plasma measurement device.

If the radiator time of interest is small compared to the time scale of the field fluctuations (HWHM$^{-1} \ll \nu_{dyn}^{-1}$, for example), the ion field varies little over the radiator time of interest and may be considered as a static field well characterized by a static field distribution function (quasi-static approximation). As soon as the two characteristic times are of the same order of magnitude, it becomes necessary to account for ion motion on the line profile.

Figure 1. (a) Temporal variation of the three components and correlation function (subwindow) of the perturbing ionic field; (b) temporal variation of the z-component of the rotational field (full line) and of the vibrational field (dash-dot line) at the emitter. The plasma conditions are $n = 10^{18}$ cm^{-3} and $T = 1$ eV for both (a) and (b).

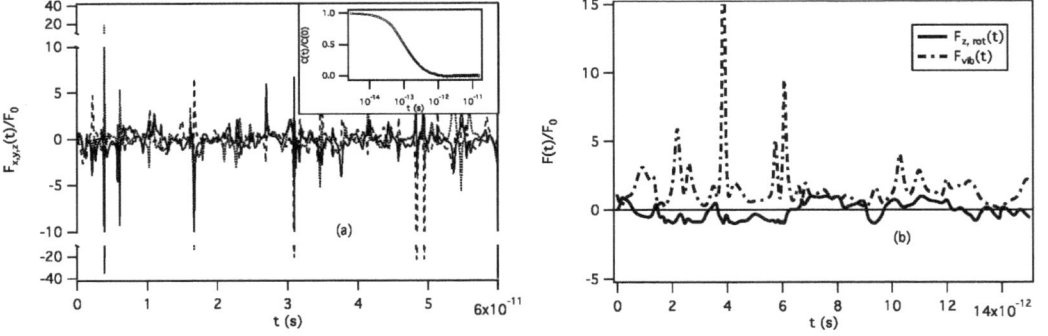

The following results have been obtained for extreme conditions that allow for a better understanding of the differences observed for the various codes and methods. Two different lines have been chosen: the hydrogen-like helium Lyman-α and -β lines. The spectral line shapes of these two lines calculated in the framework of the quasi-static approximation are plotted in Figure 2 and compared to the results accounting for ion motion obtained with a simulation code. The width of the Lyman-α line ($n = 2 - n' = 1$) is very sensitive to the ion motion effect due to the unshifted central component, whereas the width of the Lyman-β line ($n = 3 - n' = 1$), which is broader than the Lyman-alpha, due to a largest upper principal quantum number and does not have unshifted central component, is less affected.

Figure 2. Spectral line shapes of the Lyman-α (a) and Lyman-β line (b) of hydrogen-like Helium calculated in a quasi-static (dash line) and in a fluctuating (full line) ionic electric field produced by protons at $n = 10^{19}$ cm^{-3} and $T = 10$ eV.

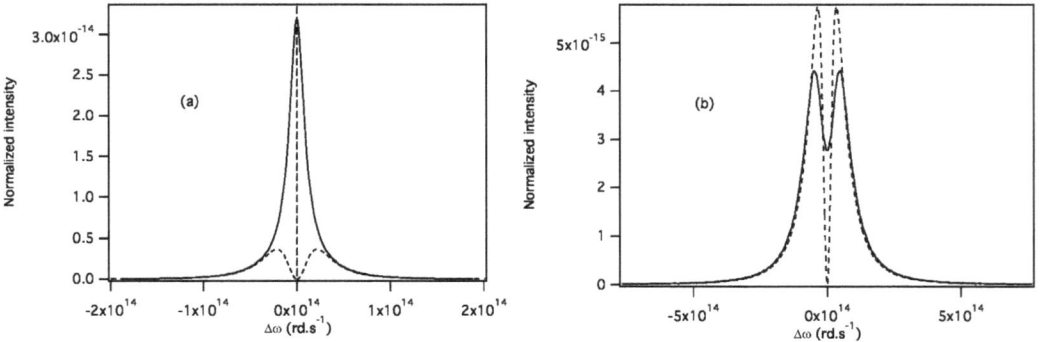

In the following, we will separately analyze how changing the direction and magnitude of the ionic electric field $\overrightarrow{F}(t)$ will influence the line shape. Comparisons between the different codes and methods are also performed.

3. Results

3.1. Generalities

Simulation results of the effects of the magnitude and direction fluctuations of the electric field on the Lyman$-\alpha$ line are plotted in Figure 3a,b, respectively, for $n = 10^{19}$ cm^{-3}. It can be seen that the vibration of the electric field with a fixed direction (Figure 3a) does not affect the central component, whereas each lateral component tends to merge around its gravity center. As expected, the wings of the line are well reproduced by the static profile (the relevant times of interest in the line wings are small compared to the characteristic times of the field fluctuations). Increasing the temperature increases the fluctuation rate, but as the static line width also depends on temperature (due to the Debye-screened potential), the ratio between these two values does not change significantly, and the effects of vibrations on the profile are very similar. Concerning the pure rotating field case (Figure 3b), the situation is completely different. As the field has a fixed magnitude, F_0, the static profile is composed of three un-broadened Stark components, one central unshifted component and two lateral ones (in Figure 3b, the components have been broadened artificially to allow for plotting). When the fluctuations of the field are taken into account, the field magnitude is unchanged, only its direction fluctuates. Here, an increase of temperature results in an increase of dynamics effects. All the Stark components are affected by the field direction changes by being broadened and merged to the center of gravity.

Figure 3. Spectral line shapes of the Lyman-α line in a pure vibrating field (**a**) and a pure rotating field (**b**) compared to the corresponding static profiles, at $n = 10^{19}$ cm^{-3} and two temperatures, $T = 1$ and $T = 10$ eV. The dynamic case results have been obtained with the BinGo simulation code.

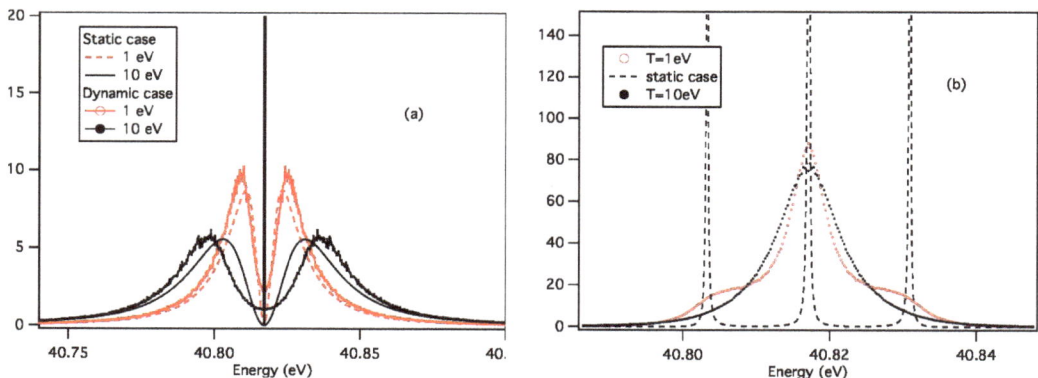

In Figure 4, the Lyman$-\beta$ line profiles calculated by numerical simulations have been plotted for the same conditions as previously. Here, again, the vibrations of the electric field affect the "blue" and "red" lateral components by merging them to their respective center of gravity. As has been already mentioned, the dynamics effects are less important in this case, because the static line-width is broader than in the previous case. In Figure 4b, the effects of the field direction changes on the line

29

profile are shown for the two different temperatures and compared to the static profile. Here, there is no unshifted component, and the effects of fluctuations are to broaden the Stark components, filling progressively the dip between the two sets of Stark components.

Figure 4. Spectral line shapes of the Lyman-β line in a pure vibrating field (**a**) and a pure rotating field (**b**) compared to the corresponding static profiles, at $n = 1 \times 10^{19}$ cm^{-3} and two temperatures, $T = 1$ and $T = 10$ eV. The dynamic case results have been obtained with the BinGo simulation code.

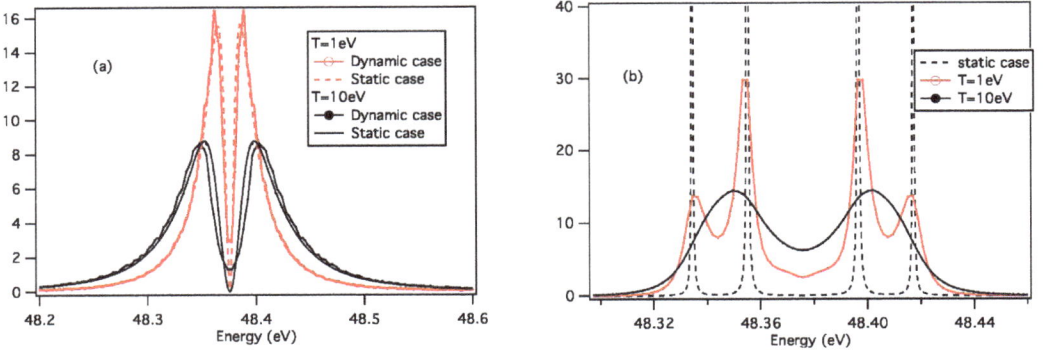

To have a better idea of the pure dynamics effect, results corresponding to various reduced masses between radiator and perturbers (μ_0, $4\mu_0$ and $16\mu_0$) have been plotted in Figure 5 for the Lyman-α line at $n = 10^{18}$ cm^{-3} and $T = 10$ eV. Here, $\mu_0 = 0.8$ is the reduced mass in the He-H$^+$ center of mass frame.

Figure 5. Reduced-mass effect on the Lyman-α line shape at $n = 10^{18}$ cm^{-3} and $T = 10$ eV in the three models of field (full (**a**); rotation (**b**) and vibration (**c**)). The color codes black, red and blue correspond to μ_0, $4\mu_0$ and $16\mu_0$, respectively. The calculations have been done with the SimU simulation code.

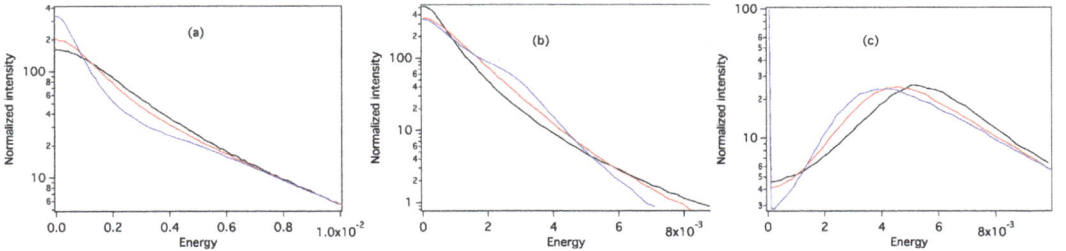

It can be seen on the full profile (Figure 5a) that for the largest value of the reduced mass, the dynamics profile still presents a structure in the wings, and when the dynamics effects increase (when the reduced mass decreases), the line broadening increases and any structure disappears. If the reduced mass were smaller, we would see a narrowing of the profile. For the rotational profile (Figure 5b), this narrowing is already seen, for the reduced mass equals μ_0. The full profile is a complex combination of both rotational and vibrational effects. The shape of the Ly-α and -β line center is determined mainly by the effects related to the rotation of the electric field. Therefore, proper consideration of these effects in the models seems to be one of the keys of the ion dynamics issue.

3.2. Code Comparisons

In this work, both MELS and PPP have been modified to account for vibration only.

The MELS code uses a stochastic microfield description of the plasma for the ion dynamics effect. The model obtains an exact solution for the line profile assuming an idealized stochastic process. It is possible to constrain the model by preserving known important properties of the actual problem by adjusting free parameters in the theory. For the present calculations, the free parameter, the fluctuation rate, was chosen to satisfy the ion impact limit, which, in principle, is equivalent to reproducing the short time limit of the momentum auto correlation function [24]. The rotational field fluctuations were computed by performing the integrations over the electric field directions as before [24], but at a single value of the field magnitude. The vibrational fluctuations were computed by restricting the angle average to a single field direction in the z-axis.

In the PPP code, the ion dynamics effects are treated by means of the frequency fluctuation model (FFM) [16,17]. The FFM is based on the premise that a quantum system perturbed by an electric microfield behaves like a set of field-dressed two-level transitions, the Stark dressed transitions. If the microfield is time varying, the transitions are subject to a collision-type mixing process—a Markov process—induced by the field fluctuations. In practice, the FFM line-shape is the result of intensity exchanges between different spectral domains of the static line-shape with an exchange rate given by ν_{dyn} (Table 1). Owing to the fact that a pure vibrational field will mix neither the central component with the lateral components nor the lateral components together, the code has been adapted in order to mix only the Stark components that must be mixed. For the rotational model, the static profile has been obtained by setting the electric field to $F_0 \vec{n}_z$, and the usual "ion-dynamics" models, FFM, have been applied. It is the mixing of the static Stark components all together that mimics the direction changes of the electric field. The chosen mixing process is a Markov process, suggesting that the cause of the change in states is so violent, that in its final states, the system has no memory of its initial state.

3.2.1. Full Cases

In this example, the line profiles are calculated for the time-dependent field created by the protons interacting through an electron-screened Coulomb (*i.e.*, Debye–Yukawa) potential. As a first observation, the three codes, BinGo, DM-simulation and SimU, give always very similar results (*cf.* Figure 6). The BinGo and DM-simulation codes are similar and account for all of the interactions between particles, whereas the SimU code simulates electron Debye-screened quasi-particles, which interact only with the charged emitter. A few minor differences appear in the center of the lines, within the uncertainties of the different methods. For instance, looking at the Lyman$-\alpha$ line

(Figure 6a), the deviation maximum appears for $n = 10^{19}$ cm^{-3} and $T = 1$ eV and represents about 6% on the peak value. The numerical simulations, BinGo, DM-simulation and SimU, will then be considered as equivalent laboratory numerical experiments, and except for the line-width comparisons, results corresponding to these three simulations will be plotted only once, for clarity.

The results obtained for both lines, with the different codes, have been compared and plotted in Figure 7 for $n = 10^{19}$ cm^{-3} and $T = 1$ eV. It can be seen that the lines calculated by the ER-simulation code are too broad. In this code, the fields measured at the emitter are screened by electrons, but no particle interactions are taken into account (the particles move on straight-path trajectories); thus, the static field distribution function is shifted toward the large fields. This leads to broader static profiles and to changing the ratio between static and dynamic effects. The Stark profile is, in this case, less affected by ion dynamics effects. This behavior is more pronounced at low temperature when the plasma coupling parameter increases.

Figure 6. Spectral line shapes of the Lyman-α (**a**) and -β (**b**) line in a time-dependent field created by interacting protons. Comparisons of the results obtained by the three simulation codes: SimU (yellow), BinGo (black) and DM-simulation (purple).

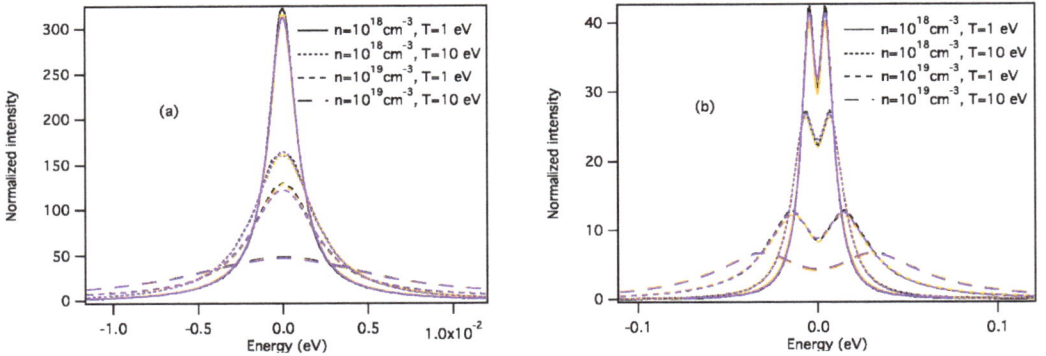

Figure 7. Spectral line shapes of the Lyman-α (**a**) and -β (**b**) line in a time-dependent field created by interacting protons at $n = 10^{19}$ cm^{-3} and $T = 1$ eV. Comparisons of the results obtained by the different codes.

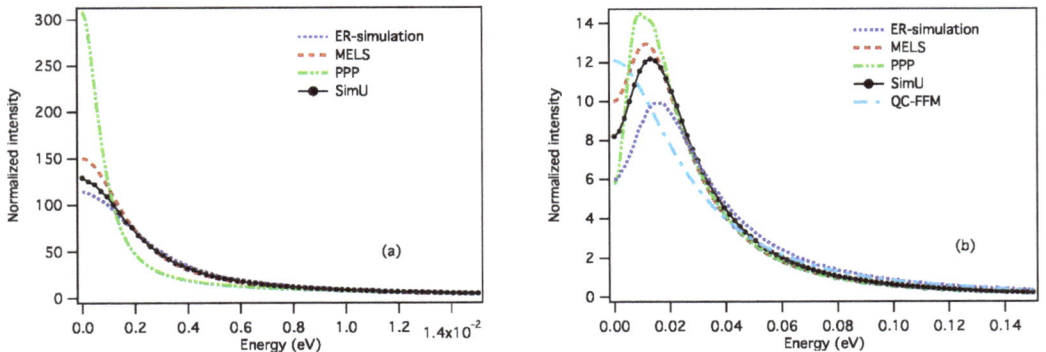

The PPP code qualitatively reproduces the same tendencies as simulation, but it underestimates the ion dynamics at the center of the line, whereas the wings are well reproduced. This could be related to a poor accounting for field direction fluctuation effects.

By construction, the QC-FFM does not show any dip at the center of the Lyman-beta line, but the wings are well represented and the line widths are consistent with those obtained using the FFM in the PPP code. Concerning the MELS code, the ion dynamics effects are well reproduced at $T = 1$ eV, but they are overestimated at $T = 10$ eV (see Figure 8).

Both codes, MELS and PPP, give similar answer to ion dynamics if their fluctuation rates ν are the same [24]. When ν increases, the line width is first increased; then, when ν becomes larger than the static line width, the line shape begins to narrow down and tends to the fast-fluctuation limit, which corresponds to $\nu = \infty$. In this work, the PPP fluctuation rates, equal to ν_{dyn} (*cf.* Table 1), are always much smaller than the MELS ones. This is the reason why the same observation in Figure 8 (too small linewidth) can be due to different underlying causes (the dynamics effects are either too weak or too strong).

Figure 8. Full line widths at half maximum (FWHM) of the Lyman-α (**a**) and -β (**b**) line in a time-dependent field created by interacting protons *versus* temperature at $n = 10^{18}$ cm^{-3} (full lines) and $n = 10^{19}$ cm^{-3} (dashed lines). Comparisons of the results obtained by the different codes: BinGo (black), Euler–Rodrigues (ER)-simulation (blue), MELS (red), PPP (green), QC-frequency fluctuation model (FFM) (turquoise blue), SimU (yellow) and DM-simulation (purple).

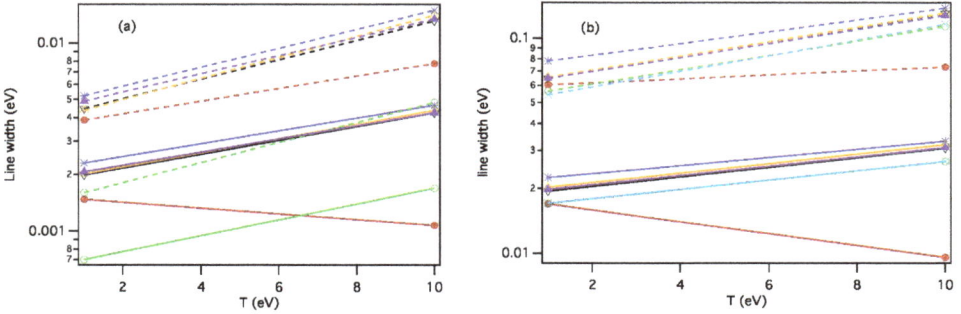

3.2.2. Vibration Case

In this section, calculations of the spectral profile have been performed assuming a pure vibrating field. Results for the Lyman-α and -β lines are plotted in Figures 9 and 10, respectively.

At $T = 1$ eV, the results obtained with ER-simulation are noticeably different from other simulation results, because in these conditions, the correlation effects are important and are not taken into account in the model. PPP compares well with the simulations of reference. At $T = 10$ eV, the correlation effects are less pronounced and ER-simulation gives results very close to those of the reference simulations. The line profiles obtained with the PPP code are slightly too narrow, especially at low density. In all of the cases, MELS shows results slightly too narrow for the Lyman-β line and results radically different for the Lyman-α line. In MELS, if the fluctuation rate, ν, is set to ν_{dyn}, the Lyman-α results compare very well with simulations, making abstraction of the central part of the lines. The central part of the lines presents a broadening, which could be attributed to the impact limit present in the BID model. For small ν, this broadening is small (leading to numerical noise on the profiles), and for larger ν, this broadening dominates.

Figure 9. Spectral line shapes of the Lyman-α line in a pure vibrating field. Comparisons of the results obtained by the different codes.

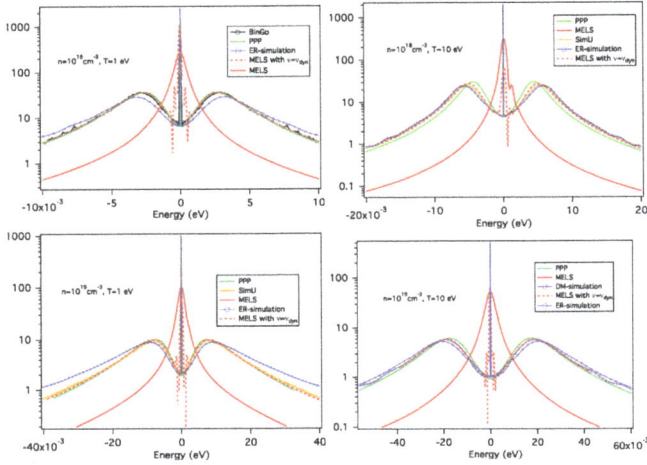

Figure 10. Spectral line shapes of the Lyman-β line in a pure vibrating field. Comparisons of the results obtained by different codes.

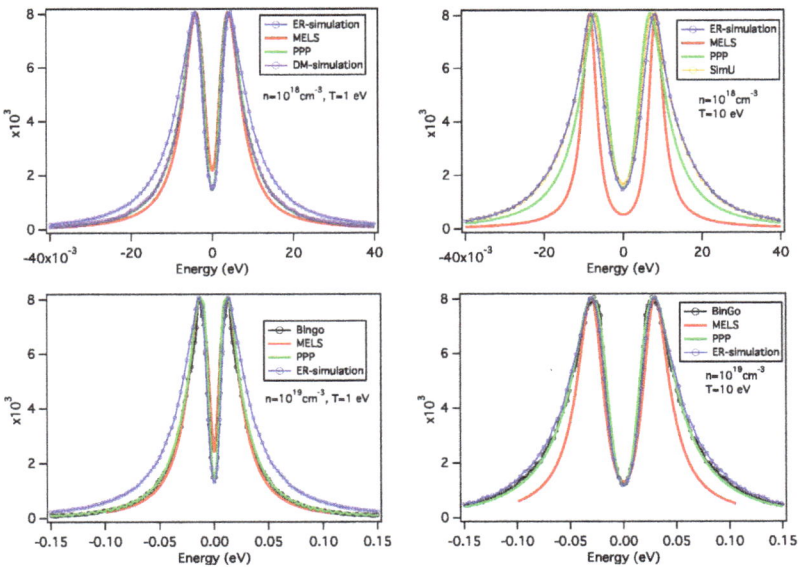

According to these remarks, it appears that the FFM and BID with the same fluctuation rates mimic correctly the magnitude fluctuation effects on the profiles. The usual BID fluctuation rate is too large and leads to too strong field-fluctuation effects.

3.2.3. Rotation Case

The Lyman-α and -β line profiles calculated in a pure rotating field for different densities and temperatures have been plotted in Figures 11 and 12, respectively. As the first result, it can be noticed that the ER-simulation code gives similar results as the reference simulations. Particle interactions have negligible effects on the line profiles in these cases. As an overall tendency, the PPP results appear to be too static and the MELS ones too dynamic. It is shown in Figures 11a and 12a that with the same fluctuation rate, PPP and MELS give similar results. It has been checked that no fluctuation rate allows the reproducing of the simulation results. Except for the Lyman-α line at $n = 10^{18}$ cm^{-3} and $T = 10$ eV (cf. Figure 11b) where the rotational profile is well reproduced with the PPP fluctuation rate, PPP and MELS results will be too structured for a small fluctuation rate, and they will be too strongly merged to the line center for a large one. In these two codes, the implemented models to account for ion dynamics effects are both stochastic models: one describes the time-dependent field fluctuations (BID) and the other assumes a Markovian process for the fluctuations of emission radiation (FFM). If the BID fluctuation rate is constant, the two codes will give similar results. Even though they both give good approximation of the line shapes in realistic conditions (when all of the broadening effects are accounting for), both of them are unable to reproduce the rotational profiles in the framework of this study. It seems that the stochastic processes had to be chosen differently.

Figure 11. Spectral line shapes of the Lyman-α line in a pure rotating field. Comparisons of the results obtained by the different codes.

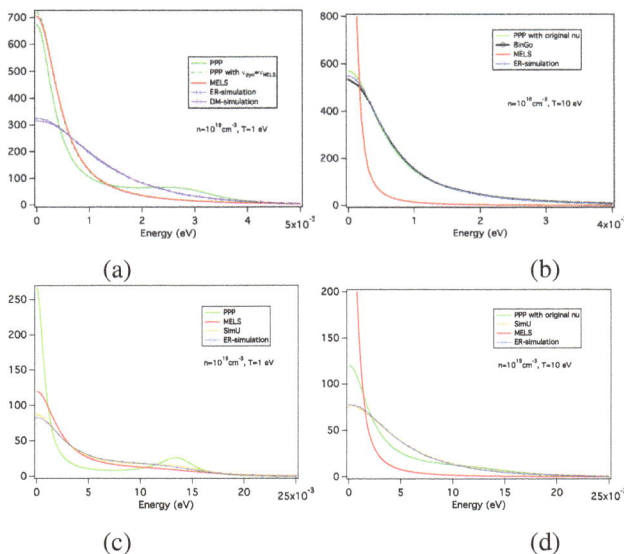

(a)

(b)

(c)

(d)

Figure 12. Spectral line shapes of the Lyman-β line in a pure rotating field. Comparisons of the results obtained by the different codes.

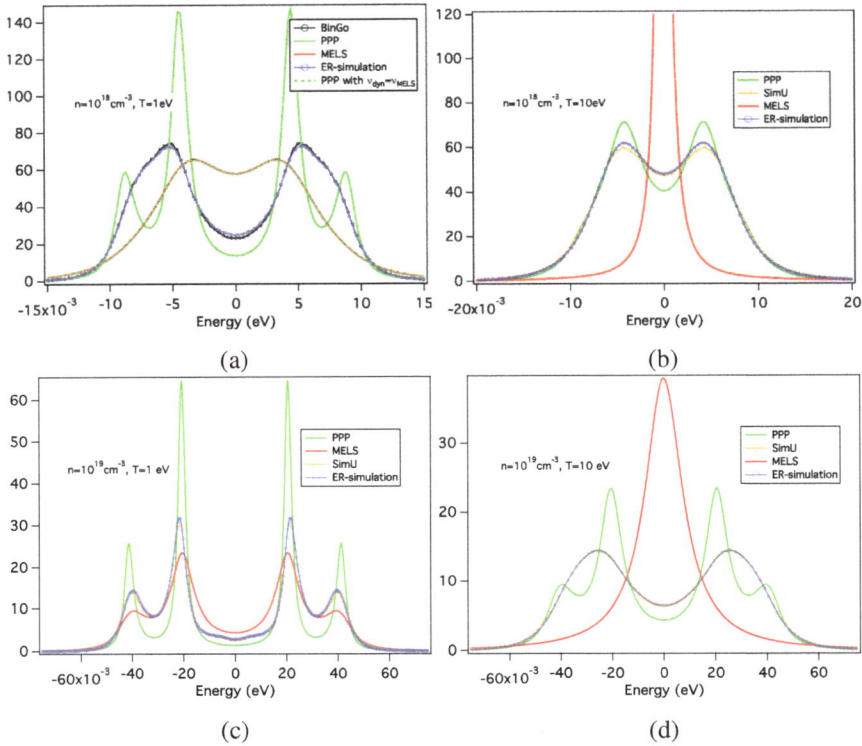

(a)

(b)

(c)

(d)

4. Discussion

Our goal in this paper has been to investigate the reasons of the big differences, which appeared in spectral line shape models intended to account for ionic field fluctuation effects (well-known as ion dynamics effects) in the framework of the first SLSP workshop held in Vienna in 2012. For this purpose, a separate study of the magnitude and direction fluctuation effects on two different spectral lines in plasma conditions chosen in order to perform severe tests on models has been performed. All the simulation codes accounting for interactions between particles or using quasi-particles interacting with the emitter give similar results. They have been taken as benchmarks. These simulation results show that the central part of the line profiles is mainly affected by rotational effects, and a good representation of these effects seems to be the clue of the ion dynamics problem. The simulation codes using quasi-particle on straight-path trajectory models, which are very economical in terms of computing time, give the right answer concerning the rotation effects, but the full profiles are too broad, due to a too large static Stark splitting for both cases at $T = 1\mathrm{eV}$ ($\Gamma = 0.23$ and $\Gamma = 0.5$, respectively), where the effects of interactions between particles are non-negligible. In this context, the intermediate approach employed by SimU in the present study, *i.e.*, accounting only for the radiator-perturber interactions, appears to provide good accuracy with only a minor (relative to the straight-path-trajectory model) increase of the computational time. Evidently, such an approach becomes questionable for strongly-coupled plasmas ($\Gamma \gtrsim 1$).

Figure 13. Autocorrelation functions of the total (black line), rotational (green line) and vibrational (red line) fields compared to the BID model correlation function (dashed blue line) and the exponential function corresponding to the FFM model (short dashed blue line).

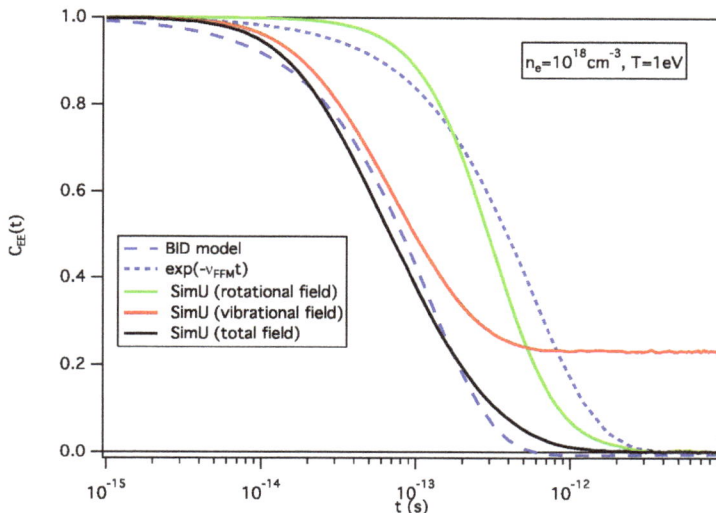

The codes based on stochastic models do not reproduce correctly the rotation effects, even though they are known to give a good approximation of the profiles in realistic conditions [28]. Figure 7 shows a rather good agreement between MELS and simulation results; the fluctuation rate used in PPP is too small and results in too static profiles. PPP would give similar results as MELS if the MELS fluctuation rate were used. We recall that the PPP and MELS codes were both developed for complex atomic systems, such as multielectron ions found in hot, dense laser-produced plasma. Thus, they have to find a satisfactory compromise between accuracy and efficiency. For a better understanding of the differences with simulations, a comparison of the different field autocorrelation functions has been plotted in Figure 13 for $n = 10^{18}$ cm^{-3} and $T = 1$ eV. The total, rotational and vibrational field autocorrelation functions obtained with the SimU simulation code have been plotted together with the total field correlation function corresponding to the BID model (*cf.* Appendix A) and an exponential function with a decay rate equal to ν_{dyn}. The absolute value of the rotational-field autocorrelation function is, by construction, F_0^2, while that of the vibrational- and total-field ones is $3/\Gamma$-times larger [29]. The vibrational field correlation function shows a plateau at the long time limit due to a non-null value of the mean vibrational field. Even though the correlation function models seem to be rather good approximations, the corresponding profiles do not coincide with the simulations. It has been verified that changing the fluctuation rate is not relevant.

It seems that the underlying stochastic models intended to account for ion dynamics are inappropriate for mimicking the pure rotating field. Other choices could be made. For instance, instead of working in the framework of strong collision models with constant fluctuation rates, a model with a frequency-dependent fluctuation rate or a diffusion model on the basis of what has been done for Doppler profiles [30] could be tested.

Acknowledgments

The organizational and financial support from the International Atomic Energy Agency is highly appreciated.

Author Contributions

The present article is based on comparisons of various codes and models developed by the different authors. All of them have been involved in all aspects of the present work.

Appendix A: Field Correlation Function in the BID Model

Consider a system of ions with electric charge Ze and mass m interacting through a screened Coulomb potential with number density n and T the temperature in energy units. The transform of the field auto-correlation function at an ion for this system is approximated in the BID model by [15]:

$$\widetilde{C}_{FF}(\omega) = \int_0^\infty dt e^{i\omega t} < \overrightarrow{F} \cdot \overrightarrow{F}(t) > \approx \frac{3mT}{Z^2 e^2} \frac{\omega \lambda(\omega)}{\omega + i\lambda(\omega)} \tag{A1}$$

The function $\lambda(\omega)$ is given by:

$$- i\lambda(\omega) = \frac{Z^2 e^2}{3mT} \int_0^\infty d\epsilon P(\epsilon) \frac{\epsilon^2}{\omega + i\nu} = \frac{\omega_p^2 \gamma}{3\nu} \frac{1}{\omega + i\nu} \tag{A2}$$

where ν is the free parameter in the BID formulation here assumed a constant, $P(\epsilon)$ is the probability distribution for the field magnitude at an ion, ω_p is the ion plasma frequency and [25,26,29]:

$$< F^2 > = \gamma < F^2 >_{OCP} = 4\pi n T \gamma \tag{A3}$$

defines the correction γ, to the field distribution second moment compared to that of the system of ions interacting through a pure rather than a screened Coulomb potential. Substitution of these results into the inverse transform yields:

$$C_{FF}(t) = < \overrightarrow{F} \cdot \overrightarrow{F}(t) > = < F^2 > \left\{ \frac{\omega_+ e^{-\omega_+ t} - \omega_- e^{-\omega_- t}}{\omega_+ - \omega_-} \right\} \tag{A4}$$

where:

$$\omega_\pm = \frac{\nu}{2} \left[1 \pm \sqrt{1 - \frac{4\gamma \omega_p^2}{3\nu^2}} \right] \tag{A5}$$

Note that:

$$\int_0^\infty dt C_{FF}(t) = 0 \tag{A6}$$

satisfying an exact property [15].

Conflicts of Interest

The authors declare no conflicts of interest.

References

1. Dufty, J. Ion motion in plasma line broadening. *Phys. Rev. A* **1970**, *2*, 534–541.
2. Frisch, U. ; Brissaud, A. Theory of Stark broadening—I soluble scalar model as a test. *J. Quant. Spectrosc. Radiat. Transf.* **1971**, *11*, 1753–1766.
3. Brissaud, A. ; Frisch, U. Theory of Stark broadening—II exact line profile with model microfield. *J. Quant. Spectrosc. Radiat. Transf.* **1971**, *11*, 1767–1783.
4. Hey, J.D.; Griem, H.R. Central structure of low-n Balmer lines in dense plasmas. *Phys. Rev. A* **1975**, *12*, 169–185.
5. Demura, A.V.; Lisitsa, V.S.; Sholin, G.V. Effect of reduced mass in Stark broadening of hydrogen lines. *Sov. Phys. J. Exp. Theor. Phys.* **1977**, *46*, 209–215.
6. Kelleher, D.E.; Wiese, W.L. Observation of ion motion in hydrogen Stark profiles. *Phys. Rev. Lett.* **1973**, *31*, 1431–1434.
7. Wiese, W.L.; Kelleher, D.E.; Helbig, V. Variations in Balmer-line Stark profiles with atom-ion reduced mass. *Phys. Rev. A* **1975**, *11*, 1854–1864.
8. Stamm, R.; Voslamber, D. On the role of ion dynamics in the stark broadening of hydrogen lines. *J. Quant. Spectrosc. Radiat. Transf.* **1979**, *22*, 599–609.
9. Stambulchik, E. Review of the 1st spectral line shapes in plasmas code comparison workshop. *High Energy Density Phys.* **2013**, *9*, 528–534.
10. Spectral Line Shapes in Plasmas code comparison workshop. Available online: http://plasma-gate.weizmann.ac.il/slsp/ (accessed on 30 May 2014).
11. Stambulchik, E.; Maron, Y. A study of ion-dynamics and correlation effects for spectral line broadening in plasma: K-shell lines. *J. Quant. Spectrosc. Radiat. Transf.* **2006**, *99*, 730–749.
12. Talin, B.; Dufour, E.; Calisti, A.; Gigosos, M.A.; González, M.A.; del Río Gaztelurrutia, T.; Dufty, J.W. Molecular dynamics simulation for modelling plasma spectroscopy. *J. Phys. A Math. Gen.* **2003**, *36*, 6049.
13. Gigosos, M.A.; Cardeñoso, V. New plasma diagnosis tables of hydrogen Stark broadening including ion dynamics. *J. Phys. B: At. Mol. Opt.* **1996**, *29*, 4795–4838.
14. Lara N. Cálculo de espectros Stark de plasmas fuertemente acoplados mediante simulación de dinámica molecular. Ph.D. Thesis, University of Valladolid, Valladolid, Spain, 2013.
15. Boercker, D.B.; Iglesias, C.A.; Dufty, J.W. Radiative and transport properties of ions in strongly coupled plasmas. *Phys. Rev. A* **1987**, *36*, 2254–2264.
16. Talin, B.; Calisti, A.; Godbert, L.; Stamm, R.; Lee, R.W.; Klein, L. Frequency-fluctuation model for line-shape calculations in plasma spectroscopy. *Phys. Rev. A* **1995**, *51*, 1918–1928.

17. Calisti, A.; Mossé, C.; Ferri, S.; Talin, B.; Rosmej, F.; Bureyeva, L.A.; Lisitsa, V.S. Dynamic Stark broadening as the Dicke narrowing effect. *Phys. Rev. E* **2010**, *81*, 016406:1–016406:6.

18. Stambulchik, E.; Maron, Y. Stark effect of high-n hydrogen-like transitions: quasi-contiguous approximation. *J. Phys. B: At. Mol. Opt.* **2008**, *41*, 095703.

19. Stambulchik, E.; Maron, Y. Quasicontiguous frequency-fluctuation model for calculation of hydrogen and hydrogenlike Stark broadened line shapes in plasmas. *Phys. Rev. E* **2013**, *87*, 053108:1–053108:8.

20. Baranger, M. *Atomic and Molecular Processes*; Bates, D.R., Ed.; Academic: New York, NY, USA, 1964.

21. Fano, U. Pressure Broadening as a Prototype of Relaxation. *Phys. Rev.* **1963**, *131*, 259–268.

22. Griem, H. *Principles of Plasma Spectroscopy*; Cambridge University Press: Cambridge, UK, 1997.

23. Griem, H.R. *Spectral Line Broadening by Plasmas*; Academic: New York, NY, USA, 1974.

24. Iglesias, C.A. Efficient algorithms for stochastic Stark-profile calculations. *High Energy Density Phys.* **2013**, *9*, 209–221.

25. Iglesias, C.A.; DeWitt, H.E.; Lebowitz, J.L.; MacGowan, D.; Hubbard, W.B. Low-frequency electric microfield distributions in plasmas. *Phys. Rev. A* **1985**, *31*, 1698–1702.

26. Iglesias, C.A.; Rogers F.J.; Shepherd R.; Bar-Shalom A.; Murillo M.S.; Kilcrease D.P.; Calisti A.; Lee R.W. Fast electric microfield distribution calculations in extreme matter conditions. *J. Quant. Spectrosc. Radiat. Transf.* **2000**, *65*, 303–315.

27. Calisti, A.; Khelfaoui, F.; Stamm, R.; Talin, B.; Lee, R.W. Model for the line shapes of complex ions in hot and dense plasmas. *Phys. Rev. A* **1990**, *42*, 5433–5440.

28. Ferri, S.; Calisti A.; Mossé C.; Rosato J.; Talin B.; Alexiou S.; Gigosos M.A.; González M.A.; González D.; Lara N.; Gomez T.; Iglesias C.; Lorenzen S.; Mancini R.; Stambulchik E. Ion dynamics effect on Stark broadened line shapes: A cross comparison of various models. *Atoms* **2014**, *2*, 1–20.

29. Iglesias, C.A.; Lebowitz, J.L.; MacGowan, D. Electric microfield distributions in strongly coupled plasmas. *Phys. Rev. A* **1983**, *28*, 1667–1672.

30. Rautian, S.G.; Sobel'man, I.I. The effect of collisions on the Doppler broadening of spectral lines. *Soviet Phys. Uspekhi* **1967**, *9*, 701–716.

Reprinted from *Atoms*. Cite as: Demura, A.V.; Stambulchik, E. Spectral-Kinetic Coupling and Effect of Microfield Rotation on Stark Broadening in Plasmas. *Atoms* **2014**, *2*, 334-356.

Article

Spectral-Kinetic Coupling and Effect of Microfield Rotation on Stark Broadening in Plasmas

Alexander V. Demura [1],* and Evgeny Stambulchik [2]

[1] National Research Centre "Kurchatov institute", Kurchatov Square 1, Moscow 123182, Russia
[2] Faculty of Physics, Weizmann Institute of Science, Rehovot 7610001, Israel;
E-Mail: Evgeny.Stambulchik@weizmann.ac.il

* Author to whom correspondence should be addressed; E-Mail: Demura_AV@nrcki.ru;
Tel.: +7-499-196-7334; Fax: +7-499-943-0073.

Received: 13 May 2014; in revised form: 17 June 2014 / Accepted: 2 July 2014 / Published: 30 July 2014

Abstract: The study deals with two conceptual problems in the theory of Stark broadening by plasmas. One problem is the assumption of the density matrix diagonality in the calculation of spectral line profiles. This assumption is closely related to the definition of zero wave functions basis within which the density matrix is assumed to be diagonal, and obviously violated under the basis change. A consistent use of density matrix in the theoretical scheme inevitably leads to interdependence of atomic kinetics, describing the population of atomic states with the Stark profiles of spectral lines, *i.e.*, to spectral-kinetic coupling. The other problem is connected with the study of the influence of microfield fluctuations on Stark profiles. Here the main results of the perturbative approach to ion dynamics, called the theory of thermal corrections (TTC), are presented, within which the main contribution to effects of ion dynamics is due to microfield fluctuations caused by rotations. In the present study the qualitative behavior of the Stark profiles in the line center within predictions of TTC is confirmed, using non-perturbative computer simulations.

Keywords: foundations of Stark broadening theory; density matrix; coupling between population and spectral distribution; microfield fluctuations caused by its rotations; MD simulations

In Memory of Professor of Moscow Physical and Engineering Institute—Vladimir Il'ich Kogan (11 July 1923–7 December 2013)—passionate outstanding scientist, pioneer of plasma physics and nuclear fusion research at Kurchatov Institute of Atomic Energy, eminent enthusiastic lecturer, great kind teacher and famous witty connoisseur of courtly linguistics and many more...

1. Introduction

The study of Stark broadening in experiments, theory and simulations has up to now achieved significant progress [1–78]. This allowed the beginning of profound detailed comparisons of various computer codes developed for calculation of spectral lines profiles in plasmas, which was the main purpose of the first two SLSP workshops [74]. However, the understanding and comparison of realizations of most successful contemporary codes on a wide set of physical cases (see [74–77] and other articles of this issue) give good reason to once more discuss, check, and revise the physical notions and ideas which form the foundation of the contemporary theory of spectral line broadening by plasmas. The present article pursues these aims.

In fact, this article deals only with two questions from the list of conceptual problems in the theory of Stark broadening by plasmas [1–77]. The first one is the construction of the spectral line broadening theory without assumption of the density matrix diagonality. Very often this assumption cannot be validated when there is an interaction mixing of states whose energy splitting is less than or comparable with the magnitude of interaction [42–45,50,53]. A consistent introduction of density matrix inevitably leads to interdependence of atomic kinetics, describing the population of atomic states with the Stark profiles of spectral lines that evidently could be defined as spectral-kinetic coupling [42–45,50,53]. Usually, this is also interrelated with the appearance of interference effects [42–45,50,53]. The other problem is related to the attempts to separate the influence of microfield fluctuations on Stark profiles into two components, perpendicular and parallel to the microfield direction [5,6], in order to separate the contribution from microfield rotation effects [12,23–25,30,33], predicted to be dominating in the central part of the line [23–25]. The necessity of this discussion arose when it was recently shown that existing approaches to accounting for ion dynamics give results that differ from one another [74–76]. This inspired attempts to describe ion dynamics effects in terms of physical mechanisms, instead of practically tacit conventional numerical comparison of complicated simulations that began with the first works done using the Model Microfield Method (MMM) in the 1970s [14,15,28,29]. The basics of the theory of thermal corrections for Stark profiles along with the results of [7–9,12,23–25] are given in Section 3. Later, the predictions, given in [12,23–25] about significance of microfield rotation effects in ion dynamics, were confirmed within other approaches [30,33], but until now it has not been explicitly shown in the results of computer simulations (see also [75]). In the present work, simultaneously with [75], the approximate way to separate microfield rotation effects in the profile of Ly-alpha within MD simulations is realized and described in Section 4. However, this separation could not in fact be achieved rigorously, due to the existence of correlations between statistical characteristics and dynamics of atomic systems that we call statistical-dynamical coupling. The specially designed numerical experiments allowed for confirming the qualitative predictions of behavior of the ion dynamical Stark profiles for Ly-alpha, given earlier in [23–25]. This includes: (1) predominance of microfield rotation in the formation of the central part of the Stark profile in plasmas for lines with the central components; (2) a specific spectral behavior of the Stark profiles near the line center as a function of the plasma temperature and reduced mass of the perturber-radiator pair; (3) a universal spectral behavior of the difference Stark profiles for two different reduced mass of the perturber-radiator pair.

For the purpose of discussion a short review of theoretical approaches and methods developed for and applied to the study of the Stark broadening by plasmas is entwined into the argumentation of each section.

2. Spectral-Kinetic Coupling

Let us consider broadening of the hydrogen atom in the well-known setting of the so-called standard theory (ST), related to formulation given in [17,18,21,51]. Then the total Stark profile of spectral line $I(\Delta\omega)$ is represented as the convolution of the Stark subprofiles, corresponding to the transitions between upper and lower substates being broadened by electrons in the fixed value of ion microfield, and integrated over ion microfield values with the microfield probability distribution function as a weight:

$$I(\Delta\omega) = \frac{1}{I_0}\Re\left\langle\langle\alpha',\beta'|\hat{\rho}\hat{\vec{d}}_n\left[i\left(\Delta\omega-\hat{C}_{n,n'}F\right)+\hat{\Phi}_{nn'}\right]^{-1}\hat{\vec{d}}_{n'}|\alpha,\beta\rangle\right\rangle \tag{1}$$

In Equation (1) $\Delta\omega=\omega-\omega_0$ is the detuning of cyclic frequency from the line center, the outer angle brackets correspond as usual to an averaging over the microfield values F (F is the absolute value of microfield), $\hat{\rho}$ is the operator of density matrix, $\hat{\vec{d}}$ is the operator of dipole moment, $\left[i\left(\Delta\omega-\hat{C}_{n,n'}F\right)+\hat{\Phi}_{nn'}\right]^{-1}$ is the operator of resolvent, $\hat{C}_{n,n'}$ is the operator of the linear Stark shift between sublevels of the upper n and lower level n' in the line space (the direct product of subspaces of the upper and lower energy levels with the principal quantum numbers n and n'), $\hat{\Phi}_{nn'}$ is the electron broadening operator, indexes $\alpha\beta$ and $\alpha'\beta'$ designate quantum states of the upper and lower levels in the bra and ket vectors of the line space, respectively.

Conventionally assuming that density matrix is diagonal in Equation (1), it could be factored out since in the region of small values of microfield all sublevels are degenerate and thus equally populated [17,18,21]. For now, the fine structure splitting is neglected. In this region, the spherical quantum functions (labeled by the quantum numbers n, l, m) form a natural zero basis of the problem due to the spherical symmetry [17,18,21]. Suppose now that the value of the ion microfield is increased, thus the sublevels are split due to the linear Stark effect in the ion electric microfield, and the degeneracy is partially removed. Now the parabolic quantum functions (labeled by the parabolic quantum numbers n_1, n_2, m) form the natural zero basis of the problem due to the cylindrical symmetry [17,18]. For sufficiently large values of the microfield, the impacts of electrons could not equate populations of sublevels [17,18], and thus the assumption of population equipartition [17,18,21], made as an initial condition, becomes violated. Evidently, since the density matrix could not have the diagonal equipartition form in the two different bases, the initial assumption of the theory [17,18,21] becomes invalid (see [42–45]). This simple example thus shows that a more consistent approach should simultaneously consider atomic kinetics and formation of spectral line profiles, that just signifies the spectral-kinetic coupling [42–45,50,53]. Those drawbacks in constructing Stark profiles under assumption of diagonal form of density matrix are weakened partially, if one introduces the dependence of electron impact broadening operator on the Stark levels splitting in the ion microfield [17,18]. *Then the non-diagonal matrix elements will be the next order corrections in comparison with the impact widths for large values of the ion microfield, and terms responsible for the line mixing will drop down more rapidly in the line wings* [17,18]. On the other hand, for small values of the ion microfield, for which the Stark splitting is of the order of the non-diagonal matrix elements of the electron-impact-broadening operator, the Stark components collapse to the center, thus becoming effectively degenerate [16–18], as it should be due to the straightforward physical reasoning (it is worth reminding that it was academician V.M. Galitsky who pointed out this effect to the authors of [16]). This collapse phenomenon is characterized by the appearance of the dependence of the decay constants and intensities of the "redefined" inside the collapse region "Stark components" on the microfield, while the energy splitting of these "Stark components" disappears. *These "redefined" Stark components appear in the process of solution of the secular equation for the resolvent operator in the line space [13–15]*. At $F = 0$, the intensity of one of the "redefined" components becomes equal to zero, while the other one gives contribution to the center of the profile identical to the contribution of the two symmetrical lateral Stark components without their redefinition during the solution of secular equation for the diagonalization of resolvent [14–16]. Therefore the collapse phenomenon of Stark components signifies the necessity to change the wave functions basis from the parabolic to spherical wave functions, or *vice versa*, depending on whether ion microfield value decreases or increases. Simultaneously, of course, this means the region of singularity for the ST assumption of the density matrix diagonality [21,51]. *From this consideration it is obvious that the collapse phenomenon has the kinetic character and in fact is one of the examples of spectral-kinetic coupling*. Thus the existence of the collapse phenomenon of Stark components at the same time means the necessity of more complete consideration of the Stark broadening within the formalism of kinetic equations for the density matrix [42–45,50,55], or in other terms, the necessity of application of the kinetic theory of Stark broadening.

The spectral-kinetic coupling discussed above is a common thing in laser physics [42]. Indeed, the lasing condition is directly connected with the difference of populations that, in turn, is proportional to the non-diagonal matrix elements of the density matrix (called coherences), describing the mixing of the upper and lower levels due to the interaction with the radiation field [42]. In the density matrix formalism it is necessary to solve the kinetic equation that leads to the system of much larger rank in comparison with conventional amplitude approach that significantly complicates finding the solution [42–45,50,55]. Moreover, the construction of terms, describing sinks and sources, is not straightforward as during their derivation it is necessary to average over a subset of variables taking account of specific physical conditions [37,42–45]. So, as a rule, these terms could be derived in a more or less general form only in the impact limit, and their concrete expressions are rather arbitrary [37,42–45,50,55]. Moreover, even the formula (1) should be changed to a more general and complex expression for the power that is absorbed or emitted by the medium [42–45,50,55].

3. Ion Dynamics in Statistical and Spectral Characteristics of Stark Profiles

Within the assumptions of ST the plasma ions are considered as static [17,18,21], and hence the resultant Stark profiles are called static or quasistatic. We now consider the ion dynamics effects, *i.e.*, the deviations from the static Stark profiles due to the thermal motion (see, for example, [7–9,12,17,18,20,21,23–36,74–76]). If these deviations are small enough (called in earlier works "thermal corrections" [7–9,12,21,23–25]), then it is possible to express them through the second moments of ion microfield time derivatives of the joint distribution functions $W(\vec{F}, \dot{\vec{F}}, \ddot{\vec{F}})$ of the ion electric microfield strength vector $\vec{F}$ and its first $\dot{\vec{F}}$ and second $\ddot{\vec{F}}$ time derivatives [5–9,12,21,23–25,48,67]. The basic idea of Markov construction of these joint distributions is that $\vec{F}, \dot{\vec{F}}$ and $\ddot{\vec{F}}$ are stochastic independent variables, formed by summation of electric fields or its derivatives over all individual ions of the medium [1,5,6]. So, these probability distribution functions are the many body objects [5–9,12,21,23–25,48,67]. However, those joint distributions possess nonzero constraint moments over, for example, microfield time derivatives, the value and direction of electric ion microfield strength vector being fixed [5–9,12,21,23–25,48,67]. So, each value of the ion microfield corresponds, for example, in fact to a nonzero "mean" value of the square of its derivative. In other words, it means that by fixing one of the initially independent stochastic variables, the mean values of the other ones become nonzero and functionally dependent on the value of the fixed variable [5,6]. So, the direct correlations between fixed stochastic variables with the moments over the other ones under this condition are evident. Another kind of correlations appears if one considers large values of ion microfields, which are produced by the nearest particle (so called "nearest neighbor approximation") [5–9,12,21,23–25,48,67]. In this case there is a direct proportionality between the value of the ion electric field and its time derivative, where the stochasticity is involved due to another stochastic variable—particle velocity. Indeed, the mean square value of particle velocities is a necessary factor in the second moments over the microfield time derivatives [5–9,12,21,23–25,48,67].

Consider now the components of the microfield time derivative that are perpendicular and parallel to the direction of the ion microfield strength vector [5–9,12,21,23–25,48,67]. By calculating the second moments of the perpendicular and parallel components of time derivative, it is possible to establish relations between them in the limits of small and large reduced microfields values $\beta = F / F_0$ (F_0 is the Holtsmark normal field value [1,5–9,12,21,23–25,48,67] and F is the current microfield value), assuming for simplicity (but without loss of generality) that ions produce the Coulomb electric field. In the case of $\beta \ll 1$, the ion microfield is formed by many distant ions and due to isotropy the following relation takes place [5,6,24,25]:

$$2 \langle \dot{\vec{F}}_{\parallel}^2 \rangle_F \sim \langle \dot{\vec{F}}_{\perp}^2 \rangle_F = \langle \dot{\vec{F}}_x^2 \rangle_F + \langle \dot{\vec{F}}_y^2 \rangle_F \tag{2}$$

On the other hand, in the case of $\beta \gg 1$, the ion microfield is formed by the nearest neighbor and the corresponding relation shows preferential direction of the microfield fluctuations along the microfield vector [5,6,24,25]:

$$\langle \dot{\vec{F}}_{\parallel}^2 \rangle_F \sim 2 \langle \dot{\vec{F}}_{\perp}^2 \rangle_F = 2 \cdot \left(\langle \dot{\vec{F}}_x^2 \rangle_F + \langle \dot{\vec{F}}_y^2 \rangle_F \right) \tag{3}$$

The general expression for $\langle \dot{\vec{F}}_{\perp}^2 \rangle_F$ in the case of a Coulomb electric field of point charges is

$$
\begin{aligned}
H(\beta) &\cdot \frac{\langle \dot{\vec{F}}_{\perp}^2 \rangle_F}{F_0^2} \\
&= q \left\{ \left[\frac{N^{2/3} \langle Z^{1/2} v_i^2 \rangle}{(\langle Z^{3/2} \rangle)^{1/3}} + \frac{N^{2/3} \langle Z^{1/2} \rangle \cdot \langle v_a^2 \rangle}{(\langle Z^{3/2} \rangle)^{1/3}} \right] \cdot \beta^{1/2} \cdot [G(\beta) - \tilde{I}(\beta)] \right. \\
&\quad \left. + p \cdot (N^{2/3} \langle v_a^2 \rangle) \left(\frac{1}{3} H(\beta) + \frac{K(\beta)}{\beta} \right) \right\}, \quad F_0 = e \cdot \kappa \cdot (\langle Z^{3/2} \rangle)^{2/3} \cdot N^{2/3}
\end{aligned}
\tag{4}
$$

$$q = \frac{45}{8} \cdot \kappa, \quad \kappa = 2\pi \left(\frac{4}{15} \right)^{2/3}, \quad p = \frac{5}{12\pi} \cdot \frac{(\langle Z \rangle)^2}{(\langle Z^{3/2} \rangle)^{4/3}}$$

Here e is the elementary charge and $\langle v_a^2 \rangle$ designates the mean of the radiator velocity square. The complex ionization composition is accounted for in Equation (4) (note the generalization of the F_0 definition). The definitions of the mean values for composition of various ion species "s" with the charges Z_s and thermal velocity $v_{i,s}$ are given by the following relations:

$$
N = \sum_s N_s, < Z^{3/2} > = \frac{1}{N} \cdot \sum_s N_s \cdot Z_s^{3/2}, \quad < Z^{1/2} v_i^2 > = \frac{1}{N} \cdot \sum_s N_s \cdot Z_s^{1/2} \cdot < v_{i,s}^2 >
$$
$$
< Z^{1/2} > = \frac{1}{N} \cdot \sum_s N_s \cdot Z_s^{1/2}, \quad < Z^2 > = \frac{1}{N} \cdot \sum_s N_s \cdot Z_s^2
\tag{5}
$$

Similarly to Equation (4), an expression for the parallel component $\langle \dot{\vec{F}}_{\parallel}^2 \rangle_F$ is

$$H(\beta) \frac{\langle \dot{\vec{F}}_{\parallel}^2 \rangle_F}{F_0^2} \tag{6}$$

$$= q \left\{ \left[\frac{N^{2/3} \langle Z^{1/2} v_i^2 \rangle}{(\langle Z^{3/2} \rangle)^{1/3}} + \frac{N^{2/3} \langle Z^{1/2} \rangle \cdot \langle v_a^2 \rangle}{(\langle Z^{3/2} \rangle)^{1/3}} \right] \cdot \beta^{1/2} \cdot \tilde{I}(\beta) \right.$$
$$\left. + p \cdot \left(N^{2/3} \langle v_a^2 \rangle \right) \cdot \left(\frac{2}{3} H(\beta) - \frac{K(\beta)}{\beta} \right) \right\}$$

In Equations (4) and (6), $H(\beta)$ is the Holtsmark function (1). The other functions are related to it through integral or differential equations (see [5,6])

$$G(\beta) = \frac{2}{\pi} \int_0^\infty dy \cdot y^{-1/2} \sin y \cdot \exp\left[-\left(\frac{y}{\beta} \right)^{3/2} \right],$$

$$\tilde{I}(\beta) = \frac{2}{\pi} \int_0^\infty dy \cdot y^{-5/2} \left(\sin y - y \cos y \right) \cdot \exp\left[-\left(\frac{y}{\beta} \right)^{3/2} \right],$$

$$H(\beta) = \frac{2}{\pi\beta} \int_0^\infty dy \cdot y \cdot \sin y \cdot \exp\left[-\left(\frac{y}{\beta} \right)^{3/2} \right],$$

$$K(\beta) = \int_0^\beta dy \cdot H(y).$$

(7)

The properties of asymptotics for these functions are given by the relations [3]

$$\beta \ll 1 \begin{cases} G(\beta) \sim \dfrac{4}{3\pi} \beta^{3/2}, \\[2mm] \tilde{I}(\beta) \sim \dfrac{4}{9\pi} \beta^{3/2}, \\[2mm] K(\beta) \sim \dfrac{4}{9\pi} \beta^3, \\[2mm] H(\beta) \sim \dfrac{4}{3\pi} \beta^2. \end{cases} \qquad , \qquad \beta \gg 1 \begin{cases} G(\beta) \sim \sqrt{\dfrac{2}{\pi}}, \\[2mm] \tilde{I}(\beta) \sim \dfrac{2}{3} \sqrt{\dfrac{2}{\pi}}, \\[2mm] K(\beta) \sim 1, \\[2mm] H(\beta) \sim \dfrac{15}{8\pi} \sqrt{\dfrac{2}{\pi}} \cdot \beta^{-5/2}. \end{cases}$$

(8)

The results of [5,6] for joint distribution functions and their moments for the Coulomb potential were generalized within the Baranger–Mozer scheme, accounting for the electron Debye screening and the ion-ion correlations [38,48,67]. The quantities $\langle \vec{F}_\parallel^2 \rangle_F$ and $\langle \vec{F}_\perp^2 \rangle_F$ characterize the statistical properties of plasma microfield and play a key role within the idea of thermal corrections to Stark profiles [7–9,12,24,25]. The terms proportional to the mean of radiator velocity square in Equations (4) and (6), in square brackets, describe the *fluctuations of microfield, induced by the relative thermal motion of the radiator atoms*. The other terms, *proportional to the mean of radiator velocity square in Equations (4) and (6) besides the factor p are due to the effects of ion-dynamics friction on the radiator motion*. As one can see, the latter terms could not be made proportional to the reduced mass of the ion-radiator pair. However, the cases when the influence of the effects of ion-dynamics friction on Stark profiles is significant have not been revealed up to now, since the corresponding deviations of profiles turn out to be small. And indeed, the full scale Molecular Dynamics (MD) simulations confirmed that the effects of ion dynamics could be well described by a so-called "reduced mass" model (RM), where the motion of radiator is neglected for moderately coupled plasma with the ion coupling parameter $\Gamma_i \leq 1$ [34,61,75]. This greatly facilitates the study of ion dynamics in simulations, since the consistent consideration of radiator motion effects in MD is quite time-consuming. The expressions for the fluctuation rates, Equations (4) and (6), also show that for plasmas with complex ion composition there could be some deviations from the RM model, caused by peculiar distributions of ion charges. As the main precision experiments have, up to now, been conducted for simple charge distributions, the expressions in Equation (5) and Equation (6) could be greatly simplified and the terms, corresponding to the ion friction, could be omitted. Hence the charge distribution is neglected below.

By analyzing the difference between profiles formed for two different reduced masses of the plasma ions, it is seen that within the idea of thermal corrections they are proportional to the second moments of parallel $\langle \dot{\vec{F}}_\parallel^2 \rangle_F$ and perpendicular $\langle \dot{\vec{F}}_\perp^2 \rangle_F$ components of the ion microfield fluctuations. The general analysis, performed in different approximations [7–9,12,24,25], has shown that the ion dynamical perturbations of Stark profiles are controlled by three main mechanisms. The first mechanism is due to the amplitude modulation, induced by the rotation of the atomic dipole along with the rotating ion microfield [12]. Due to the amplitude modulation, the projections of the atomic dipole on the coordinate axis, being at rest, are changing (or "modulated"), while the atomic dipole rotates together with the rotation of the electric microfield strength vector. The second mechanism exists due to the atomic dipole inertia with respect to the microfield rotation, and results in nonadiabatic transitions between states defined in the frame with the quantization axis along the rotating field direction [12]. The third mechanism (historically considered the first) is the phase modulation related to changes in the microfield magnitude [7–9,12,24,25]. Only this mechanism was taken into account in the earlier works (in the 1950s) [7–9], where the Stark broadening by ions was analyzed in the adiabatic approximation, i.e., only within the framework of phase modulation or frequency Stark shift [21]. As it was demonstrated in the works [24,25] within the approach of thermal corrections, the amplitude modulation gives the largest contribution to the Stark contour deformation due to ion dynamics in the vicinity of the line center. The general ideas of amplitude modulation, non-adiabatic effects and usage of the electron broadening for extending the theory of thermal corrections to the line center were proposed by Gennadii V. Sholin.

Figure 1. Function $M_\perp(x)$.

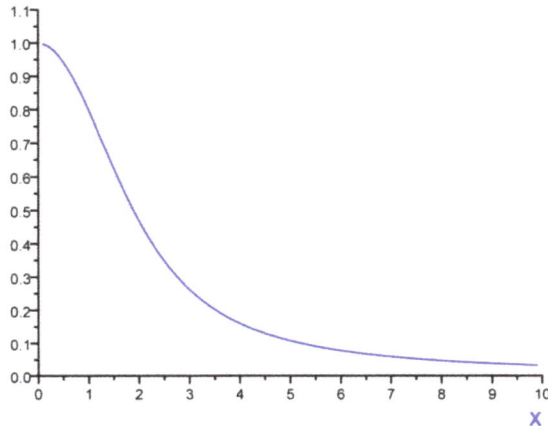

48

Recall that the thermal corrections are defined as a difference between the total profile, calculated accounting for perturbations due to ion dynamics, and the Stark profile within the ST approach [24,25]. In the case of equal temperatures of plasma ions and radiators, as well as electrons, this gives for Ly-alpha (compare with [24,25]) in the approximation of isolated individual Stark components (*i.e.*, neglecting non-diagonal elements of the electron impact broadening operator)

$$\Delta I^{(th)}_{Ly-\alpha}(\Delta\omega) = \frac{10\kappa}{\pi} \cdot \frac{(T_i/\mu)N^{2/3}}{\gamma^2} \cdot \frac{1}{CF_0}\left[\frac{CF_0}{\gamma}f_{1-\alpha}\left(\frac{\Delta\omega}{\gamma}\right) + M_\perp\left(\frac{\gamma}{CF_0}\sqrt{\left(\frac{\Delta\omega}{\gamma}\right)^2+4}\right)\cdot f_{2-\alpha}\left(\frac{\Delta\omega}{\gamma}\right)\right] \quad (9)$$

The $f_{1-\alpha}(x)$ function describes the central component contribution to the Stark profile due to amplitude modulation, while the $f_{2-\alpha}(x)$ function describes the contribution of lateral components to the Stark profile, related to the combined action of the amplitude modulation and non-adiabatic effects. Below, the explicit expressions for $f_{1-\alpha}(x), f_{2-\alpha}(x)$ are given ($\Gamma(z)$ is the gamma function):

$$f_{1-\alpha}(x) = \frac{\Gamma(1/3)}{8}\cdot\frac{3x^2-1}{(x^2+1)^3}, \quad f_{2-\alpha}(x) = \frac{1}{12}\left[2\frac{x^2-1}{(x^2+1)^2}+\frac{(x^2-2)}{(x^2+1)(x^2+4)}\right],$$

$$\int_0^\infty M_\perp(\beta)\cdot d\beta = \frac{9\pi}{8}\cdot\int_0^\infty \frac{(G(\beta)-\tilde{I}(\beta))}{\beta^{3/2}}\cdot d\beta = \frac{9\pi}{8}\left(\frac{1}{3}\cdot\Gamma\left(\frac{1}{3}\right)\right) = \frac{3\pi}{8}\cdot\Gamma\left(\frac{1}{3}\right),$$ (10)

$$M_\perp\left(x\cdot\frac{\gamma}{CF_0}\right)\propto\cdot H(\beta)\cdot\frac{<\dot{\vec{F}}_\perp^2>_F(\beta)}{\beta^2F_0^2}\bigg|_{\beta=x\frac{\gamma}{CF_0}}, \quad x\frac{\gamma}{CF_0}\sim x\cdot h_e^{1/3},\quad h_e = N\left(\frac{eC}{v_e}\right)^3 \ll 1$$

In (9) and (10) the $M_\perp\left(\frac{\gamma}{CF_0}\cdot\sqrt{x^2+4}\right)$ dimensionless function is introduced (see Figure 1),

proportional to the second moment of the microfield time derivative component, perpendicular to the microfield direction according to Equation (4) and defining in fact the mean square of microfield rotation frequency. It is defined in such a way, that $M_\perp(0)=1$, while the corresponding constant in the limit of $\beta \ll 1$ was included in the definition of the $f_{2-\alpha}(x)$ function. The behavior of $f_{1-\alpha}(x)$ and $f_{2-\alpha}(x)$ is presented in the Figures 2 and 3, respectively. The parameter γ in Equations (9) and (10) is the impact electron width of the central component in the parabolic basis. As it follows from the results of [24,25], the corrections due to the ion dynamics effects are negative in the center of a line with the central components, corresponding to decreasing of the intensity in the line center due to the ion dynamics effects and its increasing in the shoulders (the transient region between approximately the half width and the nearest line wings). As the thermal corrections have perturbative character, functions $f_{1-\alpha}(x)$ and $f_{2-\alpha}(x)$ have zero integrals. So, due to ion dynamics effects, the intensity is redistributed from the line center, increasing the total width of the lines with central Stark components [24,25]. These general features are confirmed below in the next Section 4 using MD simulations (see also [75]), that are believed to be not limited by applicability conditions of the perturbation approach [24,25]. It should be noted that within the approach of [24,25], *an exact analytical expression due to the ion-dynamics corrections for Ly-alpha, accounting for the collapse of the lateral Stark components* [13–15], was also derived [24]. It has a rather complex structure and is not presented here, but the comparison of its functional behavior with the approximation of isolated individual Stark components $f_{2-a}(x)$ is shown in Figure 3.

Figure 2. Function $f_{1-a}(x)$.

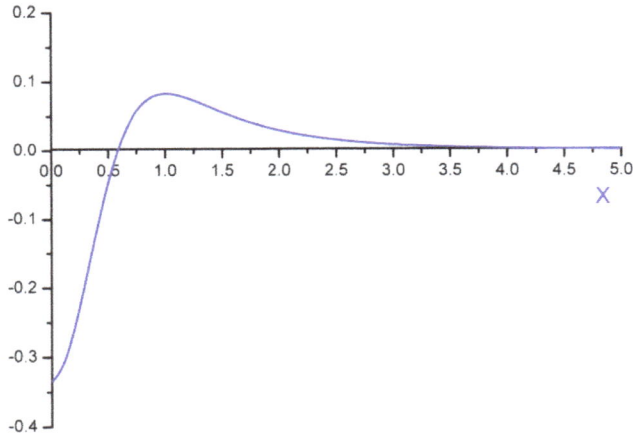

Figure 3. Function $f_{2-a}(x)$—dashed line. The solid line is the behavior of $f_{2-a}(x; \varepsilon = 1)$, obtained in numerical calculations accounting for the collapse effect [24].

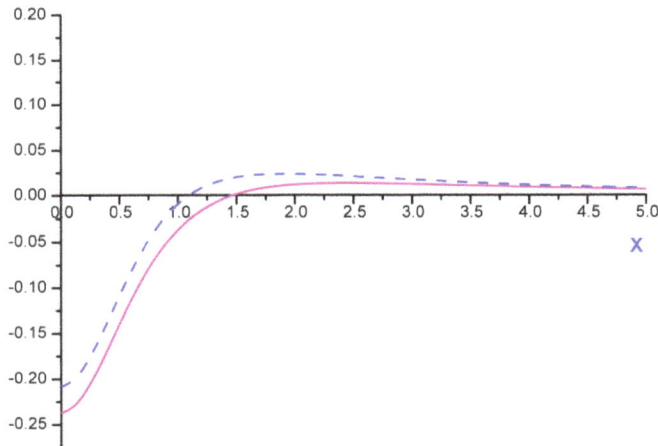

It is interesting to note several properties of the above-mentioned complex function $f_{2-\alpha}(x; \varepsilon)$, which takes into account the collapse effect of lateral Stark components (where ε is the ratio of the non-diagonal matrix element of the electron impact broadening operator to the electron impact width of the central component of Ly-alpha). Neglecting dependence of the non-diagonal matrix element of the electron impact broadening operator on the value of the ion microfield F according to [17,18], which mainly is important for large F for transition from the overlapping to isolated broadening regime of the Stark components (since $\varepsilon(\beta) \to 0$ for $\beta \to \infty$), corresponds to $\varepsilon = 1$. The ratio of the lateral component electron impact width to the central component electron impact width is equal to 2 for Ly-alpha in parabolic basis [17,18]. This is reflected in the argument of $M_\perp\left(\dfrac{\gamma}{CF_0}\cdot\sqrt{x^2+4}\right)$,

whose value is taken in the pole of resolvent $\beta=\dfrac{1}{CF_0}(\Delta\omega+i2\gamma)$, corresponding to the one lateral

component. So, remembering that the central component is more intense than the lateral ones, its strong influence on the Ly-alpha Stark shape becomes obvious. Comparing $f_{2-\alpha}(x;\varepsilon=0)$ with $f_{2-\alpha}(x;\varepsilon=1)$ at $x=0$, their ratio comes out to be about 1.26 [24]. At first glance, putting $\varepsilon=0$ in $f_{2-\alpha}(x;\varepsilon)$ allows obtaining the limit of isolated Stark components, but it turns out that $f_{2-\alpha}(x;\varepsilon=0)\neq f_{2-\alpha}(x)$. This means that there is no commutativity in the sequence of performed mathematical operations, since the $f_{2-\alpha}(x;\varepsilon)$ function is obtained in the solution of secular problem and inverting the resolvent. It is seen in Figure 3, that the difference between the approximation of isolated components ($f_{2-\alpha}(x)$) and the exact solution accounting for the collapse effect ($f_{2-\alpha}(x;\varepsilon=1)$) at $x=0$ is noticeably smaller than that between $f_{2-\alpha}(0;\varepsilon=1)$ and $f_{2-\alpha}(0;\varepsilon=0)$, as their ratio $f_{2-\alpha}(0;\varepsilon=1)/f_{2-\alpha}(0)$ is only about ~1.14. Thus this comparison demonstrates an acceptable accuracy of the approximation of isolated individual Stark components [18,19] for calculations of the thermal corrections to the Ly-alpha Stark profile.

Figure 4. The function $f_{2-\beta}(x)$.

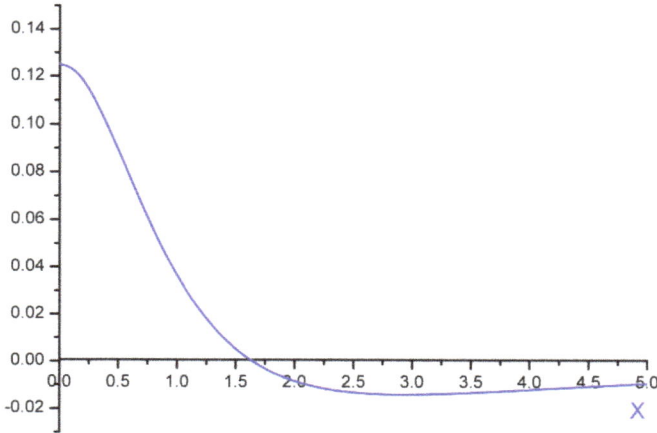

Similar to Equations (9) and (10), result for Ly-beta is [24,25].

$$\Delta I^{(th)}_{Ly-\beta}(\Delta\omega)=\dfrac{10\kappa}{\pi}\cdot\dfrac{(T_i/\mu)N^{2/3}}{w^2}\cdot\dfrac{1}{C_1F_0}\cdot f_{2-\beta}\left(\dfrac{\Delta\omega}{w}\right),\quad C_1=9ea_0/\hbar \qquad (11)$$

where $f_{2-\beta}(x)$ describes the lateral components contribution to the Stark profile, comprising to the action of only non-adiabatic effects in the case of lines without the central component [24,25],

$$f_{2-\beta}(x) = \frac{1}{6}\left[\frac{4-x^2}{(x^2+4)^2} + \frac{2-x^2}{(x^2+4)(x^2+1)}\right] \tag{12}$$

and "w" is the electron impact width of the Stark sublevel (002), designated by parabolic quantum numbers. The graph of $f_{2-\beta}(x)$ is presented in the Figure 4. The case of Lyman-beta illustrates that the ion dynamics effect increases the intensity in the center of the line without the central components and slightly decreases its width due to the lowering intensity in the nearest wings that is clearly seen in Figure 4.

Expressions (9) and (12) are derived assuming the concrete numerical values of the Stark shifts and dipole matrix elements, calculated in the parabolic basis for corresponding Stark sublevels and components of considered transitions [17,18]. The above results are obtained analytically by the perturbation theory for non-Hermitian operators and with the analytical continuation of the microfield distribution function and the second moments of its time derivatives that were shown to possess analytical properties in the upper complex plane (see [24,25]). As it follows from the validity conditions of this approach, the main contribution to integrals, describing the amplitude modulation of lines with central components, give regions of detunings near the line center, where the argument of the universal functions, describing the second moments of derivatives, is small. Moreover, the principal term of the expansion, corresponding to the amplitude modulation of the central component, does not depend on the microfield, that allows for integrating over the microfield distribution analytically [24,25]. On the other hand, the principal terms of the expansion for the lateral components related to the amplitude modulation and non-adiabatic effects, are obtained analytically via integration in the upper half of the complex plane [24,25]. As it follows from the asymptotic properties of these functions in the region of small values of argument, the effective frequency of rotations is practically constant. Due to this, in [24,25] the value of the dimensionless function $M_\perp(z)$ in Equations (9) and (11) was substituted for small values of argument near zero, corresponding to the line center. Moreover, as $M_\perp(z)$ varies very slowly on the characteristic frequency scale in the line center, in [24,25] its variation in the numerical results of Equations (9) and

(11) was neglected. It allowed to neglect the difference of values of $M_\perp\left(\dfrac{w_{\alpha\beta}}{C_{\alpha\beta}F_0}\cdot\sqrt{\left(\dfrac{\Delta\omega}{w_{\alpha\beta}}\right)^2+1}\right)$

standing beside the terms, obtained due to residue in the various poles, and equate them in fact to the common constant due to the smallness of argument. Then summation of terms due to perturbation expansion leads to more simple formulas, which are expressed as the functions f, and finalized by introduction of some general scales for the Stark constants C and impact widths w. Moreover, the significant simplification of the result also is due to the constant output of the $M_\perp(z)$ function in the region near the line center, where its argument is small. Then all the derivatives of $M_\perp(z)$ turn out to be zero (or could be considered as higher order terms of expansion), thus one is left only with the derivatives of the dispersion functions in perturbation series (see [24,25]). *However, from the principal point of view, it is important that functional behavior of $M_\perp(z)$ is proportional to the* *fluctuation of microfield component perpendicular to the microfield direction in Equations (4) and (10)*. It could be kept in the final result which would then look more cumbersome in this case than expressions (11) and (12). Indeed, the result would contain the sum of contributions of each Stark component, determined by its values of the electron impact widths and Stark constants, being multiplied by $M_\perp(z)$ from the different arguments, as explained just above (see explicit formulas for perturbation expansion, presented in [24,25]).

In the case of a line without the central component, the ion dynamics corrections are positive in the line center (see Figure 4). Thus, the intensity in the center increases, while it is decreases in the nearest wings (see Figure 4). The results of [24,25] qualitatively confirm the experimental patterns, observed in [20,22,27,31]. Also in [24,25] the difference profiles, corresponding to two different values of the reduced mass, were considered and compared with the results of experiments [20,22] for the Balmer-beta line. It was shown that relative behavior of the thermal corrections *versus* wavelength detuning from the line center $\Delta\lambda$ describes sufficiently accurately [24,25] the experimental results [20,22]. In [24,25] the difference profiles $\delta_R(\Delta\lambda)$, corresponding to the two different values of the reduced mass, were also considered. Within the notion of thermal corrections this difference is

$$\delta_R(\Delta\lambda) = \frac{1}{I_{max}^{(0)}}\left(I_{\mu_1}^{(th)}(\Delta\lambda) - I_{\mu_2}^{(th)}(\Delta\lambda)\right) \propto \frac{T_i T_e}{N^{4/3}} \cdot \frac{\mu_1 - \mu_2}{\mu_1 \mu_2} \cdot f\left(\frac{\Delta\lambda}{\Delta\lambda_0}\right) \tag{13}$$

In Equation (13) $f(x)$ is the relative behavior of the difference profile in the line center. In [24,25] it was shown, that the dependences *versus* frequency detunings from the line center, spanned by the profile difference (10), coincide well with the corresponding experimental data for the Balmer-beta line, given in [20,22].

So, according to the results and ideas of [24,25] and the discussion above, the difference profiles are proportional to the statistical characteristics of microfield and, more precisely, to the characteristics of the microfield fluctuations, related to the microfield rotations (compare with [24,25]). That is why this property could in principal be used to study microfield statistics in experiments and simulations.

In short time the ideas of [12,24,25] were accepted and the notion of dominating effect of microfield rotation into ion dynamics became widely spread [30,33].

Nowadays, the computer simulation technique has become a powerful tool for studying the physics of various non-stationary processes, and particularly plasma microfield ion dynamics effects. However, computer simulations are rather time-consuming and at present impractical for large-scale calculations. Thus, simultaneously with the computer simulations the development of model approaches, that accounted for the ion dynamical effects in an approximate manner, were carried on independently [32–37,39,41,46,49,53,54,56,59–64,68–76]. The first such model (actually, predating computer simulations) was MMM [14,15,28,29,52], later followed by various applications of the BID [37,71] and Frequency Fluctuation Model (FFM) [46,56,66] methods. Notably, neither of these models explicitly accounts for the effects of microfield rotation [24,25,30,33,53].

It is necessary now to consider the formal conditions of validity of thermal corrections approach [7–9,12,24,25]. The results of reference [12] are applicable only for the lateral Stark components and quasistatic ions, when

$$h_i = N\left(\frac{eC}{v_i}\right)^3 \gg 1 \qquad (14)$$

and in the spectral region of detunings from the line center $\Delta\omega$, corresponding to the line wings

$$\left(\frac{\Delta\omega}{CF_0}\right) \gg 1 \qquad (15)$$

On the other hand, the spectral region of applicability of the theoretical approach in [24,25] is expanded till the line center only due to the additional inclusion into consideration, besides the quasistatic ions -Equation (14), of the electron impact effect, that allowed to analyze the central Stark components, too. However, the applicability criteria of [24,25] are rather complicated and depend on the spectral region under consideration. For example, in the line center for the central Stark components the criterion of validity of [24] has the form

$$h_i^{-2/3} h_e^{-1} \Lambda^{-3} \ll 1, \quad h_i \gg 1, \quad h_e \ll 1, \quad \Lambda = \ln(\rho_D/\rho_W) \qquad (16)$$

where ρ_D and ρ_W designate the Debye and Weisskopf radii, respectively (see [21,45,58]). It is seen that condition (16) (and other criteria from [24,25]) is difficult to fulfill, which somewhat limits the practical applicability of the theory. For the line wings, the results of [24,25] reproduce the results of the earlier work [12] under the criterion for the separate Stark components

$$h_i^{-2/3} I^{(th)}(\Delta\omega/CF_0) \ll H(\Delta\omega/CF_0) \qquad (17)$$

where $I^{(th)}(\Delta\omega/CF_0)$ is the thermal correction profile that represents, within the assumptions of [18,19], a sum of contributions from the amplitude modulation, non-adiabatic effects, and phase modulation, and $H(\Delta\omega/CF_0)$ is a microfield distribution function.

The results of [12,24,25] proved the numerical predominance of amplitude modulation and non-adiabatic effects contributions over the phase modulation one in the line wings, where the perturbation approach of [12,24,25] is applicable practically for any plasma parameter. The magnitudes of amplitude modulation and non-adiabatic effects contributions are of the same order in this region of Stark profiles [12]. Moreover, the amplitude modulation and non-adiabatic effects contributions have the same sign, which is opposite to the sign of the phase modulation correction [12]. So, the cancellation of non-adiabatic and atom reorientation effects (amplitude modulation) does not take place, as was proposed earlier by Spitzer in his very instructive papers [2–4], and they play the dominant role in the line wings.

4. Ion Dynamics Modeling and Statistical-Dynamical Coupling

The work during the preparation of SLSP workshops and along with their conduction revealed unexpected spread of results of various computational models, done for *ion perturbers only* (see for example [74–76]).

In this respect, the study of directionality correlations, presented at SLSP-1 [65], inspired the authors to test whether the rotation effects really are responsible for a dominating contribution according to the predictions of [24,25].

To this end the Ly-α profile was calculated using a computer simulation (CS) method [57]. A one-component plasma (OCP) was assumed, consisting only of one type of ions. Furthermore, to avoid effects of plasma non-ideality such as the Debye screening, the ions were assumed moving along straight path trajectories. Time histories of the electric field $\vec{F}(t)$, formed by the ions, were stored, to be used as an input when numerically solving the time-dependent Schrödinger equation of the hydrogen atom.

It is instructive to separately analyze how changing the direction, and the magnitude of the microfield influence the line shape. Let us define "rotational" and "vibrational" microfields as

$$\vec{F}_{rot}(t) = F_0 \frac{\vec{F}(t)}{F(t)}, \quad F(t) = \left|\vec{F}(t)\right|, \tag{18}$$

and

$$\vec{F}_{vib}(t) = \vec{n}_z F(t), \tag{19}$$

respectively (compare with [75]), where F_0 is again the Holtsmark normal field for singly charged ions [1].

The effect of varying reduced mass was modeled by enabling time dilation of the field histories. Evidently, the field that is changing slower by a factor of s, corresponds to that, formed by particles moving s times more slowly, i.e., with an s^2-times larger reduced mass. We note that by reusing the field histories generated only once, any possible inaccuracy due to a finite statistical quality of the simulations, such as a deviation from the Holtsmark distribution of the field magnitudes [1], should be present in all calculations and thus, cancel out when the difference profiles are evaluated. The parameters of the base run ($s = 1$) were selected to correspond to protons (i.e., $\mu_0 = 0.5$) with the particle density $N = 10^{17}$ cm^3 and temperature $T = 1$ eV, while the additional runs with $s = 2, 4$, and 8 corresponded to $\mu = 2, 8$, and 32, respectively.

The resulting Ly-α profiles are presented in Figure 5. It is seen that the rotational microfield component has a significantly more pronounced effect on the total line shapes, while changing only the magnitude of the field while keeping its direction constant (the "vibrational" component) has only a minor effect on the shape of the lateral components. Evidently, with no change in the field direction, the central component remains the δ-function (not shown on Figure 5b). We note that the width of the central component due to the rotational microfield component increases when μ decreases. This is in a qualitative agreement with Equation (9).

The Ly-α profiles calculated with the full microfields (Figure 6a) show a resembling behavior: the line HWHM, mostly determined by the central component, scales approximately inversely with s within the range of parameters assumed, while the shape of the lateral components remains mostly unchanged. We note that varying s may alternatively be interpreted as scaling the temperature according to $T = T_0/s^2$. The observed dependence is, thus, qualitatively similar to the Ly-α T-dependence inferred in an ion-dynamics study [76]. We note that the shape of the central component is practically Lorentzian.

We now turn to analyzing the difference profiles defined in the spirit of Equation (13). However, the theory of thermal corrections [24,25] was derived perturbatively with the zero-order broadening due to the electron impact effect, while in the present CS calculations no electron broadening was included.

Therefore, it is the ion-dynamical broadening itself that fulfills this role, and one should expect to find a self-similar solution. We checked it by calculating difference between the line shapes calculated with the time-dilation factors s and $s' = s + \delta s$, keeping the $\delta s/s$ ratio constant, and normalizing the frequency axis of the resulting difference profiles to the line width. In other words, the ion-induced HWHM w_i plays the role of the electron impact one in Equation (9).

Figure 5. CS Ly-α profiles, broadened by an OCP, assuming $N = 10^{17}$ cm^{-3} and $T = 1$ eV. The ion radiator reduced mass $\mu = s^2 \mu_0$, where $\mu_0 = 0.5$. a) Line shapes influenced by the rotational field component (18); b) Line shapes influenced by the vibrational field component (19).

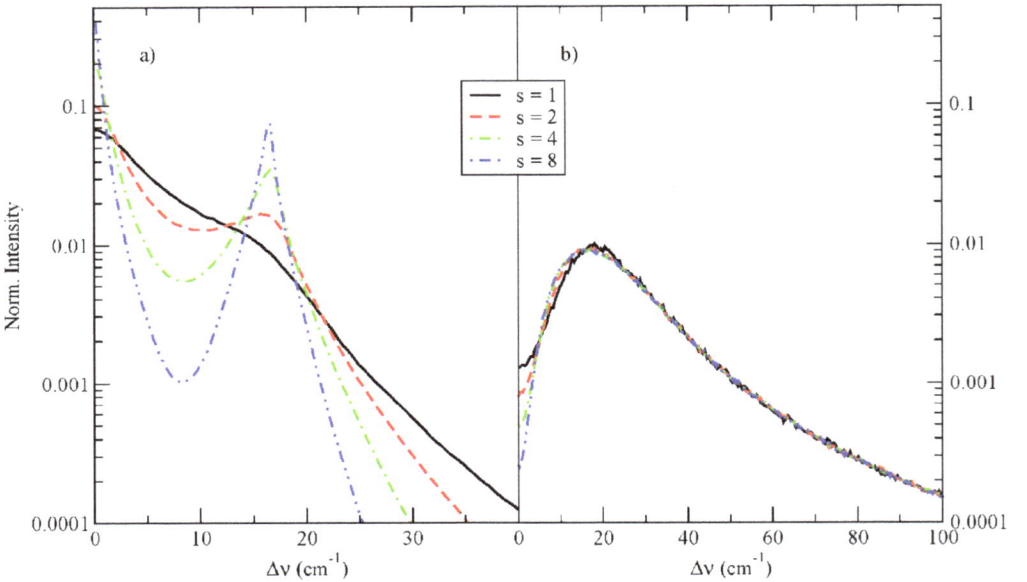

Although it is desirable to keep $\delta s/s$ as "infinitesimal" as possible, in practice too small a ratio results in rather noisy profiles due to a finite accuracy of the simulations; for this reason, $\delta s/s = 1/4$, corresponding to $\delta\mu/\mu = 9/16$, was used. The results are shown in Figure 6b. Indeed, the normalized difference profiles remain practically the same over the 64-fold variation of μ tested. Furthermore, the profiles in the central region are qualitatively similar to the prediction of the theory of thermal corrections [24,25] (cf. Figure 2). It appears, however, that the functional form is rather close to

$$\propto \frac{x^2 - 1}{\left(x^2 + 1\right)^2} \tag{20}$$

also shown in Figure 6b. It is easy to see that such a functional form corresponds to a difference between two Lorentzians, confirming the shape of the Ly-α central component inferred from our CS calculations.

Figure 6. a) CS full Ly-α profiles, broadened by an OCP, assuming $N = 10^{17}$ cm^{-3} and $T = 1$ eV. The ion radiator reduced mass $\mu = s^2\mu_0$, where $\mu_0 = 0.5$; b) Profiles differences between line shapes, calculated with s and $s' = s + \delta s = 5/4s$ (*i.e.*, $\delta\mu/\mu = 9/16$). The profile differences are scaled to the lineshape HWHM w_i.

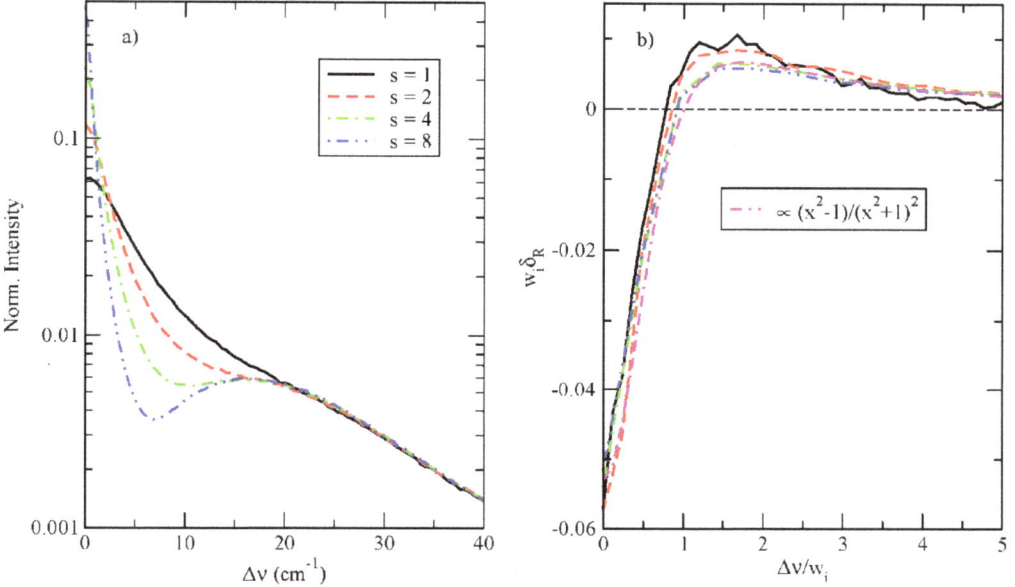

5. Discussion

Let us discuss the separation of rotational and vibrational (phase modulation) effects of ion dynamics (used also in [75]). It is assumed that the mean magnitude of the field is equal to the normal Holtsmark field value. On the other hand, the solution for a fixed angular velocity and a fixed magnitude of the electric field for hydrogen is known exactly [10,13]. Furthermore, the solution of the Schrödinger equation for this problem and hence the profile strongly depends on the microfield value, which is set equal to F_0. These profiles are characterized by well distinguished properties that are consequences of atom dynamics in the rotating microfield [10,13,75]. *When the profile patterns are averaged over all microfield directional histories, the fixed value of the microfield leads to a statistical-dynamical coupling through specifics of solutions of the Schrödinger equation, i.e., coupling between the microfield statistics and specific dynamics of the atomic system* [10,13]. This is illustrated by the instructive detailed patterns presented in [75] for the rotational contribution of ion dynamics effects. Let us now consider the proposed separation of phase modulation or vibrational effects. Here only the microfield orientation is assumed to be constant, while its magnitude as a function of time is preserved. The solution of the Schrödinger equation in this case reflects the specifics of a fixed orientation, and after averaging over microfield histories this imposes very characteristic features of the Stark profiles [7,8,75]. *These profiles are also subject to statistical-dynamical coupling caused only by the fixed microfield orientation.*

The present CS results show that the convolution of the separated rotational and vibrational profiles does not equal to the one, obtained by using the full microfield histories. This is due to constrains, involved in the separation of the rotational and vibrational effects, and is clearly a consequence of the statistical-dynamical coupling. In fact, it has the same origin as the coupling between the fixed microfield values and the means of their derivatives [5,6], discussed in the beginning of Section 3.

The analysis performed in the previous section has confirmed the formation of the central part of the hydrogen line with the central component, predominantly broadened under the action of the amplitude modulation. While in the quasistatic region of the Stark broadening the natural scaling is defined by $CF_0 \sim n^2 N_i^{2/3}$, and in the impact regime it is proportional to $\sim n^4 (T_i/\mu)^{-1/2} N_i$, from the dimension consideration it follows that in the regime of Stark broadening controlled by the ion dynamics the characteristic scale w_i could be just proportional to $(T_i/\mu)^{1/2} N_i^{1/3}$ as there could be no dependence on the dipole atomic moment or microfield. On the right hand side in Figure 6b the difference profiles for the same set of artificial values of reduced mass are plotted. Their qualitative behavior is similar to the ones discussed in the Section 2 functions $f_{1-\alpha}(x)$, $f_{2-\alpha}(x)$. From the performed analysis in the previous section it could be deduced that the characteristic scale of HWHM is proportional to $(2T_i/\mu)^{1/2}$, and at the same time the analysis, given in [67], has shown that in this range of parameters ($N_e \sim 10^{17}$ cm^{-3}, $T \sim 1$ eV) HWHM is proportional to $N_i^{1/3}$. Combining these two dependences lead to the conclusion that, *for the chosen plasma parameters*, the HWHM needs to be proportional to the typical ion microfield frequency: $w_i \sim (2T_i/\mu)^{1/2} N_i^{1/3}$.

These properties discovered in the process of CS could be in fact treated in a quite simple manner. Indeed, the main contribution to the broadening is due to the central component. The existence of the field and its orientation define those Stark sublevels that give rise to the central Stark component, but the microfield does not affect those states, and they do not depend on the microfield value since they do not possess a dipole moment. The microfield rotations change the quantization of quantum states, which can be considered as their *decay* or determination of their *life time*. Earlier, the formal model of this type with the decay rate, depending on the microfield value and based on equations like Equation (4), was suggested in [17], but was not thoroughly studied. From this idea it follows that any interaction should lead to a finite life time of the system. This hypothesis is supported by the observation of a nearly Lorentzian profile of the central component in the simulations with $w_i \propto v_i N_i^{1/3}$.

6. Conclusions

The existence of spectral-kinetic coupling in the formation of Stark profiles is stated. It arises since the spectral profiles and balance equations could not, in general, be considered separately. A consistent approach should include balance equations as well as spectral line profiles in one system of equations for density matrix.

MD simulations qualitatively confirm the results, obtained within the notion of thermal corrections, namely that the formation of the Stark profile center is mainly due to the microfield rotation, while the wings are affected by the phase modulation. Here, it is worth mentioning that *in the line wings, where the theory of thermal corrections is practically always valid, the ion dynamics contributions of the amplitude modulation and the non-adiabatic effects have the same sign and significantly exceed numerically the contribution of the phase modulation, which has the opposite sign.*

The existence of the statistical-dynamical coupling between the plasma microfield statistics and the dynamics of atomic system of radiators, applied to the averaging of the dynamic solution over samples of histories of microfield evolution in plasma (which is a special case of time dependence in quantum mechanics induced by the environment [78]), may be the cause that prevents the convolution of separate contributions to the Stark profile from the rotational and vibrational effects to be equal to the Stark profile, obtained under the full microfield evolution.

The difference profiles, obtained by a subtraction of experimental or simulated profiles, corresponding to two different reduced masses of perturber-radiator pair, could be used as a tool for studying the statistical properties of microfields.

It is pointed out that results of MD simulations of ion dynamics could be treated by a hypothetical model of the quantum states decay caused by the changes of quantization axes due to the microfield rotations. This may allow for explaining the success of a variety of models that consider neither the microfield rotation, nor the detailed evolution of the microfields.

We hope that this work will inspire further studies of the ion dynamics effects on Stark profiles.

Acknowledgments

We wish to thank A. Calisti, S. Ferri, M.A. Gigosos, M.A. Gonzalez, C.A. Iglesias and V.S. Lisitsa for many fruitful discussions on the subject.

The work of A.V.D. was partially supported by the Russian Foundation for Basic Research (project No. 13-02-00812) and by the Council of the President of the Russian Federation for Support of Young Scientists and Leading Scientific Schools (project No. NSh-3328.2014.2). The work of E.S. was partially supported by Israel Science Foundation and the Cornell University Excellence Center.

A.V.D. greatly appreciates the invitations and support of the Organizing Committee of the SLSP-1&2 workshops and the International Atomic Energy Agency that made possible his participation in these meetings.

Author Contributions

Sections 1–3 were prepared by A.V.D. The computer simulations, described in Section 4, were performed by E.S. Both authors contributed equally to the rest of this work.

Conflicts of Interest

The authors declare no conflict of interest.

59

References

1. Holtsmark, J. Über die Verbreiterung von Spektrallinien. *Ann. Phys. (Leipz.)* **1919**, *58*, 577–630.
2. Spitzer, L. Stark-Effect broadening of hydrogen lines. I Single encounters. *Phys. Rev.* **1939**, *55*, 699–708.
3. Spitzer, L. II. Observable profiles. *Phys. Rev.* **1939**, *56*, 39–47.
4. Spitzer, L. Impact broadening of spectral lines. *Phys. Rev.* **1940**, *58*, 348–357.
5. Chandrasekhar, S.; von Neumann, J. The statistics of gravitational field arising from random distributions of stars. I. The speed of fluctuations. *Astrophys. J.* **1942**, *95*, 489–531.
6. Chandrasekhar, S.; von Neumann, J. II. The speed of fluctuations; dynamical friction; spatial correlations. *Astrophys. J.* **1943**, *97*, 1–27.
7. Kogan, V.I. Broadening of Spectral Lines in Hot Plasma. In *Plasma Physics and the Problem of Controlled Thermonuclear Reactions*; Leontovich, M.A., Ed.; Academy of Science USSR Press: Moscow, Russia, 1958; Volume IV, pp. 259–304.
8. Kogan, V.I. Broadening of Spectral Lines in Hot Plasma. *Plasma Physics and the Problem of Controlled Thermonuclear Reactions*; Leontovich, M.A., Ed.; Pergamon Press: London, UK, 1960; Volume IV; p. 305.
9. Wimmel, H.K. Statistical ion broadening in plasmas. *J. Quant. Spectrosc. Radiat. Transf.* **1960**, *1*, 1–29.
10. Wimmel, H.K. Erratum. *J. Quant. Spectrosc. Radiat. Transf.* **1964**, *4*, 497–499.
11. Ishimura, T. Stark effect of the Lyman alpha line by a rotating electric field. *J. Phys. Soc. Jpn.* **1967**, *23*, 422–429.
12. Kogan, V.I.; Selidovkin, A.D. On fluctuating microfield in system of charged particles. *Beitr. Aus. Plasmaphys.* **1969**, *9*, 199–216.
13. Sholin, G.V.; Lisitsa, V.S.; Kogan, V.I. Amplitude modulation and non-adiabaticity in the Stark broadening of hydrogen lines in a plasma. *Sov. Phys. JETP* **1971**, *32*, 758–765.
14. Lisitsa, V.S. Hydrogen Atom in Rotating Electric Field. *Opt. Spectrosc. USSR* **1971**, *31*, 468.
15. Frisch, U.; Brissaud, A. Theory of Stark broadening—I. Soluble scalar model as a test. *J. Quant. Spectrosc. Radiat. Transf.* **1971**, *11*, 1753–1766
16. Brissaud, A.; Frisch, U. —II. Exact line profile with model microfield. *J. Quant. Spectrosc. Radiat. Transf.* **1971**, *11*, 1767–1783.
17. Strekalov, M.L.; Burshtein, A.I. Collapse of shock-broadened multiplets. *JETP* **1972**, *34*, 53–58.
18. Sholin, G.V.; Demura, A.V.; Lisitsa, V.S. Theory of Stark broadening of hydrogen lines in plasma, *Sov. Phys. J. Exp. Theor. Phys.* **1973**, *37*, 1057–1065
19. Sholin, G.V.; Demura, A.V., Lisitsa, V.S. *Electron. Impact Broadening of Stark Sublevels of Hydrogen Atom in Plasmas*; Preprint IAE-2332, Kurchatov Institute of Atomic Energy: Moscow, Russia, 1972; pp. 1–21.
20. Vidal, C.R.; Cooper, J.; Smith, E.W. Hydrogen Stark-broadening tables. *Astrophys. J. Suppl. Ser.* **1973**, *25*, 37–136.
21. Kelleher, D.E.; Wiese, W.L. Observation of ion motion in hydrogen Stark profiles. *Phys. Rev. Lett.* **1973**, *31*, 1431–1434.

22. Griem, H.R. *Spectral Line Broadening by Plasmas*; Academic Press: New York, NY, USA, 1974.

23. Wiese, W.L.; Kelleher, D.E.; Helbig, V. Variation in Balmer-line Stark profiles with atom-ion reduced mass. *Phys. Rev. A* **1975**, *11*, 1854–1864.

24. Demura, A.V.; Lisitsa, V.S.; Sholin, G.V. On the ion motion effect in Stark profiles of hydrogen lines in a plasma. In Proceedings of the XIIth ICPIG, Eindhoven, The Netherlands, 1975; p. 37.

25. Demura, A.V.; Lisitsa, V.S.; Sholin, G.V. *Theory of Thermal Corrections to Stark Profiles of Hydrogen Spectral Lines*; Preprint IAE-2672; Kurchatov Institute of Atomic Energy: Moscow, Russia, 1976; pp. 1–47.

26. Demura, A.V.; Lisitsa, V.S.; Sholin, G.V. Effect of reduced mass in Stark broadening of hydrogen lines. *Sov. Phys. J. Exp. Theor. Phys.* **1977**, *46*, 209–215.

27. Gruntzmacher, K.; Wende, B. Discrepancies between the Stark broadening theories for hydrogen and measurements of Ly-α Stark profiles in a dense equilibrium plasma. *Phys. Rev. A* **1977**, *16*, 243–246.

28. Seidel, J. Hydrogen Stark broadening by model electronic microfields. *Z. Naturforschung* **1977**, *32*, 1195–1206

29. Seidel, J. Effects of ion motion on hydrogen Stark profiles. *Z. Naturforschung* **1977**, *32*, 1207–1214.

30. Voslamber, D. Effect of emitter-ion dynamics on the line core of Lyman-α. *Phys. Lett. A* **1977**, *61*, 27–29.

31. Gruntzmacher, K.; Wende, B. Stark broadening of the hydrogen resonance line L_β in a dense equilibrium plasma. *Phys. Rev. A* **1978**, *18*, 2140–2149.

32. Stamm, R.; Voslamber, D. On the role of ion dynamics in the Stark broadening of hydrogen lines. *J. Quant. Spectrosc. Radiat. Transf.* **1979**, *22*, 599–609.

33. Voslamber, D.; Stamm, R. Influence of different ion dynamical effects on Lyman lines. In *Spectral Line Shapes Volume 1*; Wende, B., Ed.; Walter de Gruyter & Co.: Berlin, Germany, 1981; pp. 63–72.

34. Seidel, J.; Stamm, R. Effects of radiator motion on plasma-broadened hydrogen Lyman-β. *J. Quant. Spectrosc. Radiat. Transf.* **1982**, *27*, 499–503.

35. Stamm, R.; Talin, B.; Pollock, E.L.; Iglesias, C.A. Ion-dynamics effects on the line shapes of hydrogenic emitters in plasmas. *Phys. Rev. A* **1986**, *34*, 4144–4152.

36. Calisti, A.; Stamm, R.; Talin, B. Effect of the ion microfield fluctuations on the Lyman-α fine-structure doublet of hydrogenic ions in dense plasmas. *Europhys. Lett.* **1987**, *4*, 1003–1008.

37. Boercker, D.B.; Dufty, J.W.; Iglesias, C.A. Radiative and transport properties of ions in strongly coupled plasmas. *Phys. Rev. A* **1987**, *36*, 2254.

38. Demura, A.V. *Theory of Joint Distribution Functions of Ion. Microfield and Its Space and Time Derivatives in Plasma with Complex. Ionization Composition*; Preprint IAE-4632/6. Kurchatov Institute of Atomic Energy: Moscow, Russia, 1988; pp. 1–17.

39. Calisti, A.; Stamm, R.; Talin, B. Simulation calculation of the ion-dynamic effect on overlapping neutral helium lines. *Phys. Rev. A* **1988**, *38*, 4883–4886.

40. Demura, A.V. Microfield Fluctuations in Plasma with Low Frequency Oscillations. In *XIXth ICPIG Contributed Papers*; ed. J.M. Labat; Faculty of Physics, University of Belgrade: Belgrade, 1990; Volume 2, pp. 352–353.

41. Calisti, A.; Khelfaoui, F.; Stamm, R.; Talin, B.; Lee, R.W. Model for the line shapes of complex ions in hot and dense plasmas. *Phys. Rev. A* **1990**, *42*, 5433–5440.

42. Rautian, S.G.; Shalagin, A.M. *Kinetic Problems of Nonlinear Spectroscopy*; North Holland: New York, NY, USA, 1991.

43. Anufrienko, A.V.; Godunov, A.L.; Demura, A.V.; Zemtsov, Y.K.; Lisitsa, V.S.; Starostin, A.N.; Taran, M.D.; Shchipakov, V.A. Nonlinear interference effects in Stark broadening of ion lines in a dense plasma. *Sov. Phys. J. Exp. Theor. Phys* **1990**, *71*, 728–741.

44. Anufrienko, A.V.; Bulyshev, A.E.; Godunov, A.L.; Demura, A.V.; Zemtsov, Y.K.; Lisitsa, V.S.; Starostin, A.N. Nonlinear interference effects and ion dynamics in the kinetic theory of Stark broadening of the spectral lines of multicharged ions in a dense plasma. *JETP* **1993**, *76*, 219–228.

45. Sobelman, I.I.; Vainstein, L.A.; Yukov, E.A. *Excitation of Atoms and Broadening of Spectral Lines*; Springer: Heidelberg, Germany; New York, NY, USA, 1995.

46. Talin, B.; Calisti, A.; Godbert, L.; Stamm, R.; Lee, R.W.; Klein, L. Frequency-fluctuation model for line-shape calculations in plasma spectroscopy. *Phys. Rev. A* **1995**, *51*, 1918–1928.

47. Gigosos, M.A.; Cardenoso, V. New plasma diagnosis tables of hydrogen Stark broadening including ion dynamics. *J. Phys. B* **1996**, *29*, 4795–4838.

48. Demura, A.V. Instantaneous joint distribution of ion microfield and its time derivatives and effects of dynamical friction in plasmas. *J. Exp. Theor. Phys.* **1996**, *83*, 60–72.

49. Alexiou, S.; Calisti, A.; Gautier, P.; Klein, L.; Leboucher-Dalimier, E.; Lee, R.W.; Stamm, R.; Talin, B. Aspects of plasma spectroscopy: Recent advances. *J. Quant. Spectrosc. Radiat. Transf.* **1997**, *58*, 399–413.

50. Kosarev, I.N.; Stehle, C.; Feautrier, N.; Demura, A.V.; Lisitsa, V.S. Interference of radiating states and ion dynamics in spectral line broadening. *J. Phys. B* **1997**, *30*, 215–236.

51. Griem, H. *Principles of Plasma Spectroscopy*; Cambridge University Press: Cambridge, UK, 1997.

52. Stehle, C.; Hutcheon, R. Extensive tabulation of Stark broadened hydrogen line profiles. *Astron. Astrophys. Suppl. Ser.* **1999**, *140*, 93–97.

53. Barbes, A.; Gigosos, M.A.; Gonzalez, M.A. Analysis of the coupling between impact and quasistatic field mechanisms in Stark broadening. *J. Quant. Spectrosc. Radiat. Transf.* **2001**, *68*, 679–688.

54. Gigosos, M.A.; Gonzalez, M.A.; Cardenoso, V. Computer simulated Balmer-alpha, -beta and -gamma Stark line profiles for non-equilibrium plasma diagnostics. *Spectrochim. Acta B* **2003**, *58*, 1489–1504.

55. Demura, A.V.; Rosmej, F.B.; Stamm, R. Density Matrix Approach to Description of Doubly Excited States in Dense Plasmas. In *Spectral Line Shapes* (18th International Conference on

Spectral Line Shapes, Auburn, Alabama 4-9 June 2006); E. Oks, M. Pindzola, Eds.; AIP Conference Proceedings vol. 874; AIP: Melville, NY, USA, 2006; pp. 112–126.

56. Calisti, A.; Ferri, S.; Mosse, C.; Talin, B. Modélisation des profils de raie dans les plasmas: PPP—Nouvelle version. *J. Phys. IV Fr.* **2006**, *138*, 95–103.

57. Stambulchik, E.; Maron, Y. A study of ion-dynamics and correlation effects for spectral line broadening in plasma: K-shell lines. *J. Quant. Spectrosc. Radiat. Transf.* **2006**, *99*, 730–749.

58. Oks, E.A. *Stark Broadening of Hydrogen and Hydrogenlike. Spectral Lines in Plasmas. The Physical Insight*; Alpha Science International Ltd.: Oxford, UK, 2006.

59. Calisti, A.; Ferri, S.; Mosse, C.; Talin, B.; Lisitsa, V.; Bureyeva, L.; Gigosos, M.A.; Gonzalez, M.A.; del Rio Gaztelurrutia, T.; Dufty, J.W. Slow and fast micro-field components in warm and dense hydrogen plasmas. *ArXiv e-prints* **2007**, [arXiv:physics.plasm-ph/0710.2091].

60. Calisti, A.; del Rio Gaztelurrutia, T.; Talin, B. Classical molecular dynamics model for coupled two component plasma. *High Energy Density Phys.* **2007**, *3*, 52–56.

61. Ferri, S.; Calisti, A.; Mosse, C.; Talin, B.; Gigosos, M.A.; Gonzalez, M.A. Line shape modeling in warm and dense hydrogen plasma. *High Energy Density Phys.* **2007**, *3*, 81–85.

62. Stambulchik, E.; Alexiou, S.; Griem, H.; Kepple, P.C. Stark broadening of high principal quantum number hydrogen Balmer lines in low-density laboratory plasmas. *Phys. Rev. A* **2007**, *75*, 016401.

63. Calisti, A.; Ferri, S.; Talin, B. Classical molecular dynamics model for coupled two component plasma. *High Energy Density Phys.* **2009**, *5*, 307–311.

64. Godbert-Mouret, L.; Rosato, J.; Capes, H.; Marandet, Y.; Ferri, S.; Koubiti, M.; Stamm, R.; Gonzalez, M.A.; Gigosos, M.A. Zeeman-Stark line shape codes including ion dynamics. *High Energy Density Phys.* **2009**, *5*, 162–165.

65. Stambulchik, E.; Maron, Y. Plasma line broadening and computer simulations: A mini-review. *High Energy Density Phys.* **2010**, *6*, 9–14.

66. Calisti, A.; Mosse, C.; Ferri, S.; Talin, B.; Rosmej, F.; Bureyeva, L.A.; Lisitsa, V.S. Dynamic Stark broadening as the Dicke narrowing effect. *Phys. Rev. E* **2010**, *81*, 016406.

67. Demura, A.V. Physical Models of Plasma Microfield. *Int. J. Spectrosc.* **2010**, 671073:1–671073:42.

68. Calisti, A.; Talin, B. Classical Molecular Dynamics Model for Coupled Two-Component Plasmas—Ionization Balance and Time Considerations. *Contrib. Plasma Phys.* **2011**, *51*, 524–528.

69. Calisti, A.; Ferri, S.; Mosse, C.; Talin, B.; Gigosos, M.A.; Gonzalez, M.A. Microfields in hot dense hydrogen plasmas. *High Energy Density Phys.* **2011**, *7*, 197–202.

70. Ferri, S.; Calisti, A.; Mosse, C.; Mouret, L.; Talin, B.; Gigosos, M.A.; Gonzalez, M.A.; Lisitsa, V. Frequency-fluctuation model applied to Stark-Zeeman spectral line shapes in plasmas. *Phys. Rev. E* **2011**, *84*, 026407.

71. Mancini, R.C.; Iglesias, C.A.; Calisti, A.; Ferri, S.; Florido, R. The effect of improved satellite line shapes on the argon Heβ spectral feature. *High Energy Density Phys.* **2013**, *9*, 731–736.

72. Iglesias, C.A. Efficient algorithms for stochastic Stark-profile calculations. *High Energy Density Phys.* **2013**, *9*, 209–221

73. Iglesias, C.A. Efficient algorithms for Stark-Zeeman spectral line shape calculations. *High Energy Density Phys.* **2013**, *9*, 737–744.

74. Stambulchik, E. Review of the 1st Spectral Line Shapes in Plasmas code comparison workshop. *High Energy Density Phys.* **2013**, *9*, 528–534.

75. Calisti; A.; Demura, A.; Gigosos, M.; Gonzalez-Herrero, D.; Iglesias, C.; Lisitsa, V.; Stambulchik, E. Influence of micro-field directionality on line shapes. *Atoms* **2014**, *2*, 259–276.

76. Ferri, S.; Calisti, A.; Mossé, C.; Rosato, J.; Talin, B.; Alexiou, S.; Gigosos, M.A.; González, M.A.; González-Herrero, D.; Lara, N.; *et al.* Ion Dynamics Effect on Stark-Broadened Line Shapes: A Cross-Comparison of Various Models. Atoms **2014**, *2*, 299–318.

77. Alexiou, S.; Dimitrijević, M.S.; Sahal-Brechot, S.; Stambulchik, E.; Duan, B.; Gonzalez-Herrero, D.; Gigosos, M.A. The second woorkshop on lineshape comparison: Isolated lines. *Atoms* **2014**, *2*, 157–177.

78. Briggs, J.S.; Rost, J.M. Time dependence in quantum mechanics. *Eur. Phys. J. D* **2000**, *10*, 311–318.

Reprinted from *Atoms*. Cite as: Rosato,J.; Capes, H.; Stamm, R. Ideal Coulomb Plasma Approximation in Line Shape Models: Problematic Issues. *Atoms* **2014**, *2*, 253-258.

Article

Ideal Coulomb Plasma Approximation in Line Shape Models: Problematic Issues

Joel Rosato *, Hubert Capes and Roland Stamm

Aix Marseille Université, CNRS, PIIM UMR 7345, Marseille 13397, France;
E-Mails: h.capes@laposte.net (C.H.); roland.stamm@univ-amu.fr (S.R.)

* Author to whom correspondence should be addressed; E-Mail: joel.rosato@univ-amu.fr;
 Tel.: +33-49128-8624.

Received: 23 March 2014; in revised form: 11 June 2014 / Accepted: 13 June 2014 / Published: 19 June 2014

Abstract: In weakly coupled plasmas, it is common to describe the microfield using a Debye model. We examine here an "artificial" ideal one-component plasma with an infinite Debye length, which has been used for the test of line shape codes. We show that the infinite Debye length assumption can lead to a misinterpretation of numerical simulations results, in particular regarding the convergence of calculations. Our discussion is done within an analytical collision operator model developed for hydrogen line shapes in near-impact regimes. When properly employed, this model can serve as a reference for testing the convergence of simulations.

Keywords: Stark broadening; hydrogen plasmas; numerical simulations; collision operators

1. Introduction

The Spectral Line Shapes in Plamas Code Comparison Workshop (SLSP) [1] focuses on a set of standardized physical problems to be addressed using codes from different research groups/labs. Amongst these problems is the description of Stark line shapes with ion dynamics effects, referred to as the cases "1" and "2" in the first (2012) and second (2013) editions of the Workshop. In order to get a simple interpretation of what the codes effectively calculate, a set of idealizing assumptions has been considered. For example, the electrons and the ions are assumed to move along straight path trajectories and they produce unscreened Coulomb potentials (ideal plasma approximation). The purpose of this paper is to show that the latter assumption raises a problem of consistency in the interpretation of "*ab initio*" numerical simulations, *i.e.*, simulations that are free from physical approximations in the evaluation of the plasma microfield and in the calculation of the atomic dipole autocorrelation function. They commonly serve as a reference for testing other models. Although very convenient in practice (in particular for programming purposes), the simulations can take a long time and become useless in the case where a large amount of particles is required in the evaluation of the microfield. This occurs for weakly coupled plasmas, when the Debye length is much larger than the mean interparticle distance. A relevant strategy when performing a simulation is to take a box (either of a cubic, spherical, or more complex shape) of characteristic size of the order of the Debye length (typically larger by a factor of several units), which means that the number of particles can be very large in weakly coupled plasma conditions. If the plasma is so weakly coupled that a calculation cannot be performed on a reasonable time scale, the box size is reduced and the corresponding number of particles is adjusted in such a way that a relevant statistical quantity (like the microfield probability density function or the microfield autocorrelation function) is well reproduced within a few percents error bars. We suggest that this procedure does not suffice to obtain reference profiles in the case of infinite Debye length. The neglect of far perturbers may result in a significant underestimate of the line broadening if the microfield is dynamic. Our discussion is based on the use of an analytical model for Stark broadening in regimes such that the impact approximation is not far from being satisfied by the perturbers under consideration (ions or electrons) [2,3]. We consider the Lyman α line broadened due to the electrons only at $N = 10^{17}$ cm^{-3} and $T = 100$ eV (subcase "1.1.3.1.1" of the 2013 SLSP), and we assume an unscreened Coulomb electric field as required in the Workshop statement of cases.

2. A Collision Operator Model for the $\lambda_D \to \infty$ Limit

The plasma conditions yield a value of about 4×10^{-3} for the ratio b_W/r_0 between the Weisskopf radius ($b_W = \hbar n^2/m_e v$, where $n = 2$ is the upper principal quantum number and $v = \sqrt{2T/m_e}$ is the electrons' thermal velocity) and the mean interparticle distance ($r_0 = N^{-1/3}$), which suggests that a collision operator may be used. The standard impact models for hydrogen (e.g., Griem *et al.* [4,5]) and their extensions (e.g., the Lewis cut-off [6] and the unified theory [7,8]) account for particle correlations within a Debye screening model and, hence, are not compatible with the unscreened electric field assumption ($\lambda_D \to \infty$) done in the subcase 1.1.3.1.1. This "artificial" setting of an infinite Debye length requires a careful reconsideration of the role of far perturbers (weak collisions), in a more general framework than that involved in the standard collision operator models. Recently, it has been shown that the perturbers that effectively contribute to the line broadening are those which are located at a distance smaller than $v/\bar{\gamma}$, where $\bar{\gamma}$ is the line's characteristic width, because of the finite lifetime of the emitter (see the discussions in [2,3]). This length may be interpreted as an upper cut-off in place of the Debye length for our "artificial" Coulomb plasma. Applying the model reported in [2] to the subcase 1.1.3.1.1 yields the following formula for the line shape

$$I(\Delta\omega) = \frac{1}{\pi}\text{Re}\frac{1}{-i\Delta\omega + \gamma(\Delta\omega)} \tag{1}$$

where $\Delta\omega$ is the frequency detuning and where $\gamma(\Delta\omega)$ is the collision operator taken in the bracket $\langle 2,1,0|...|2,1,0\rangle$ (spherical base). The latter is given by {see [2], Equation (A4), setting $\lambda_D \to \infty$}

$$\gamma(\Delta\omega) = \frac{9g}{8\sqrt{\pi}}\int_0^\infty du e^{-u}\left[\sqrt{\frac{\pi u}{-i\widetilde{\Delta\omega}}}\text{erf}\left(\sqrt{\frac{-i\widetilde{\Delta\omega}}{u}}\right) + E_1\left(\frac{-i\widetilde{\Delta\omega}}{u}\right)\right] \tag{2}$$

We have used here a system of units such that the Weisskopf radius b_W and the thermal velocity v are equal to unity; $g = 4\pi N b_W^3/3$ is the number of particles in the Weisskopf sphere, which is much smaller than unity (namely, of the order of 2×10^{-7} here), and the quantity $\widetilde{\Delta\omega}$ is defined as $\Delta\omega + i\theta\gamma_0$ where $\gamma_0 \equiv \gamma(\Delta\omega = 0)$ and θ is an atomic physics factor of the order of unity. In our calculations we have set $\theta = 1$. Like standard collision operators, $\gamma(\Delta\omega)$ exhibits a decomposition into a strong collision part (the first term of the right-hand side) and a weak collision part (the second term). It is worth noting that Equation (2) does not provide a closed expression for $\gamma(\Delta\omega)$, because of the presence of γ_0 (through the $\widetilde{\Delta\omega}$ term) in the error and exponential integral functions; instead, it implies an equation of the form $\gamma_0 = F(\gamma_0)$ where F is a nonlinear function. In [2], this equation was solved numerically by iterations and it was shown that this method is quickly convergent.

3. Questioning the Validity of Simulations

Our Equation (1) applied to the subcase 1.1.3.1.1 leads to a line shape about twice larger than that obtained from the simulation results that have been presented during the Workshop. These simulations assume a finite box that contains several thousands of particles. According to the model, we suggest that this deviation stems from the neglect of the far perturbers (i.e., those inside the $v/\bar{\gamma}$ sphere but outside the simulation box) in the simulations. It is quite difficult to test this argument by enlarging the simulation box up to $v/\bar{\gamma}$ because this would imply a very large number of particles, up to several billions. Instead, we have modified our model [2] in such a way to account for a finite plasma size R. The general collision operator formula accounting for finite Debye length {[2], Equation (A4)} has been used with the formal substitution $\lambda_D \equiv R$.

Figure 1 shows an application to a cubic box of size $R = 3400b_W$. The three plots correspond to (i) the result of the model assuming finite R (referred to as "UTPP with box cut-off"); (ii) the model assuming infinite R ("UTPP"); and (iii) a numerical simulation performed with our code [9]. The simulation has been performed using about 2000 particles, which corresponds to the setting of $R = 3400b_W$ in the plasma conditions associated with the subcase 1.1.3.1.1. The dipole autocorrelation function has been evaluated assuming 5000 histories, with a time step of about $0.08\ r_0/v$. The setting of a smaller time grid yields no significant modification in the results. In order to get a dipole autocorrelation function sufficiently close to zero at large times, we have taken 10^6 time steps. Our simulation (not presented at the Workshop 2013 edition) is in agreement with the other simulation codes presented at the Workshop [10]. These codes also assumed several thousands of particles. As can be seen in the figure, the model accounting for the box size yields a line shape that coincides with the simulation result (up to the statistical noise), whereas the model that assumes infinite R yields a line shape about twice larger. The setting of several thousands of particles in the simulations presented at the Workshop was supported by the reproducibility of plasma statistical quantities (such as the microfield probability density function). However, this does not mean that all of the line broadening is taken into account. The line width has a logarithmic dependence on R up to a threshold of the order of $v/\bar{\gamma}$ (see Figure 2), which is much larger than the box size considered in the simulations. This suggests that the simulations performed at the workshop have not converged in the plasma conditions (subcase "1.1.3.1.1") considered in our investigation. Note, we cannot provide such a strong statement in other cases where our collision operator model is not applicable; a (straightforward) convergence test should involve a plot of the simulated line width in terms of R.

Figure 1. Hydrogen Lyman α line shape calculated according to the subcase 1.1.3.1.1. The three plots correspond to (i) the result of the model assuming a finite box size R (dashed line); (ii) the model assuming infinite R (solid line); and (iii) a numerical simulation (circles). The second plot (solid line) serves as a reference. Keeping finite R results in an underestimate of the line width.

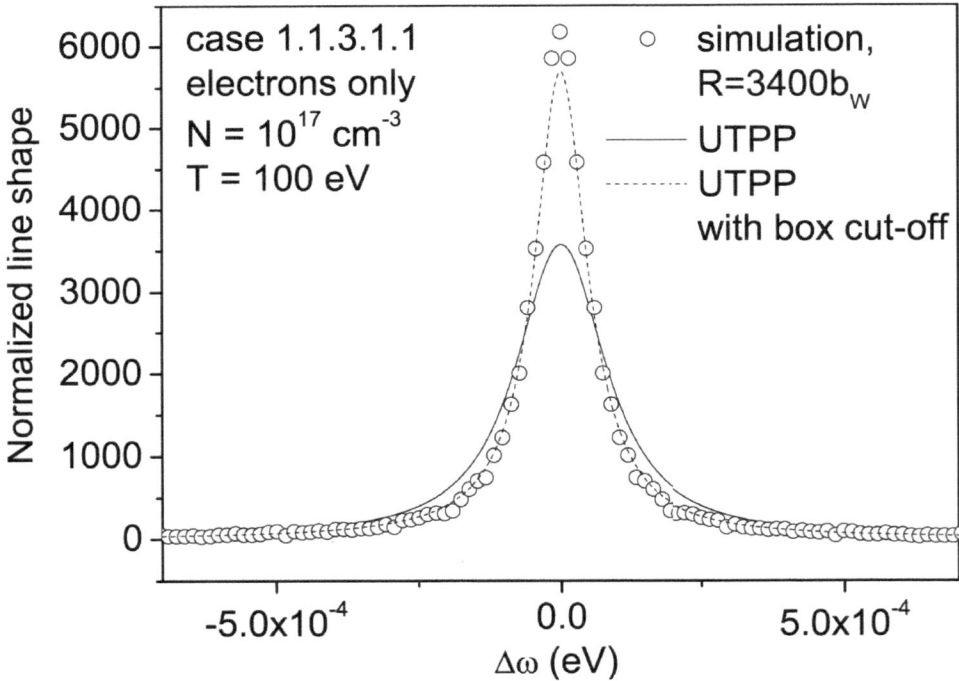

Figure 2. The collision operator at the line center ($\Delta\omega = 0$) is a weakly increasing function of the box size R and saturates when R is of the order of 10^8. This roughly corresponds to the value of $v/\bar{\gamma}$ (see text).

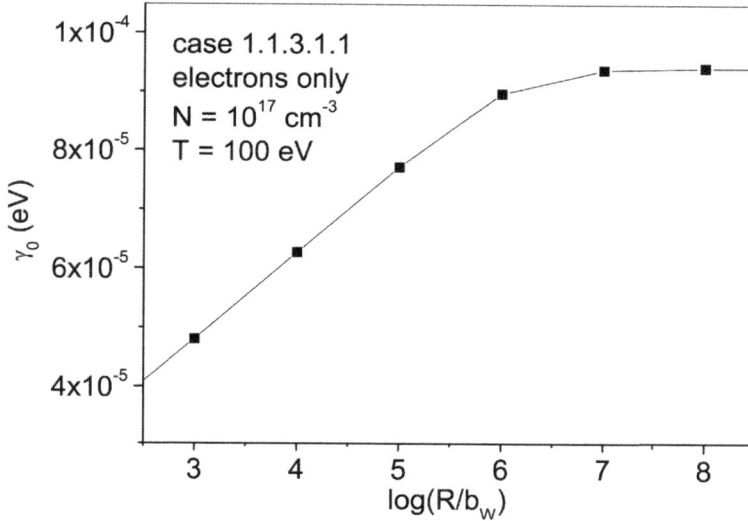

case 1.1.3.1.1
electrons only
$N = 10^{17}$ cm^{-3}
$T = 100$ eV

y-axis: γ_0 (eV)

x-axis: $\log(R/b_w)$

4. Conclusions

In conclusion, we have shown that an artificial setting of an infinite Debye length in numerical simulations requires a careful interpretation of the results. With an analytical model, we have shown that the simulations may neglect a considerable amount of perturbers that effectively contribute to the broadening; the deviations between the simulations and the analytical model are sufficiently large (up to a factor of 2) so that they must deserve a special analysis. The ideal plasma conditions considered in this work were motivated by the need of a physical model sufficiently simple so that line shape codes results can be easily compared (SLSP Workshop). Our results suggest that this physical model is not appropriate; a more relevant one should retain particle correlations within a Debye field. The analytical line shape model used in our work is restricted to weakly coupled plasmas in conditions such that the microfield evolves at a time scale much shorter than the time of interest. A further investigation should focus on an extension able to account for more general cases, e.g., when the emitter suffers simultaneous strong collisions (namely, when the ratio b_W/r_0 is significant).

Acknowledgments

This work was carried out within the framework of the European Fusion Development Agreement and the French Research Federation for Fusion Studies. It is supported by the European Communities under the contract of Association between Euratom and CEA. The views and opinions expressed herein do not necessarily reflect those of the European Commission.

69

Conflicts of Interest

The authors declare no conflict of interest.

References

1. Stambulchik, E. Review of the 1st Spectral Line Shapes in Plasmas code comparison workshop. *High Energy Density Phys.* **2013**, *9*, 528–534.
2. Rosato, J.; Capes, H.; Stamm, R. Influence of correlated collisions on Stark-broadened lines in plasmas. *Phys. Rev. E* **2012**, *86*, 046407:1–046407:8.
3. Rosato, J.; Capes, H.; Stamm, R. Divergence of the Stark collision operator at large impact parameters in plasma spectroscopy models. *Phys. Rev. E* **2013**, *88*, 035101:1–035101:3.
4. Griem, H.R.; Kolb, A.C.; Shen, K.Y. Stark broadening of hydrogen lines in a plasma. *Phys. Rev.* **1959**, *116*, 4–16.
5. Griem, H.R.; Blaha, M.; Kepple, P.C. Stark-profiles calculations for Lyman-series lines of one-electron ions in dense plasmas. *Phys. Rev. A* **1979**, *19*, 2421–2432.
6. Lewis, M. Stark broadening of spectral lines by high-velocity charged particles. *Phys. Rev.* **1961**, *121*, 501–505.
7. Voslamber, D. Unified model for Stark broadening. *Z. Naturforsch.* **1969**, *24a*, 1458–1472.
8. Smith, E.W.; Cooper, J.; Vidal, C.R. Unified classical-path treatment of Stark broadening in plasmas. *Phys. Rev.* **1969**, *185*, 140–151.
9. Rosato, J.; Marandet, Y.; Capes, H.; Ferri, S.; Mossé, C.; Godbert-Mouret, L; Koubiti, M; Stamm, R. Stark broadening of hydrogen lines in low-density magnetized plasmas. *Phys. Rev. E* **2009**, *79*, 046408:1–046408:7.
10. Ferri S.; Calisti A.; Mossé C.; Rosato J.; Talin B.; Alexiou S.; Gigosos M.A.; González M. A.; González D. Ion dynamics effect on Stark broadened line shapes: A cross comparison of various models *Atoms*, submitted.

Reprinted from *Atoms*. Cite as: Alexiou, S.; Dimitrijević, M.S.; Sahal-Brechot, S.; Stambulchik, E.; Duan, B.; González-Herrero, D.; Gigosos, M.A. The Second Workshop on Lineshape Code Comparison: Isolated Lines. *Atoms* **2014**, 2, 157-177.

Article

The Second Workshop on Lineshape Code Comparison: Isolated Lines

Spiros Alexiou [1,*], **Milan S. Dimitrijević** [2], **Sylvie Sahal-Brechot** [3], **Evgeny Stambulchik** [4], **Bin Duan** [5], **Diego González-Herrero** [6] and **Marco A. Gigosos** [6]

[1] TETY, University of Crete, 71409 Heraklion, TK2208, Greece

[2] Astronomical Observatory, Volgina 7, Belgrade 11060, Serbia; E-Mail: mdimitrijevic@aob.rs

[3] LERMA-UMR CNRS 8112 and UPMC, Observatoire Paris Meudon, 5 Pl Jules Janssen, Meudon 92195, France; E-Mail: sylvie.sahal-brechot@obspm.fr

[4] Faculty of Physics, Weizmann Institute of Science, Rehovot 76100, Israel; E-Mail: evgeny.stambulchik@weizmann.ac.il

[5] Institute of Applied Physics and Computation Mathematics, Beijing 100088, China; E-Mail: alexduan1967@hotmail.com

[6] Departamento de Optica, Universidad de Valladolid, Valladolid 47071, Spain; E-Mails: diegohe@opt.uva.es (D.G.-H.); gigosos@coyanza.opt.cie.uva.es (M.A.G.)

* Author to whom correspondence should be addressed; E-Mail: moka1@otenet.gr; Tel.+306937545547.

Received: 14 March 2014; in revised form: 9 April 2014 / Accepted: 25 April 2014 / Published: 12 May 2014

Abstract: In this work, we briefly summarize the theoretical aspects of isolated line broadening. We present and discuss test run comparisons from different participating lineshape codes for the 2s-2p transition for LiI, B III and NV.

Keywords: Stark broadening; isolated lines; impact approximation

1. The Importance of Level Spacings

In a plasma, free electrons and ions interact and perturb atomic states, resulting in line broadening and shift [1,2]. Each free plasma particle coming appreciably close to the emitter will contribute to the broadening and, in principle, also, the shift. This contribution will depend not only on the proximity of the emitter and perturber, but also on the duration of the interaction, i.e., the time interval during which the interaction is appreciable. Isolated lines are lines where for the upper and lower levels, the closest perturbing level is energetically much further away than the inverse collision duration. For instance, for the 2s-2p lines considered, the 2s-level states are perturbed by the 2p states in the sense that the important broadening process is:

$$|2s\rangle \rightarrow (via\ a\ collision)\ |2p\rangle \rightarrow (via\ a\ collision)\ |2s\rangle \tag{1}$$

In general, we consider the broadening process:

$$|\alpha\rangle \rightarrow (via\ a\ collision)\ |\alpha'\rangle \rightarrow (via\ a\ collision)\ |\alpha\rangle \tag{2}$$

For isolated lines, the relevant energy spacings, $\omega_{\alpha\alpha'}$, between the collisionally connected states, α and α', is of paramount importance. These energy spacings effectively reduce the interaction time and the effective impact parameters. For example, the relevant plasma-related quantity is, for perturbative collisions and a long-range dipole interaction [3]:

$$\phi = \frac{ne^2}{\hbar^2} \int_{-\infty}^{\infty} dt_1 \int_{-\infty}^{t_1} dt_2 \langle \mathbf{E}(t_1) \cdot \mathbf{E}(t_2) \rangle e^{\imath\omega_{\alpha\alpha'}(t_1-t_2)} \tag{3}$$

where

$$\langle \mathbf{E}(t_1) \cdot \mathbf{E}(t_2) \rangle = \frac{2\pi}{3} \int_0^{\infty} v f(v) dv \int_0^{\rho_{max}} \rho d\rho \mathbf{E}(t_1) \cdot \mathbf{E}(t_2) \tag{4}$$

Here, $\mathbf{E}(t)$ is the electric field at the emitter, due to the perturbing particle with impact parameter ρ and velocity (at an infinite distance from the emitter) v. We note the decreasing width with increasing temperature, e.g., $\frac{lnT}{\sqrt{T}}$ [4]; the width is dominated by a $T^{-1/2}$ decay [5] for straight line trajectories.

The imaginary exponential has the following important effects compared to the collisionally degenerate (hydrogenic, $\omega_{\alpha\alpha'} = 0$) case:

1. Only a part of the collision duration is effective, i.e., the effective collision duration is shortened, not the entire collision.
2. Since the impact parameter affects the collision duration ($\propto \rho/v$), the effective impact parameters are also smaller.
3. The effective velocities increase (slow collisions are adiabatic).
4. As a corollary to all the above, broadening is decreased.

For strong collisions, we also have (canceling) higher order terms in the Dyson expansion. However, it is still true that the collision duration is effectively shortened by the imaginary exponentials.

In all calculations, only the 2s and 2p levels were considered. The only exception was the fully quantum-mechanical code [6], which used 13 nonrelativistic configurations $1s^2nl(n = 2, 3, 4, 5, l \leq 4)$ for B III and N V.

2. Electron *vs.* Ion Broadening

For hydrogen, given a velocity and impact parameter, a straight line trajectory gives the same contributions for any *single* electron and singly-charged ion.

For a hydrogen-like emitter, electrons are attracted; hence, they come closer, but also are accelerated, hence spending less time in the vicinity of the emitter. Ions are repelled, hence staying further away, but also slowing down; hence, they stay longer in the vicinity of the emitter. The action is $\int V(t)dt$, with $V(t)$ the emitter-perturber interaction; hence, the net effect is that the same velocity electron and singly charged ion contribute the same to broadening [3]. Of course, ion perturbers win overall, due to their velocity distribution favoring smaller velocities, both for hydrogen and hydrogen-like emitters.

For isolated lines, the action is $\int V(t)e^{i\omega_{\alpha\alpha'}t}dt$. Therefore, only times of the order $1/\omega_{\alpha\alpha'}$ contribute. For electrons, this is not too bad: electrons rely on coming close, and since they are accelerated, their effective collision duration is small anyway. For ions, however, the consequences are far greater: ions were kept away and relied on a longer duration interaction to match the electrons. Now that a longer duration is negated by the shrinking of the effective collision duration to $\propto 1/\omega_{\alpha\alpha'}$, the ions are ineffective [3].

The bottom line is that, for isolated lines, ions normally cannot compete with electrons, with two exceptions: (A) when $\omega_{\alpha\alpha'}$ is small compared to kT, in which case, the collision duration was very short anyway; (B) ions can compete in non-dipole interactions. In contrast to the dipole case, we can have the channel

$$|\alpha\rangle \xrightarrow{\text{via a quadrupole collision}} |\alpha\rangle \xrightarrow{\text{via another quadrupole collision}})|\alpha\rangle$$

as result of a quadrupole excitation/deexcitation. Non-dipole interactions are, however, outside the scope of the workshop comparison and were explicitly neglected in all calculations, except the fully quantal one.

As a result, isolated lines allow a test of electron broadening alone. For the calculations submitted to the workshop and considered in the present paper, ion broadening was explicitly neglected for both widths and shifts. The electron densities used were $n_e = 10^{17}$ e/cm^3 for LiI, $n_e = 10^{18}$ e/cm^3 for B III and $n_e = 10^{19}$ e/cm^3 for NV. However, the results are presented, normalized to a density of 10^{17} e/cm^3, that is, the B III results were divided by 10 and the NV results by 100.

3. Penetrating Collisions

These are collisions where the perturbing particle penetrates the emitter wavefunction extent for the states involved. As already discussed above, since smaller impact parameters gain in importance, penetration is more important for isolated lines:

For instance, for neutrals, the collision duration is $\propto \rho/v$, so that the impact parameters, ρ, and velocities, v, that contribute significantly are such that $\rho\omega_{\alpha\alpha'} \leq v$, e.g., small ρ. Hence, close collisions, which typically dominate for isolated lines, are significantly affected by penetration [7], which softens these collisions. This effect can be quite important and typically produces a factor of a two-width decrease [8].

Thus, if penetration is accounted for, perturbation theory is typically valid. This is convenient, since by softening the interaction, penetration makes those important close collisions perturbative; hence, much easier to solve. On the downside, there is no one-size fits all formula, i.e., instead of universal broadening functions, now we have per species and line broadening functions, although universal formulas that account for penetration have been given for hydrogen [10] and are possible for H-like ions. Furthermore, wavefunction information is needed, and oscillator strengths are no longer adequate [8].

4. Quantal Calculations

In contrast to semiclassical calculations, which assume the perturbing particles to be classical particles moving in well-defined trajectories, quantal calculations model the perturbing particles in terms of their wavefunctions and consider quantal effects, such as temporary electron capture and resonances. Only one fully quantal calculation participated in the workshop (DARC).

Fully quantal calculations have been performed by a number of different methods, e.g., Coulomb–Born (CB), distorted waves (DW), up to close coupling (CC) (R-matrix [11] and convergent close coupling (CCC) [12]). CB agreed very well with CC, which is understandable, due to the weakness of the interaction. Comparisons of quantal and semiclassical calculations have been done and summarized [8,13]. Good agreement is found if the penetration is taken into account in semiclassical calculations. What is puzzling is a recent DW calculation by Elabidi *et al.* [14] citing the agreement with the experiments; these results are a factor of two larger than the aforementioned CC calculations. The present K-matrix CC calculations of DARC also agree with experiments and show the factor of two disagreements with the previous CC calculations. It is not clear why the DW [14] and DARC show such a large difference with R-matrix and CCC.

4.1. Work to be Done on Quantal Calculations

This is an open issue, and work is needed to pinpoint the origin of these differences. Comparing partial wave results and/or cross-sections may be helpful. We have not resolved the discrepancy with experiments. A code comparison may be needed to assess the relative importance of various factors in quantal calculations.

4.2. Work to be Done with Semiclassical Calculations

If penetration and non-perturbative aspects are accounted for, the remaining differences between the quantal and semiclassical approach are due to [8,13]:

a. The demarcation line between what may and what may not be treated semiclassically.

b. Back-reaction issues, i.e., in classical terminology, deviations from the (straight line or hyperbolic) trajectory, due to the energy transfer between the perturber and emitter; e.g., for collisional excitation/deexcitation, in our case, between the 2s and 2p levels, given that these energy transfers are comparable to the thermal velocities.

Answers to these questions may be given by Feynman path calculations. Such calculations have started appearing, but only for the simplest cases [15].

5. (Dynamic) Shifts

Shifts are caused by the finite collision duration. The same considerations apply as for widths: Ion (dipole) shifts are typically negligible; penetration is also an important factor. In contrast to widths, where one is looking for the guaranteed nonnegative deviation of $Re(I - S_a S_b^*)$ of a quantity, possibly small, from unity, with shifts, one is looking for the deviation of a possibly small quantity, $Im(I - S_a S_b^*)$ from zero. Furthermore, canceling opposite sign contributions arise [16], and the end result is that a small error in the calculation or deviation from thermal distribution functions can change the shift sign. This study was repeated [17] and confirmed the results for $\Delta n = 0$, but found consistent red shifts for $\Delta n \neq 0$. The recommendation of [17] is that $\Delta n \neq 0$ shifts are reliable, while $\Delta n = 0$ are not.

6. The Codes

Three types of codes participated in the workshop: DARC [6], a fully quantal (K-matrix) close-coupling code that, in principle, should be the benchmark, but uses a different physical model from the rest, full simulation codes (SimU [18] and Simulation [19]), which, in principle, may also account for perturber–perturber interactions, though for the cases presented in this comparison, perturbing electrons were modeled as non-interacting quasiparticles interacting with the emitter via Debye-shielded fields and semiclassical impact codes (SCP [20–23], a perturbative impact code and Starcode [8]). Starcode is a non-perturbative impact code, which solves the Schrödinger equation if it considers that the perturbative calculation has too large calculation uncertainties (and with the current tolerance, it invariably does). Although Starcode is a single code, it has a number of options, and in the present paper, we refer to two Starcode versions: one version, referred to as Starcode, which accounts for penetrating collisions, and one, called Starcode-NP, which does not. In practice, this is controlled by a single option in the input file.

The problem specification was that only dipole electron broadening should be considered. Except for DARC, which includes non-dipole interactions, all other codes have adhered, either by default or by selection of the appropriate options, to this specification. However, judging from Starcode calculations with monopoles and quadrupoles, the inclusion of monopole and quadrupole interactions makes too small a difference to explain the differences with DARC. Previous quantal close-coupling calculations have also produced widths that are a factor of two smaller than the ones quoted by DARC, as discussed above.

In addition, the problem specifications were to neglect fine structure, but only the simulation codes adhered to this specification. Among the codes, only SimU uses a Debye-shielded electron-emitter interaction, while all others use either a pure Coulomb interaction (DARC, Simulation) or a cutoff length of the order of the Debye length. In view of the remarks on short impact parameters being most important, these differences are not expected to be significant. For example, for SimU, the difference in the calculations using pure Coulomb or Debye-shielded fields is of the order of 0.1%. Small differences when using Coulomb or Debye-shielded fields have also been confirmed in tests using Simulation.

Atomic data are discussed in the Appendix.

7. Strong Collisions

The term "strong collisions" has been used in the literature to denote either "collisions that may not be treated by perturbation theory" or "collisions that may not be treated within the model used", e.g., collisions for which the semiclassical approximation is not valid or collisions for which the normally employed long-range approximation is not valid. Note that the two definitions may be in conflict. For example, penetrating collisions may be non-perturbative within the long-range approximation and perturbative if the long-range approximation is not used. Impact codes SCP and Starcode use the second definition and estimate the strong collision contribution based on unitarity. Starcode, which is non-perturbative, actually computes the non-perturbative contribution exactly within the impact approximation, but excludes from the computation the phase space for which its model is physically not valid. Simulation and SimU make no distinction between strong and weak collisions, as they are treated in exactly the same way. The simulation codes compute strong collisions (even if the model used is physically invalid due to quantal effects and/or the long-range dipole approximation), as well as weak ones. However, we consider simulation codes to employ the first definition when we wish to consider the differences between simulation codes and perturbative results. Hence, in principle, differences between the simulation codes and Starcode-NP should be due to either non-impact (overlapping) collisions or the phase space part that is excluded from Starcode and included in the simulation. For DARC, this distinction is rather meaningless.

Generally speaking, good agreement is expected between all codes for the non-strong (in either meaning) contributions, while things are not as clear for strong collisions in either meaning.

8. Calculational Uncertainties

With respect to calculational uncertainties, the codes use quite different ways of estimating them.

- DARC. Calculation uncertainties were estimated by increasing the orbital angular momentum of the colliding electron up to 25. They turned out to be $\ll 5\%$ [6].
- SCP. For simple spectra, an error bar of 20% was adopted, based on the average agreement with the experimental results for He I lines [24] and Si-like ions [25], namely 17% for widths and 20% for shifts.

- SimU. This is a full simulation code that computes the line profile directly instead of first computing the autocorrelation function. It is a semiclassical, long-range dipole code that handles non-perturbative aspects, as well as non-impact effects rigorously. SimU calculates $|\vec{D}(\omega)|^2$, where $\vec{D}(t)$ is the dipole operator and $\vec{D}(\omega)$ is its Fourier transform. The procedure is repeated N times ($N \gg 1$) and averaged, which corresponds to an averaging over a statistically representative ensemble of radiators. The uncertainty due to a finite statistical sampling is $\propto 1/\sqrt{N}$, and the proportionality coefficient is inferred by observing convergence over the course of simulations. For details, see [26]. Another important factor affecting the calculational accuracy is the finite step of the frequency grid, equal to $t_{\max}^{-1}$, where $t_{\max}$ is the time over which $\vec{D}(t)$ is calculated in each run prior to averaging. Contrary to the first (statistical) factor, the latter always results in an overestimate of the line width by $\sim t_{\max}^{-1}$. For the shift values, this factor contributes $\pm \sim t_{\max}^{-1}/2$. The total uncertainties were within 10% for the cases considered.

- Simulation. The spectra is obtained through a Fourier transform of the autocorrelation function, which has been obtained by computer simulation and averaged over a large enough number of samples. The error is estimated using several sets of configurations to average the autocorrelation function and comparing the results obtained in each case. These fluctuations are about 3% using sets of 10,000 runs.

- Starcode. Starcode has two sources of errors: numerical errors, e.g., from integrations or solutions of the Schrödinger equation, which are controlled through appropriate tolerances and are generally negligible, and errors arising from the model used, which are the ones quoted here. Starcode recognizes that there is a part of the phase space for which the model used (e.g., semiclassical trajectories and, in the version without penetration, trajectories that penetrate the wavefunction extent) is not reliable. However, the contribution of that part of the phase space may be bound by unitarity. Thus, one first defines $v_0 = \frac{\hbar}{m\rho_{max}}$. This is a velocity below which all collisions have a de Broglie wavelength larger than the screening length, ρ_{max}, and, hence, are definitely non-classical. The corresponding phase space makes a contribution:

$$E_1 = 2\pi n \int_0^{v_0} v f(v) dv \int_0^{\rho_{max}} \rho d\rho Re[I - S_a(\rho, v)S_b^*(\rho, v)] \tag{5}$$
$$= \pi n \alpha \sqrt{\tfrac{8m}{\pi kT}} \rho_{max}^2 (1 - (1 + w_0)e^{-w_0})$$

with n the electron density, $f(v)$ the velocity distribution, S_a and S_b the S-matrices for the upper and lower levels, respectively, $w_0 = \frac{mv_0^2}{2kT}$ and α is an estimate for the average $Re[I - S_a(\rho, v)S_b^*(\rho, v)]$, which is between zero and two, due to unitarity. In practice, this contribution is negligible in our examples.

For $v > v_0$, the semiclassical approach is not valid for $\rho < \frac{b\hbar}{mv}$, where b is a parameter, taken to be one in the present calculations, which specifies how much larger than the de Broglie wavelength an impact parameter must be in order to be treatable semiclassically. Hence, if penetration is allowed, this non-semiclassical phase space provides the dominant contribution:

$$E_2 = 2\pi n \int_{v_0}^{\infty} v f(v) dv \int_0^{\frac{b\hbar}{mv}} \rho d\rho Re[I - S_a(\rho, v)S_b^*(\rho, v)] = 2\pi n \alpha (\frac{b\hbar}{m})^2 \sqrt{\frac{2m}{\pi kT}} e^{-w_0} \tag{6}$$

If penetration is not allowed, then the upper limit of the ρ integral is $max(R, \frac{b\hbar}{mv})$ with R the relevant wavefunction extent, e.g., the maximum spatial extent of the 2s and 2p levels in our case. Hence, E_2 in the case with penetration is replaced by $E_{2NP} + E_3$ with:

$$E_{2NP} = 2\pi n \int_{v_0}^{v_1} v f(v) dv \int_0^{\frac{b\hbar}{mv}} \rho d\rho Re[I - S_a(\rho,v)S_b^*(\rho,v)] \qquad (7)$$
$$= 2\pi n \alpha (\frac{b\hbar}{m})^2 \sqrt{\frac{2m}{\pi kT}} (e^{-w_0} - e^{-w_1})$$

with $v_1 = \frac{\hbar}{mR}$ and $w_1 = \frac{mv_1^2}{2kT}$ and

$$E_3 = 2\pi n \int_{v_1}^{\infty} v f(v) dv \int_0^R \rho d\rho Re[I - S_a(\rho,v)S_b^*(\rho,v)] = \pi n \alpha \sqrt{\frac{8m}{\pi kT}} R^2 (1 + w_1) e^{-w_1} \qquad (8)$$

This ensures that, provided one trusts the choices of b and R, only truly semiclassical paths are computed, and a rigorous, unitarity-based error bar may be returned for the part of the phase space that is not treatable semiclassically. Of course, other sources of error, e.g., the back reaction, are not accounted for.

For the cases treated here with penetration not allowed, the wavefunction extent was dominant for all temperatures in the LiI calculations, for $T = 15$ and 50 eV for B III and $T = 50$ eV for NV. The de Broglie cutoff was dominant for the remaining cases. Hence, if the de Broglie cutoff is dominant, the Starcode and Starcode-NP calculations should have the same error bar, and Starcode-NP should have a larger width, due to the stronger interaction. If R is the dominant cutoff, then the error bar of Starcode-NP should be significantly larger and the computed width of Starcode might even exceed (with the error bar not accounted for) Starcode-NP, due to the larger phase space that is contributing. Of course, the situation will be reversed when the calculation uncertainties are accounted for. Width results are quoted as (the width from the part of phase space that is computed reliably $+\frac{1}{2}$ error bar) $\pm\frac{1}{2}$ error bar. Shifts are quoted as (the shift from the part of phase space that is computed reliably) $\pm$ error bar. All Starcode and Starcode-NP strong collision estimates assume a constant $Re(I - S_a S_b^*) = 1$ for that part of the phase space.

9. Results: Widths

Widths are shown in Table 1. All widths have been normalized to a density of 10^{17} e/cm^3.

Table 1. 2s-2p Transition width (FWHM) comparisons. All widths are in cm^{-1} and normalized to a density of 10^{17} e/cm^3.

Species	T (eV)	Model	DARC	SCP	SimU	Simulation	Starcode-NP	Starcode
LiI	5	straight		1.05	0.82	0.88	0.95	0.84
LiI	15	straight		1.10	0.87	0.96	0.87	0.72
LiI	50	straight		1.00	1.08	0.88	0.80	0.51
B III	5	hyperbolic	0.322	0.33	0.2		0.3	0.2
B III	5	straight		0.15	0.112	0.114		
B III	15	hyperbolic	0.259	0.207	0.173		0.2	0.13
B III	15	straight		0.139	0.128	0.128		
B III	50	hyperbolic	0.187	0.138	0.136		0.15	0.092
B III	50	straight		0.126	0.13	0.122		
NV	5	hyperbolic	0.199	0.197	0.1135		0.133	0.088
NV	5	straight		0.071	0.015	0.043		
NV	15	hyperbolic	0.1228	0.117	0.08		0.084	0.056
NV	15	straight		0.0535	0.0488	0.0487		
NV	50	hyperbolic	0.0812	0.07	0.0584		0.06	0.036
NV	50	straight		0.0476	0.044	0.0469		

9.1. General Remarks

1. In general, hyperbolic path calculations produced broader lines than straight line paths. This is expected, because due to the energy spacings (imaginary exponentials), only times closest to the time of the closest approach ($t = 0$) are important, but for these times, the electron is closer to the emitter if it is moving in a hyperbolic rather than a straight line path. Hence, the interaction is stronger, which results in larger widths.

2. Non-perturbative collisions (impact or not) are treated correctly in simulations and by strong collision or error estimates in perturbative impact codes. This is not an issue for non-perturbative impact codes. Non-perturbative collisions are significant if penetration is not accounted for.

3. Furthermore, in all cases, Starcode (with penetration) produces the lowest widths. This, however, is understood and expected, as Starcode with the option in question uses a different physical model that accounts for penetration, which, in turn, softens the interaction and results in smaller widths.

4. DARC consistently produces large widths, in many cases, more than twice the results of Starcode, which also accounts for penetrating collisions. This is not understood and at variance with previous fully quantum mechanical close-coupling calculations, as discussed above.

5. When investigating the differences between impact codes and simulations, particularly at low temperatures, one should keep in mind that impact codes treat all particles as impacting, while with simulation codes, the slow particles that are not impacting are treated correctly, i.e., differences could arise from non-impact effects for some particles. This would, in

principle, mean lower widths for the simulations and should be less of an issue, at least for non-perturbative calculations at higher T. If that is correct, agreement should improve with non-perturbative impact calculations at higher T. As shown below, this is indeed the case for B III and NV, while the opposite is true for LiI. Note, however, that quantitatively non-impact effects are not expected to be an issue, as discussed below.

6. Another issue is the temperature dependence, where we find some codes and cases with an increasing width with increasing temperature or a width that varies very little with temperature. Intuitively, one might expect that higher temperatures mean weaker collisions and, hence, smaller widths. We know, however, that an increase in temperature sometimes actually results in larger widths, e.g., for initially strong collisions, such as non-impact ions [27].

Summing up, a non-decreasing temperature behavior, as observed in the simulation codes, may be the result of either non-impact effects or a significant or dominant strong collision term. The first possibility seems unlikely from simple estimates, discussed below, while if non-impact effects were negligible and the second were the case, then, since Starcode-NP is also a non-perturbative impact code, the differences from Starcode-NP would be due to the non-semiclassical phase space, as well as the emitter relevant wavefunction extent. This would have to be significant enough to change the widths vs. the temperature behavior. To check this conjecture, we list in Table 2 the Starcode-NP width contributions from the phase space part where Starcode-NP is presumably valid (labeled "non-strong" in Table 2, even though it includes non-perturbative collisions), as well as the "strong" collision contribution from the non-semiclassical and wavefunction extent phase space, where Starcode-NP (or any other workshop code, except DARC) is not valid. The actual width quoted is as described above, e.g., $0.83 + 0.23/2 \pm 0.23/2$ for the lowest LiI temperature.

Table 2. 2s-2p Transition width (FWHM) strong collision importance in Starcode-NP. All widths are in cm^{-1} and normalized to a density of 10^{17} e/cm^3.

Species	T (eV)	Starcode-NP Non-Strong	Starcode-NP Strong
LiI	5	0.83	0.23
LiI	15	0.69	0.39
LiI	50	0.48	0.7
B III	5	0.286	0.042
B III	15	0.172	0.048
B III	50	0.11	0.078
NV	5	0.114	0.038
NV	15	0.072	0.025
NV	50	0.0454	0.03

Therefore, we conclude that the differences between Starcode-NP and the simulation codes, as well as the different temperature trends exhibited are within the calculation uncertainties.

9.2. LiI Results

Figure 1 displays the results from five codes. Interestingly, only Starcode, in both versions, displays a decreasing width with temperature behavior (Within the semiclassical model, threshold issues do not arise, as we have virtual excitations and deexcitations, and a higher T means weaker interactions, due to their shorter duration.) All three other calculations produce an increase in the width as T rises from five to 15 eV. Simulation varies very little with temperature. With SimU, that trend continues out to $T = 50$ eV, whereas for SCP and Simulation, it is reversed. For SCP, this may be attributed to the symmetrization, which respects the quantum properties of the S-matrix rather than the method of computing the strong collision term, which is less than half the width already at 5 eV and decreases with T. For simulations, a possible explanation would be a non-impact behavior for electrons at low T. However, simple estimates do not support this conjecture. For example, at $T = 1$ eV, $n = 10^{17}$ e/cm^3 and $L - \alpha$, electrons are well impacted [28], and it is hard to see why they would be non-impacted at 5 eV for Li with a longer lifetime (inverse HWHM), due to the smaller matrix elements and imaginary exponentials. Hence, the most likely origin is [27], the increasing width *vs.* temperature for strong collisions. As already discussed, however, within the calculational uncertainties, Starcode-NP could also be made to show an increasing width with temperature.

9.3. B III Results

Figure 2 displays the widths from five codes. General trends are confirmed, with DARC giving the largest widths. SCP and Starcode-NP exhibit fairly good agreement, but the disagreement of SimU with Starcode-NP and the agreement of SimU with Starcode at low T (which is not expected, as SimU does not account for penetrating collisions, which make the difference between Starcode and Starcode-NP) and Starcode-NP at high T (which is expected) is puzzling. As shown in Table 2, the non-strong Starcode-NP contribution alone is significantly larger than SimU's total width, and the strong contribution will only increase the width of Starcode-NP and, hence, the difference.

81

Figure 1. A comparison of LiI 2s-2p transition widths as computed by SCP (solid black), SimU (dotted red), Simulation (dashed green) and Starcode with (dash-dotted blue) and without (dash-double dotted orange) penetration accounted for.

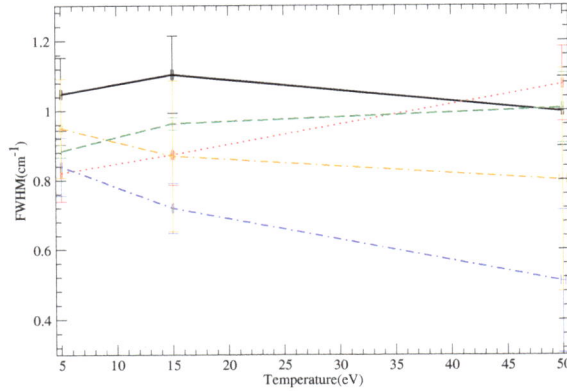

Figure 2. A comparison of hyperbolic trajectory B III 2s-2p transition widths as computed by SCP (solid black), SimU (dotted red), DARC (dashed green) and Starcode with (dash-dotted blue) and without (dot-double dash orange) penetration accounted for.

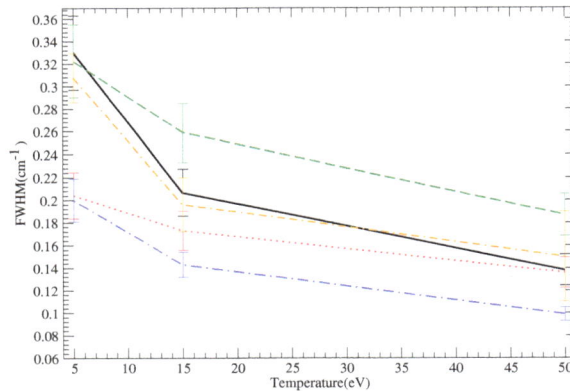

Figure 3 compares straight and hyperbolic trajectory results where applicable. Overall, the comparison demonstrates the difference, particularly for SCP, that arises from including *vs.* neglecting the attraction by the emitter. One would have expected a closer agreement between SimU and Simulation, which use the same physical model. For SimU and Simulation, we see a width increase with increasing T, as already discussed. This increase is reversed for Simulation, but not for SimU. At high T, the calculations converge, as expected.

Figure 3. A comparison of hyperbolic and straight line trajectories for the NV 2s-2p transition widths, as computed by SCP-hyperbolic (solid black), SCP-straight (dotted red), SimU-hyperbolic (dashed green), SimU-straight (dash-dotted blue) and Simulation (dash-double dotted orange).

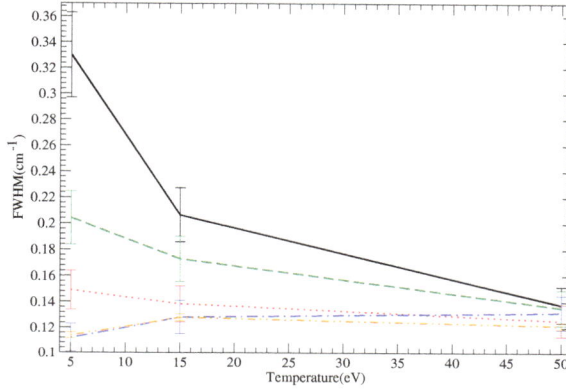

9.4. NV Results

Figure 4 displays the widths from five codes. SimU and Starcode-NP agree quite well, which is expected, at least for high temperatures, where non-impact effects should play no role. SCP and DARC also agree quite well with each other; however, their disagreement with other codes at the highest T value is not understood, since close coupling effects are expected to be negligible at this high T. All codes show the expected decrease for the widths as a function of T, as collisions weaken with increasing emitter charge Z [25].

Figure 4. A comparison of hyperbolic trajectory NV 2s-2p transition widths, as computed by SCP (solid black), SimU (dotted red), DARC (dashed green) and Starcode with (dash-dotted blue) and without (dot-double dash orange) penetration accounted for.

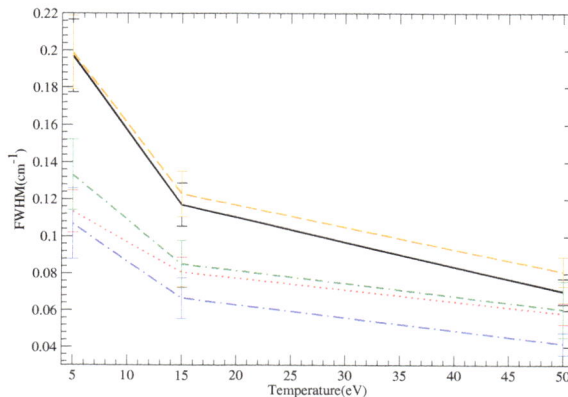

Figure 5 compares straight and hyperbolic trajectory results, where applicable. Overall, the comparison demonstrates the significant difference that arises from including *vs.* neglecting the attraction by the emitter. Due to the higher Z, it is expected that these differences would be more pronounced than the differences found for B III. The close agreement of SimU-straight and Simulation-straight is expected, as both codes use the same physical model.

Figure 5. A comparison of hyperbolic and straight line trajectories for the NV 2s-2p transition widths, as computed by SCP-hyperbolic (solid black), SCP-straight (dotted red), SimU-hyperbolic (dashed green), SimU-straight (dash-dotted blue) and Simulation (dash-double dotted orange).

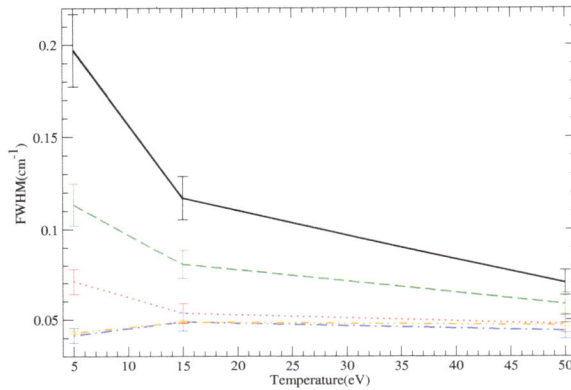

10. Results: Shifts

As discussed, shift calculations within the stated case definitions are probably inaccurate, but the comparisons may well illustrate trends and code differences. The results are summarized in Table 3. Shifts were normalized to a density of 10^{17} e/cm^3.

Table 3. 2s-2p Transition shift comparisons of all shifts are in cm^{-1} and normalized to a density of 10^{17} e/cm^3.

Species	T (eV)	Model	DARC	SCP	SimU	Simulation	Starcode-NP	Starcode
LiI	5	straight		0.551	0.498	0.457	0.21	0.2
LiI	15	straight		0.42	0.43	0.368	0.29	0.28
LiI	50	straight		0.271	0.282	0.247	0.29	0.27
B III	5	hyperbolic	0.085	0.024	0.0846		0.05	0.063
B III	5	straight		0.0751	0.065	0.062		
B III	15	hyperbolic	0.058	0.0276	0.051		0.04	0.044
B III	15	straight		0.0606	0.0515	0.0515		
B III	50	hyperbolic	0.0265	0.0241	0.0363		0.03	0.03
B III	50	straight		0.0398	0.0352	0.0355		
NV	5	hyperbolic	0.0694	0.00423	0.031		1	2
NV	5	straight		0.0284	0.0222	0.02375		
NV	15	hyperbolic	0.03796	0.0054	0.0183		0.009	0.014
NV	15	straight		0.0233	0.0199	0.0199		
NV	50	hyperbolic	0.01776	0.0065	0.01275		.008	0.01
NV	50	straight		0.01535	0.0138	0.01355		

10.1. General Remarks

1. In general, we expect that if shifts are perturbative, they should also decrease with increasing T, due to the weakening of the interaction. For the same reason, we expect a convergence of the straight line and hyperbolic trajectory results as $T \to \infty$. No such statement of decreasing shifts with increasing T can be made with certainty in the non-perturbative case.

2. Starcode, in particular (with or without penetration), often shows a very weak increase of shift with T, even in the version with penetration accounted for, which results in weak, perturbative collisions. The reason is that Starcode explicitly excludes the non-semiclassical phase space, e.g., impact parameters shorter than the de Broglie wavelength. (In addition, Starcode, without accounting for penetration, also excludes impact parameters inside the wavefunction extent, because, again, these may not be treated within a long-range approximation). When T decreases, this phase space, which may not be computed within the code's semiclassical framework, increases. Hence, the phase space that gives a shift that may presumably be reliably computed shrinks, while the phase space that may not be treated grows. For both shifts and widths, Starcode binds the contribution of the non-semiclassical phase space by unitarity and returns it as an error bar. Therefore, while the quoted shift may be fairly insensitive to T, the error bar is not. These results are to be interpreted as an uncertainty in shift calculations inherent in semiclassical calculations, i.e., the part of the phase space that may not be treated semiclassically can make a substantial shift contribution rather than a temperature-insensitive shift.

3. Straight line SimU and Simulation agreement is excellent for the charged cases (B III and NV), as expected and, surprisingly, much poorer (though still good) for LiI.

4. SCP consistently gives significantly smaller shifts for ion lines (B III and NV). Furthermore, straight-line SCP calculations give significantly larger shifts than hyperbolic path SCP calculations. This is counterintuitive, since as discussed, hyperbolic paths result in a stronger effective interaction close to the emitter and, hence, action. Furthermore, the straight-line SCP shifts are monotonically decreasing as a function of T, in contrast to the hyperbolic trajectory results. This behavior follows the semiclassical b-functions [2], which result in lower shifts for hyperbolic compared to straight-line paths and, in addition, a decreasing b-function with increasing ion charge.

10.2. LiI

Figure 6 plots the shifts computed by five codes *vs.* electron temperature T. Except for Starcode (explained above), in both versions, the shift is decreasing with T, which is expected in view of the weakening of the interaction. The differences between the other three codes are generally better than 20%, although it is not clear why SimU and Simulation, which essentially use the same model and, hence, would, in principle, be expected to give the same results, differ by that amount.

Figure 6. A comparison of LiI 2s-2p transition shifts as computed by SCP (solid black), SimU (dotted red), Simulation (dashed green) and Starcode with (dash-dotted blue) and without (dot-double dash orange) penetration accounted for.

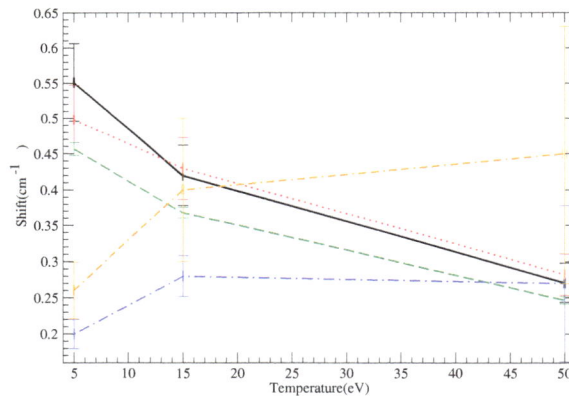

10.3. B III

Figure 7 plots the shifts computed by five codes *vs.* electron temperature T. The general trend (shifts decreasing with T) is observed, with the exception of the non-penetrative Starcode, which gives identical (within errors) results for the 15 and 50 eV cases, which has been discussed above, and SCP, which shows a small increase between five and 15 eV. SCP is significantly lower for small T, which might be attributable to the use of perturbative impact theory by SCP, as it reflects the behavior of the semiclassical b-function. This is consistent with the convergence seen for higher T.

Figure 7. A comparison of B III 2s-2p transition shifts as computed by SCP (solid black), SimU (dotted red), DARC (dashed green) and Starcode with (dash-dotted blue) and without (dot-double dash orange) penetration accounted for.

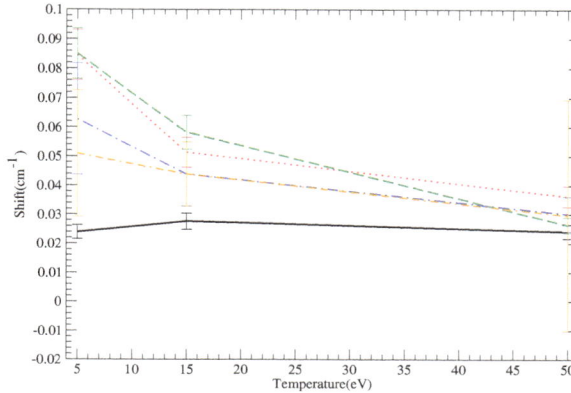

Figure 8. A comparison of hyperbolic and straight-line trajectories for the B III 2s-2p transition widths, as computed by SCP-hyperbolic (solid black), SCP-straight (dotted red), SimU-hyperbolic (dashed green), SimU-straight (dash-dotted blue) and Simulation (dash-double dotted orange).

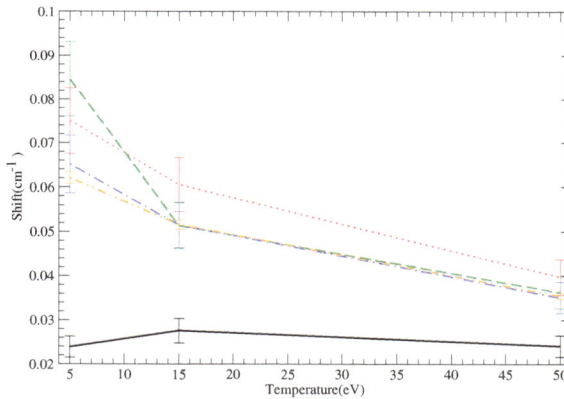

Figure 8 plots the shifts computed by five codes *vs.* electron temperature T. The straight line SimU and Simulation agreement is very good, as discussed above. The SimU hyperbolic trajectory shifts are also in very good agreement with the mentioned straight-line results, except at the lowest T. SCP is also in fair agreement.

10.4. NV

Figure 9 plots the shifts computed by five codes *vs.* electron temperature T. DARC produces significantly larger shifts at low T, with all codes converging for large T. The large difference with

Starcode seems to indicate that higher order partial waves (except the s-wave, which is viewed as an error bar in Starcode) differ substantially from the semiclassical results. SCP is lowest and exhibits insensitivity to T. SimU and Starcode without penetration also agree quite well, except at the lowest T. The lowest temperature point discrepancy between SimU and Starcode-NP could be due to the fact that Starcode does not compute the short impact parameter ($\rho < \frac{\hbar}{mv}$) contribution, which is not semiclassical, but adds an estimate to the shift, as discussed above. The calculations agree within their respective calculation uncertainties, although one might have expected a better agreement. However, attributing the difference to the phase space not included in the Starcode-NP calculation (but added as an error bar) does not seem to hold: For example, as already discussed, Starcode-NP calculations treat collisions with impact parameters larger than the de Broglie wavelength and use a unitarity-based estimate for smaller impact parameters. If instead, Starcode-NP treats all collisions with impact parameters larger than 0.1 times the de Broglie wavelength and only uses unitarity estimates for impact parameters smaller than 0.1 times the de Broglie wavelength, the shift actually decreases and is in the range of $1 = 1.5 \text{ cm}^{-1}$ at the lowest temperature, compared to a value of 3 cm^{-1} for SimU.

Figure 9. A comparison of NV 2s-2p transition shifts as computed by SCP (solid black), SimU (dotted red), DARC (dashed green) and Starcode with (dash-dotted bluw) and without (dot-double dash orange) penetration accounted for.

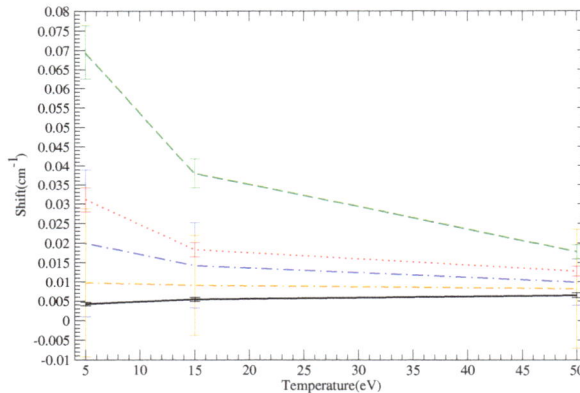

Figure 10 plots the shifts computed by five codes *vs.* electron temperature T.

The three straight line results are in fairly good agreement, with SimU and Simulation being in excellent agreement. The hyperbolic trajectory and straight line SimU results are interesting in that the hyperbolic trajectory shift is larger at the lowest T (expected, as discussed in view of the stronger effective interaction for hyperbolic rather than straight line trajectories) and slightly smaller for the other two T values; however, that difference is within the calculation uncertainties, and the two calculations converge, as expected, for high T.

Figure 10. A comparison of hyperbolic and straight line trajectories for the NV 2s-2p transition widths, as computed by SCP-hyperbolic (solid), SCP-straight (dotted), SimU-hyperbolic (dashed), SimU-straight (dash-dotted) and Simulation (dash-double dotted).

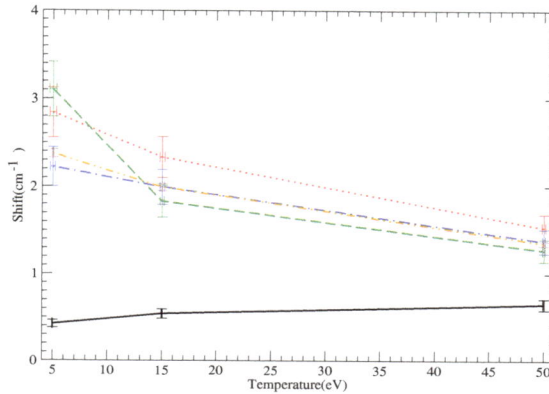

Acknowledgments

The authors would like to acknowledge the International Atomic Energy Agency (B.J. Braams and H.-K. Chung) for the organizational and financial support of this workshop.

Author Contributions

The present work is based on codes developed by all authors, who also participated in all aspects of this work.

Appendix A: Atomic Data

Atomic data were taken from the problem specification, with the 2s state taken as zero energy:

Table A1. 2s-2p Atomic Data.

Species	2p Energy (cm^{-1})	Oscillator Strength
LiI	14,903.89	0.7472
B III	48,381.07	0.3629
NV	80,635.67	0.234

Energies were averaged over fine structure components for $l > 0$, while reduced matrix elements were obtained from the respective multiplet-averaged oscillator strengths, f, via the equation:

$$|(nl|r|n'l')| = (-1)^{l+l_>}\sqrt{\frac{3f(2l'+1)}{2(E_{nl}-E_{n'l'})}} \tag{9}$$

89

with $l_> = max(l, l')$, i.e., one in our case.

Not all codes adhered to this specification. DARC, SCP and Starcode included fine structure and matrix elements used from atomic structure packages, as well as energies from these packages or experimental ones. However, in all cases, no significant differences were introduced in either the energy differences or the reduced matrix elements.

Conflicts of Interest

The authors declare no conflict of interest.

References

1. Griem, H.R. *Plasma Spectroscopy*; McGraw-Hill: New York, NY, USA, 1964.
2. Griem, H.R. *Spectral Line Broadening by Plasmas*; Academic: New York, NY, USA, 1974.
3. Alexiou, S. Collision operator for isolated ion lines in the standard Stark-broadening theory with applications to the Z scaling in the Li isoelectronic series 3P-3S transition. *Phys. Rev. A* **1994**, *49*, 106–119.
4. Elabidi, H.; Sahal-Brechot, S. Checking the dependence on the upper level ionization potential of electron impact widths using quantum calculations. *Eur. Phys. J. D* **2011**, *61*, 285–290.
5. Seidel, J. Effects of ion motion on hydrogen Stark profiles. *Z. Naturforsch.* **1977**, *32a*, 1207–1214.
6. Duan, B.; Bari, M.A.; Wu, Z.Q.; Jun, Y.; Li, Y.M. Widths and shifts of spectral lines in He II ions produced by electron impact. *Phys. Rev. A* **2012**, *86*, 052502.
7. Griem, H.; Ralchenko, Y.V.; Bray, I. Stark broadening of the B III 2s-2p lines. *Phys. Rev. E* **1997**, *56*, 7186–7192.
8. Alexiou, S.; Lee, R.W. Semiclassical calculations of line broadening in plasmas: Comparison with quantal results. *J. Quant. Spectrosc. Radiat. Transf.* **2006**, *99*, 10–20.
9. Alexiou, S. *Phys. Rev. Lett.* Problem with the standard semiclassical impact line-broadening theory. **1995**, *75*, 3406–3409.
10. Alexiou, S.; ad Poquérusse, A. Standard line broadening impact theory for hydrogen including penetrating collisions. *Phys. Rev. E* **2005**, *72*, 046404.
11. Seaton, M.J. Line-profile parameters for 42 transitions in Li-like and Be-like ions. *J. Phys. B* **1988**, *21*, 3033–3054.
12. Ralchenko, Y.V.; Griem, H.; Bray, I.; Fursa, D.V. Electron collisional broadening of 2s3s-2s3p lines in Be-like ions. *J. Quant. Spectrosc. Radiat. Transf.* **2001**, *71*, 595–607.
13. Alexiou, S.; Lee, R.W.; Glenzer, S.; Castor, J. Analysis of discrepancies between quantal and semiclassical calculations of electron impact broadening in plasmas. *J. Quant. Spectrosc. Radiat. Transf.* **2000**, *65*, 15–22.
14. Elabidi, H.; Ben Nessib, N.; Cornille, M.; Dubau, J.; Sahal-Brechot, S. Electron impact broadening of spectral lines in Be-like ions: Quantum calculations. *J. Phys. B* **2008**, *41*, 025702.

15. Bouguettaia, H.; Chihi, Is.; Chenini, K.; Meftah, M.T.; Khelfaoui, F.; Stamm, R. Application of path integral formalism in spectral line broadening: Lyman-α in hydrogenic plasma. *J. Quant. Spectrosc. Radiat. Transf.* **2005**, *94*, 335-346.

16. Alexiou, S. Problems with the use of line shifts in plasmas. *J. Quant. Spectrosc. Radiat. Transf.* **2003**, *81*, 461-471.

17. Gunderson, M.A.; Haynes, D.A.; Kilcrease, D.P. Using semiclassical models for electron broadening and line shift calculations of $\Delta n=0$ and $\Delta n \neq 0$ dipole transitions. *J. Quant. Spectrosc. Radiat. Transf.* **2006**, *99*, 255–264.

18. Stambulchik, E.; Maron, Y. A study of ion-dynamics and correlation effects for spectral line broadening in plasma: K-shell lines. *J. Quant. Spectrosc. Radiat. Transf.* **2006**, *99*, 730–749.

19. Gigosos, M.A.; Cardenoso, V. New plasma diagnosis tables of hydrogen Stark broadening including ion dynamics. *J. Phys. B: Atomic, Mol. Opt. Phys.* **1996**, *29*, 4795-4838.

20. Sahal-Brechot, S. Impact theory of the broadening and shift of spectral lines due to electrons and ions in a plasma. *Astron. Astrophys.* **1969**, *1*, 91-123.

21. Sahal-Brechot, S. Impact theory of the broadening and shift of spectral lines due to electrons and ions in a plasma. *Astron. Astrophys.* **1969**, *2*, 322–354.

22. Sahal-Brechot, S. Stark broadening of isolated lines in the impact approximation. *Astron. Astrophys.* **1974**. *35*, 319–321.

23. Fleurier, C.; Sahal-Brechot, S.; Chapelle, J. Stark profiles of some ion lines of alkaline earth elements. *J. Quant. Spectrosc. Radiat. Transf.* **1977**, *17*, 595–604.

24. Dimitrijević, M.S.; Sahal-Brechot, S. Comparison of measured and calculated Stark broadening parameters for neutral-helium. *Phys. Rev. A* **1985**, *31*, 316–320.

25. Elabidi, H.; Sahal-Brechot, S.; Ben Nessib, N. Quantum Stark broadening of 3s-3p spectral lines in Li-like ions; Z-scaling and comparison with semi-classical perturbation theory. *Eur. Phys. J. D* **2009**, *54*, 51–64.

26. Stambulchik, E.; Alexiou, S.; Griem, H.R.; Kepple, P.C. Stark broadening of high principal quantum number hydrogen Balmer lines in low-density laboratory plasmas. *Phys. Rev. E* **2007**, *75*, 016401.

27. Gigosos, M.A.; Gonzalez, M.A.; Konjevic, N. On the Stark broadening of Sr+ and Ba+ resonance lines in ultracold neutral plasmas. *Eur. Phys. J. D* **2006**, *40*, 57–63.

28. Hegerfeldt, G.C.; Kesting, V. Collision-time simulation technique for pressure-broadened spectral lines with applications to Ly-α. *Phys. Rev. A* **1988**, *37*, 1488–1496.

91

Reprinted from *Atoms*. Cite as: Sahal-Bréchot, S.; Dimitrijević, M.S.; Ben Nessib, N. Widths and Shifts of Isolated Lines of Neutral and Ionized Atoms Perturbed by Collisions With Electrons and Ions: An Outline of the Semiclassical Perturbation (SCP) Method and of the Approximations Used for the Calculations. *Atoms* **2014**, 2, 225-252.

Article

Widths and Shifts of Isolated Lines of Neutral and Ionized Atoms Perturbed by Collisions With Electrons and Ions: An Outline of the Semiclassical Perturbation (SCP) Method and of the Approximations Used for the Calculations

Sylvie Sahal-Bréchot [1,*], **Milan S. Dimitrijević** [1,2] **and Nabil Ben Nessib** [3]

[1] Laboratoire d'Étude du Rayonnement et de la Matière en Astrophysique et Atmosphères (LERMA2), Observatoire de Paris, UMR CNRS 8112, UPMC, 5 Place Jules Janssen, 92195 Meudon CEDEX, France
[2] Astronomical Observatory, Volgina 7, 11060 Belgrade, Serbia; E-mail: mdimitrijevic@aob.rs
[3] Department of Physics and Astronomy, College of Science, King Saud University. Riyadh 11451, Saudi Arabia; E-mail: nbennessib@ksu.edu.sa

* Author to whom correspondence should be addressed; E-Mail: sylvie.sahal-brechot@obspm.fr; Tel.: +331-450-774-42; Fax: +331-450-771-00.

Received: 24 April 2014; in revised form: 22 May 2014 / Accepted: 26 May 2014 / Published: 10 June 2014

Abstract: "Stark broadening" theory and calculations have been extensively developed for about 50 years. The theory can now be considered as mature for many applications, especially for accurate spectroscopic diagnostics and modeling, in astrophysics, laboratory plasma physics and technological plasmas, as well. This requires the knowledge of numerous collisional line profiles. In order to meet these needs, the "SCP" (semiclassical perturbation) method and numerical code were created and developed. The SCP code is now extensively used for the needs of spectroscopic diagnostics and modeling, and the results of the published calculations are displayed in the STARK-B database. The aim of the present paper is to introduce the main approximations leading to the impact of semiclassical perturbation method and to give formulae entering the numerical SCP code, in order to understand the validity conditions of the method and of the results; and also to understand some regularities and systematic trends. This would also allow one to compare the method and its results to those of other methods and codes. [1]

Keywords: Stark broadening; impact approximation; isolated lines; semiclassical perturbation picture

---

1. Introduction

"Stark broadening" theory and calculations have been extensively developed for about 50 years. Following the pioneering work by [1–4], the theory and calculation of collisional line broadening in the impact approximation showed a great expansion from the sixties. The theory can now be considered as mature for many applications, especially for accurate spectroscopic diagnostics and modeling, in astrophysics, laboratory plasma physics and technological plasmas, as well. This requires the knowledge of numerous collisional line profiles. Hence, calculations based on a simple method that is accurate and fast enough are necessary for obtaining numerous results. In order to meet these needs, the impact semiclassical perturbation (SCP) method and the associated numerical code were created and developed by Sahal-Bréchot for isolated lines in the sixties and seventies [5–9], then updated [10–12] in the eighties and nineties and then, again, in the present new century [13]. Since the impact approximation was assumed, the method was inspired by the developments of the theory of electron-atom and electron-ion collisions which have been rapidly expanding since the sixties. In particular [14], the semiclassical perturbation method appeared to be especially suitable, due to the speed of the numerical calculations and to the sufficient expected accuracy. Nowadays, the SCP code is extensively used for the needs of spectroscopic diagnostics and modeling, and the results of the published calculations are displayed in the STARK-B database [15].

The aim of the present paper is to introduce the main approximations leading to the impact semiclassical perturbation method and to the formulae entering the numerical SCP code. This would give an idea of the validity conditions of the method and of the results, which are discussed in Section 4. This would also allow one to compare the different codes, and this would help to understand some regularities and systematic trends observed in the experiments and in the results of calculations.

In the next section, we will recall the basic key assumptions leading to the theory of collisional line broadening in impact approximation. We will define the concepts of "complete collision" approximation and of "isolated lines". Then, we will recall the specific approximations made in the semiclassical perturbation treatment for isolated lines of neutral and ionized atoms perturbed by electrons and positive ions in the impact approximation. We will give the formulae entering the

---

[1] The content of the present paper was presented and discussed at the first Spectral Line Shapes in Plasmas (SLSP) code comparison Workshop, which was held in Vienna, Austria, April 2–5, 2012.

SCP numerical code. Then, we will discuss the approximations made and the obtained results. This would allow one to compare the results obtained by other numerical codes and to understand some regularities and systematic trends observed in the experiments and in the results of calculations.

2. Collisional Line Broadening in the Impact Approximation

Following the pioneering work by Baranger [1–4], the theory and calculation of collisional line broadening widths and shifts in the impact approximation for electrons and ion interactions is very briefly outlined in the following.

We consider a neutral or ionized atom surrounded by the perturbers, P, moving around it. We begin with the very general formula [1–4], which gives the intensity, $I(\omega)$, of a spectral line ($i \rightarrow f$) at the angular frequency, ω:

$$I(\omega) = \frac{4\omega^4}{3c^3} |\langle f |\mathbf{d}| i \rangle|^2 \tag{1}$$

$$I(\omega) = \frac{1}{2\pi} \int_{-\infty}^{+\infty} \exp\left(\mathrm{i}\omega s\right) \Phi(s) \mathrm{d}s. \tag{2}$$

Dropping the factor, $4\omega^4/3c^3$, the autocorrelation function $\Phi(s)$ reads:

$$\Phi(s) = \mathrm{Tr}\left[\mathbf{d}\, T^*(s)\, \mathbf{d}\, T(s)\, \rho\right]. \tag{3}$$

Here, ρ is the density matrix of the whole system atom (A) surrounded by the bath of particles (B: photons R and perturbers P), $\mathbf{d}$ is the dipole moment, $I(\omega)$ is the Fourier transform of the autocorrelation function, $\Phi(s)$, and $T(s)$ is the evolution operator of the whole system. Tr is the trace over the states of the whole system and c is the velocity of the light.

• With the *no-back reaction approximation*, which is the first key approximation, the bath, B, remains described by its unperturbed density operator (or its distribution function in the classical picture), irrespective of the amount of energy and polarization diffusing into it from A:

$$\rho(t) = \rho_A(t) \otimes \rho_B(t).$$

It is also assumed that the bath, B, is in a stationary state, and thus, $\rho_B(t) = \rho_B$. We also suppose that the bath of colliding perturbers P (density matrix ρ_P) is decoupled from the bath of photons R (density matrix ρ_R). Thus:

$$\rho_B = \rho_P \otimes \rho_R,$$

and we will only consider in the following the interaction with the bath of colliding particles. Hence, we can write:

$$\Phi(s) = \mathrm{Tr}_A\left[\rho_A \, \mathrm{Tr}_P[\mathbf{d}\, T^*(s)\, \mathbf{d}\, T(s)\, \rho_P\,]\right] \tag{4}$$

where Tr_A is the trace over the atomic states and Tr_P is a trace over all the perturbers. When the perturbers can be treated classically, the trace over the perturbers is replaced by a statistical average $[...]_{AV}$ over all modes of motion of the perturbers, which move on a classical path [4]:

$$\Phi(s) = \mathrm{Tr}_A \left[\rho_A \, \mathbf{d} \, T^*(s) \, \mathbf{d} \, T(s) \right]_{AV}. \tag{5}$$

• The second key approximation is the *impact approximation*: the interactions are separated in time. In other words, the atom interacts with one perturber only at a given time: the mean duration, τ, of an interaction must be much smaller than the mean interval between two collisions, ΔT. This can be expressed as:

$$\tau \ll \Delta T,$$

where $\tau \approx \frac{\rho_{typ}}{v_{typ}}$, ρ_{typ} is a mean typical impact parameter and v_{typ} a mean typical relative velocity.

ΔT is of the order of the inverse of the collisional line width, which can be very roughly written as equal to $N_P v_{typ} \rho_{typ}^2$, where N_P is the density of the perturbers.

Thus, the validity condition of the impact approximation can be written as:

$$\rho_{typ} << N_P^{-1/3}. \tag{6}$$

The "collision volume", of the order of ρ_{typ}^3, must be smaller than N_P^{-1}, the volume per perturber [1]. In other words, the perturbers are independent and their effects are additive.

• Then, we will make the *complete collision approximation*: within this approximation, atom–radiation and atom–perturber interactions are decoupled. This implies that the collision must be considered as instantaneous in comparison with the time, Γ^{-1}, characteristic of the evolution of the excited state under the effect of the interaction with the radiation. In other words, the interaction process has time to be completed before the emission of a photon.

The complete collision approximation can become invalid in line wings, even if it remains valid in the line center. Its condition of validity is:

$$\tau << 1/\Delta\omega,$$

where $\Delta\omega$ is the detuning.

Then, using Equation (4) in the quantum picture or Equation (5) in the semiclassical picture, the calculation of the line profile becomes an application of the theory of atomic collisions. This implies that we have to express and calculate the scattering S-matrix.

• In addition, in the case of *isolated lines* (*cf.* Figure 1), which implies that neighboring levels do not overlap [3], the line profile, $F(\omega)$, between the levels, i, $(\alpha_i J_i)$ and f $(\alpha_f J_f)$, is Lorentzian:

$$I(\omega) = \rho_A \left(\alpha_i J_i \right) \frac{4\omega^4}{3c^3} \, F\left(\omega\right) \tag{7}$$

$$F(\omega) = \frac{1}{\pi} \int_0^\infty e^{i(\omega - \omega_{if})s} \, \Phi\left(s\right) \mathrm{d}s \tag{8}$$

$$\Phi\left(s\right) = \exp\left(-\left(w + \mathrm{i}d\right)s\right) \tag{9}$$

and

$$F(\omega) = \frac{w}{\pi \left[(\omega - \omega_{if} - d)^2 + w^2\right]} \tag{10}$$

ρ_A is the reduced atomic density matrix at the stationary state.

Figure 1. The studied line, $i - f$ (red-bold arrow, example of an emission line): $\alpha_i J_i$ is the initial level, and $\alpha_f J_f$ is the final level. The perturbing levels are αJ, $\alpha' J'$. Black-thin arrows: the inelastic transitions in excitation and deexcitation. The levels are broadened (grey strips), but do not overlap.

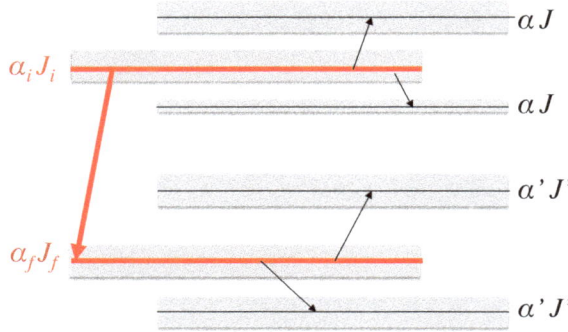

- The determination of the atomic density matrix elements (the populations of the levels) deserves special attention. In fact, Baranger [1–4] assumes LTE (local Thermodynamical equilibrium), and then, $\rho_A(\alpha_i J_i)$ is the Boltzmann factor:

$$\rho_A(\alpha_i J_i) = g_i \frac{\exp\left(-\frac{E_i}{kT}\right)}{Z(T)} \tag{11}$$

where g_i is the statistical weight of the $(\alpha_i J_i)$ level, E_i its energy, k the Boltzmann constant and $Z(T)$ the partition function at the temperature, T.

Out of LTE, it is more complicated, because the calculation of the profile and of the populations are not decoupled for obtaining the intensity of the line. In particular, (7) is no longer valid. This is the problem of the *redistribution of radiation*, which has been extensively studied for many years, especially for the study of stellar atmospheres [16–18] and some laboratory plasmas.

However, within the complete redistribution approximation, the calculation of the line profile and of the populations become decoupled, and (7) is again valid. This needs to assume that the radiation is weak (the atom-radiation interaction is treated with the second-order perturbation theory) and to make the *Markov approximation*: the time evolution of $\rho_A(t)$ does not depend on its past history and only depends on the time, t.

The coupling of the atomic density matrix and the line profile needs to go beyond the Markov approximation, but this is outside the scope of the present theory. See, for instance, [18].

Consequently, out of LTE, within the complete redistribution approximation, the statistical equilibrium equations for a collisional radiative model can be solved for obtaining the populations by ignoring the line profiles, and then, the resulting populations are multiplied to the line profile given by Equation (8) for obtaining the intensity.

We now concentrate our attention on the determination of the line profile given by (10). We will determine the half-width, w (half width at half-maximum intensity), and the shift, d, of the profile of the $i - f$ line. They are given by:

$$w + \mathrm{id} = \mathrm{Tr}_P \left(1 - S_{ii} S_{ff}^*\right) \tag{12}$$

where the trace over the perturbers, Tr_P, is given by:

$$Tr_P = N_P \int_0^\infty v f(v) dv \int_0^\infty 2\pi \rho \, \mathrm{d}\rho \oint \frac{\mathrm{d}\Omega}{4\pi}. \tag{13}$$

Equation (13) is written in the semiclassical picture, where the colliding perturbers move along a classical path of impact parameter ρ, but this is not essential. Here, v is the relative velocity of the perturber, $f(v)$ the distribution of velocities (Maxwell distribution) and N_P the density of the perturbers.

$$\oint \frac{\mathrm{d}\Omega}{4\pi}$$

is the angular average over all directions of the colliding perturbers. Hence:

$$w + \mathrm{id} = N_P \int_0^\infty v f(v) dv \int_0^\infty 2\pi \rho \, \mathrm{d}\rho \left\langle 1 - S_{ii} S_{ff}^* \right\rangle_{angular\ average} \tag{14}$$

This angular average is a linear combination of products of "3-j" coefficients and of initial and final Zeeman states of the S-matrix elements [3]. Thus, Equation (14) reads:

$$w + \mathrm{id} = N_P \int_0^\infty v f(v) dv \int_0^\infty 2\pi \rho \, \mathrm{d}\rho$$

$$\times \left[1 - \sum_{\substack{M_i M_i' \\ M_f M_f' \\ \mu}} (-1)^{2J_f + M_f + M_f'} \begin{pmatrix} J_i & 1 & J_f \\ -M_i & \mu & M_f \end{pmatrix} \begin{pmatrix} J_i & 1 & J_f \\ -M_i' & \mu & M_f' \end{pmatrix} \right. \tag{15}$$

$$\times \left. \left\langle \alpha_f J_f M_f \left| S^* \right| \alpha_f J_f M_f' \right\rangle \left\langle \alpha_i J_i M_i \left| S \right| \alpha_i J_i M_i' \right\rangle \right]$$

This is the basic "Baranger's formula" which gives the half-width and the shift of an isolated line perturbed by collisions (*i.e.*, impact and complete collision approximation). We notice that the S-matrix is symmetric, unitary and can be calculated in any reference frame. Its matrix elements have to be calculated for a given relative velocity, v, and an impact parameter, ρ, in the semiclassical description. This will appear in the following equations.

In fact, the T-matrix, where $T = 1 - S$ (and, thus, $T^*T = 2Re(T)$), is used in the SCP method. Thus, we will use it in the following.

After elementary calculations using the properties of the "3-j" coefficients, Equation (15) reads:

$$w + \mathrm{i}d = N_P \int_0^\infty v\, f(v)dv \int_0^\infty 2\pi\, \rho\, \mathrm{d}\rho$$

$$\times \left[\sum_{M_i} \tfrac{1}{2J_i+1} \langle \alpha_i J_i M_i | T(\rho, v) | \alpha_i J_i M_i \rangle + \sum_{M_f} \tfrac{1}{2J_f+1} \langle \alpha_f J_f M_f | T^*(\rho, v) | \alpha_f J_f M_f \rangle \right.$$

$$- \sum_{\substack{M_i M'_i \\ M_f M'_f \\ \mu}} (-1)^{2J_f + M_f + M'_f} \begin{pmatrix} J_i & 1 & J_f \\ -M_i & \mu & M_f \end{pmatrix} \begin{pmatrix} J_i & 1 & J_f \\ -M'_i & \mu & M'_f \end{pmatrix}$$

$$\left. \times\ \langle \alpha_i J_i M_i | T(\rho, v) | \alpha_i J_i M'_i \rangle \langle \alpha_f J_f M_f | T^*(\rho, v) | \alpha_f J_f M'_f \rangle \right] \tag{16}$$

Then, the half-width at half-maximum intensity w, and the full width at half-maximum intensity $W = 2w$ become:

$$W = 2w = N_P \int v\, f(v)\mathrm{d}v$$

$$\times \left[\left(\sum_{\alpha J} \sigma\,(\alpha_i J_i \to \alpha J, v) + \sum_{\alpha' J'} \sigma\,(\alpha_f J_f \to \alpha' J', v) \right) \right.$$

$$- 2\mathrm{Re} \int_0^\infty 2\pi\, \rho\, \mathrm{d}\rho\ \Big[\sum_{\substack{M_i M'_i \\ M_f M'_f \\ \mu}} (-1)^{2J_f + M_f + M'_f} \tag{17}$$

$$\times \begin{pmatrix} J_i & 1 & J_f \\ -M_i & \mu & M_f \end{pmatrix} \begin{pmatrix} J_i & 1 & J_f \\ -M'_i & \mu & M'_f \end{pmatrix}$$

$$\left. \times\ \langle \alpha_f J_f M_f\, | T^*(\rho, v)\, | \alpha_f J_f M'_f \rangle \langle \alpha_i J_i M_i\, | T(\rho, v)\, | \alpha_i J_i M'_i \rangle \Big] \right]$$

In Equation (17):

$$\sum_{\alpha J} \sigma\,(\alpha_i J_i \to \alpha J, v)$$

is a sum of all the inelastic cross-sections originating from the initial level towards all perturbing levels, αJ, in excitation and deexcitation (*cf.* Figure 1), with $\alpha_i J_i \neq \alpha J$, and of the elastic cross-section $\sigma(\alpha_i J_i \to \alpha_i J_i, v)$.

In fact:

$$\sigma\,(\alpha_i J_i \to \alpha J, v) = \int_0^\infty 2\pi\, \rho\, \mathrm{d}\rho\, P(\alpha_i J_i \to \alpha J,\, \rho,\, v) \tag{18}$$

where $P(\alpha_i J_i \to \alpha J,\, \rho,\, v)$ is the transition probability between the levels, $(\alpha_i J_i)$ and (αJ), for the impact parameter, ρ, and the relative velocity, v.

We have similar expressions for the inelastic cross-sections originating from the final level $\alpha_f J_f$. Then the perturbing levels are $\alpha' J'$ and the elastic cross-section is $\sigma(\alpha_f J_f \to \alpha_f J_f, v)$.

We notice that the transition probability:

$$P(\alpha_i J_i \to \alpha J,\, \rho,\, v) = \frac{1}{2J_i + 1} \sum_{M_i M} |\langle \alpha_i J_i M_i | T(\rho, v) | \alpha J M \rangle|^2 \tag{19}$$

is a sum over the final M substates and an average over the initial M_i substates of the T-matrix elements.

The third term of Equations (16) and (17) is the so-called "interference term", which is a linear combination of off-diagonal elastic elements of the initial and final elastic elements of the T-matrix. For collisions with electrons, it is often small (10% of the total width or thereabout), but not always.

Equations (16) and (17) will be used in the SCP method and numerical code.

Finally, we notice that the fine structure (and, *a fortiori*, hyperfine structure) can generally be ignored, and consequently, the fine structure components (or hyperfine components) have the same width and the same shift, which are equal to those of the multiplet. This is due to the fact that the electronic spin, S (or nuclear spin I), has no time to rotate during the collision time (of the order of ρ/v, the mean duration of the collision). This is only true in LS coupling.

3. The Semiclassical Perturbation Approximation (SCP) for Stark-Broadening Studies

We study the case of isolated lines in the impact and complete collision approximations recalled in the preceding section.

Within the semiclassical approximation, the perturbers (electrons or positive ions) are assumed to be classical particles, and they move along a classical path unperturbed by the interactions with the radiating atom. The atom is described by its quantum wave-functions and energy levels. With the perturbation approximation, the atom–perturber interaction is treated by the time-dependent perturbation theory. The long-range approximation will be made.

3.1. The Semiclassical Approximation and the Parametric Representation of the Orbits

For neutral radiating atoms, the trajectory is rectilinear, and for radiating ions, it is a hyperbola. The parametric representation of the trajectory can be found in [19] for the repulsive case and also in [6] for the attractive case and the straight path case. The radiating atom is at the origin, O, of the axes. The quantization axis, Oz, is perpendicular to the collision plane, xOy, and the velocity vector of the perturber is parallel to Ox at time $t = -\infty$. Z_A is the charge of the radiating atom ($Z_A = 0$ for a neutral), Z_P the charge of the perturber ($Z_P = 1$ for an electron), μ the reduced mass of the system atom–perturber, v the relative velocity and ρ the impact parameter.

The coordinates are given in Table 1.

Table 1. Parametric representation of the orbits (trajectories).

	Attractive Hyperbola	**Repulsive Hyperbola**	**Straight Path**
x	$a\left(\varepsilon - \cosh u\right)$	$a\left(\varepsilon + \cosh u\right)$	ρ
y	$a\sqrt{\varepsilon^2 - 1}\,\sinh u$	$a\sqrt{\varepsilon^2 - 1}\,\sinh u$	$\rho\sinh u$
t	$\frac{a}{v}\left(\varepsilon\sinh u - u\right)$	$\frac{a}{v}\left(\varepsilon\sinh u + u\right)$	$\frac{\rho}{v}\sinh u$
distance of closest approach	$a(\varepsilon - 1)$	$a(\varepsilon + 1)$	ρ

For radiating ions, $a = e^2 \frac{Z_A Z_P}{\mu v^2}$ is the semi-major axis of the hyperbola, e is the electron charge and $\varepsilon = \left(1 + \frac{\rho^2}{a^2}\right)^{\frac{1}{2}}$ is the eccentricity.

3.2. The Time-Dependent Perturbation Approximation for the Calculation of the S (or T) Matrix

We use the following expression for the S-matrix:

$$S = \mathcal{T}\left(\exp\left(\frac{1}{i\hbar}\int_{-\infty}^{+\infty}\tilde{V}(t)\,\mathrm{d}t\right)\right) \tag{20}$$

where $\mathcal{T}$ is the chronological operator and $\tilde{V}(t)$ is the atom–perturber interaction potential in interaction representation. Now, we make the second order perturbation theory: we expand the S−matrix given by Equation (20) in multipoles (the so-called Dyson series), and we retain the two first terms of the expansion:

$$S = 1 + \frac{1}{i\hbar}\int_{-\infty}^{+\infty}\tilde{V}(t)\mathrm{d}t + \frac{1}{i^2\hbar^2}\int_{-\infty}^{+\infty}\tilde{V}(t)\mathrm{d}t\int_{-\infty}^{t}\tilde{V}(t')\mathrm{d}t' \tag{21}$$

and T follows:

$$T = +\frac{i}{\hbar}\int_{-\infty}^{+\infty}\tilde{V}(t)\mathrm{d}t + \frac{1}{\hbar^2}\int_{-\infty}^{+\infty}\tilde{V}(t)\mathrm{d}t\int_{-\infty}^{t}\tilde{V}(t')\mathrm{d}t'. \tag{22}$$

3.3. The Atom-Perturber Interaction Potential

We will only study the case of an ideal plasma, which is valid if the Debye length is much larger than the mean distance between perturbers. Hence the atom–perturber interaction V is the electrostatic potential between the N atomic electrons, the nucleus of charge $(Z_A + N)$ and the perturber of charge, Z_P (with $Z_P = -1$ for an electron):

$$V = \frac{(Z_A + N)\, Z_P\, e^2}{r_P} - Z_P\, e^2 \sum_{i=1}^{N}\frac{1}{r_{iP}} \tag{23}$$

where r_P is the distance between the nucleus and the perturber and r_{iP} the distance between the ith atomic electron and the perturber. $Z_A = 0$ for a neutral atom.

Then, $1/r_{iP}$ is expanded into multipolar components. We will only retain the long range part, since we will make the long-range perturbation theory. Penetrating orbits are outside the scope of SCP method and code. The $Y_{\lambda\mu}$ denote the spherical harmonics.

$$V = \frac{Z_A Z_P\, e^2}{r_P} - \sum_{\lambda=1}^{\infty}\frac{4\pi Z_P\, e^2}{2\lambda + 1}\frac{1}{r_P^{\lambda+1}}\sum_{\mu=-\lambda}^{+\lambda}\sum_{i=1}^{N}r_i^\lambda Y_{\lambda\mu}(\hat{r}_P)Y_{\lambda\mu}^*(\hat{r}_i). \tag{24}$$

The first term of this expansion is the Coulomb term, which does not play any role in the calculation of the S-matrix, due to its spherical symmetry. The following ones have to be retained: the dipole term ($\lambda = 1$) and the quadrupole term ($\lambda = 2$).

3.4. Determination of the T-Matrix Elements Using Equations (22) and (24) and the Coordinates of the Perturber

Thanks to the preceding subsections, we are now able to obtain the semiclassical perturbation expressions of Equations (16) and (17). The $T-$matrix elements, which enter Formula (16), read:

$$\langle \alpha_i J_i M_i | T(\rho, v) | \alpha_i J_i M_i' \rangle = \frac{\mathrm{i}}{\hbar} \int_{-\infty}^{+\infty} \langle \alpha_i J_i M_i | V(\rho, v, t) | \alpha_i J_i M_i' \rangle \, \mathrm{d}t$$

$$+ \frac{1}{\hbar^2} \sum_{\alpha J M} \int_{-\infty}^{+\infty} \langle \alpha_i J_i M_i | V(\rho, v, t) | \alpha J M \rangle \, \mathrm{e}^{\mathrm{i}\omega_{ij}t} \mathrm{d}t \int_{-\infty}^{t} \langle \alpha J M | V(\rho, v, t') | \alpha_i J_i M_i' \rangle \, \mathrm{e}^{-\mathrm{i}\omega_{ij}t'} \mathrm{d}t'$$

(25)

where ω_{ij} is the angular frequency difference between the initial i ($\alpha_i J_i$) level and the perturbing j (αJ) level. We have an analogous expression for the final f ($\alpha_f J_f$) level.

The first term of Equation (16) is the first direct term, originating from the initial level. It reads:

$$\sum_{M_i} \langle \alpha_i J_i M_i | T(\rho, v) | \alpha_i J_i M_i \rangle = \frac{\mathrm{i}}{\hbar} \sum_{M_i} \int_{-\infty}^{+\infty} \langle \alpha_i J_i M_i | V(\rho, v, t) | \alpha_i J_i M_i \rangle \, \mathrm{d}t$$

$$+ \frac{1}{\hbar^2} \sum_{\alpha J M M_i} \int_{-\infty}^{+\infty} \langle \alpha_i J_i M_i | V(\rho, v, t) | \alpha J M \rangle \, \mathrm{e}^{\mathrm{i}\omega_{ij}t} \mathrm{d}t \int_{-\infty}^{t} \langle \alpha J M | V(\rho, v, t') | \alpha_i J_i M_i \rangle \, \mathrm{e}^{-\mathrm{i}\omega_{ij}t'} \mathrm{d}t'$$

(26)

The second term of Equation (16) is the second direct term, originating from the final level, αJ_f, and has an analogous expression.

The third term of Equation (16) is the interference term.

Then, we use the second order long-range expansion of the interaction potential (Equation (24)). We do not enter the details of the calculations, which can be found in Sahal-Bréchot [6].

We summarize the different steps and only give the resulting formulae in the following:

• The quadrupolar potential is taken into account only for elastic collisions and for inelastic collisions between the fine structure levels of the initial and final levels.

• The fist term of Equation (26) is taken as equal to zero:
  $-$It is exactly equal to zero for the ($\lambda = 1$)-dipolar potential contribution, due to selection rules on the "$3j$" coefficients.
  $-$It is only different from zero for the ($\lambda = 2$)-quadrupolar potential contribution in the case of $J_i \geq 1$ integer numbers, but is neglected.

• The second term of Equation (26) contains quadrupolar and dipolar elements.

• There is no interference terms between dipolar and quadrupolar contributions, which add independently.

• The Debye screening is taken into account by introducing an upper cutoff at the Debye length. The calculations are detailed in [6] for obtaining the $T-$matrix elements and in [7] for the integration over the impact parameters and the choice of the cutoffs.

The full width at half-maximum intensity, W, and the shift, d, of the $i - f$ line can be put under the form [7]:

$$W = N_P \int v f(v) dv \left(\sum_{i' \neq i} \sigma_{ii'}(v) + \sum_{f' \neq f} \sigma_{ff'}(v) + \sigma_{el} \right) \qquad (27)$$

Here, the perturbing levels are denoted as i' and f' for simplicity in the writing.

• The inelastic contribution of the upper level, i, is calculated as follows [7]. The dipolar interaction potential is taken into account:

$$\sum_{i' \neq i} \sigma_{ii'}(v) = \frac{1}{2} \pi R_1^2 + \int_{R_1}^{R_D} 2\pi \rho d\rho \sum_{i' \neq i} P_{ii'}(\rho, v) \qquad (28)$$

$$P_{ii'}(\rho, v) = \frac{1}{\hbar^2} \left| \int_{-\infty}^{+\infty} V_{ii'} \exp\left(-\frac{i}{\hbar} \Delta E_{ii'} \, t\right) \right|^2 \qquad (29)$$

with $\Delta E_{ii'} = \hbar \omega_{ii'}$.

The upper cutoff, R_D, is the Debye radius. The lower cutoff, R_1, is chosen as in [7,14]:

$$\sum_{i' \neq i} P_{ii'}(\rho, v) = \frac{1}{2}$$

or, in the case of electron collisions:

$$\sum_{i' \neq i} \sigma_{ii'}(v) = \int_{\min(\langle r_i \rangle, \langle r_i' \rangle)}^{R_D} 2\pi \rho d\rho \sum_{i' \neq i} P_{ii'}(\rho, v) \qquad (30)$$

where $\langle r_i \rangle$ is the mean radius of the i level and $\langle r_i' \rangle$ is the mean radius of the i' level. An hydrogenic approximation is sufficient for calculating these mean radii, and we retain as in [14] the smallest result between the one of Equation (29) and the one of Equation (30). This minimizes [14] the role of close collisions, which are considered as overestimated by the perturbation theory.

In the case of collisions with positive ions (cf. [20]), we retain the smallest between the result of Equation (28) and of the following one:

$$\sum_{i' \neq i} \sigma_{ii'}(v) = \pi \langle r_i \rangle^2 P_{ii'}\left(\langle r_i \rangle, v\right) + \int_{\langle r_i \rangle}^{R_D} 2\pi \rho d\rho \sum_{i' \neq i} P_{ii'}(\rho, v) \qquad (31)$$

The inelastic contribution from the lower level, f, is calculated in the same way.

• The contribution of elastic collisions, σ_{el}, is calculated as follows:

$$\sigma_{el} = 2\pi R_2^2 + \int_{R_2}^{R_D} 2\pi \rho d\rho \sin^2 \varphi + \sigma_r \qquad (32)$$

with:

$$\varphi = (\varphi_p^2 + \varphi_q^2)^{\frac{1}{2}} \qquad (33)$$

and:

$$\varphi_p = \sum_{i' \neq i} \varphi_{ii'} - \sum_{f' \neq f} \varphi_{ff'} \qquad (34)$$

The phase shifts, φ_p and φ_q, are due, respectively, to the dipolar potential (the polarization potential in the adiabatic approximation) and to the quadrupolar potential. The contribution of the dipolar interaction to the width is often denoted as the quadratic contribution. Their expressions will be detailed hereafter. The lower cutoff, R_2, is chosen as $\varphi(R_2, v) = 1$ [7]. The interference term is taken into account in φ_p and φ_q.

Here, σ_r is the contribution of the Feshbach resonances [9], which concerns only ionized radiating atoms colliding with electrons. It is an extrapolation of the excitation collision strengths (and not the cross-sections) under the threshold by means of the semiclassical limit of the Gailitis approximation (see [9] for details of the calculations).

• The shift, d, is given by (the dipolar interaction potential is the only one to be taken into account):

$$d = N_P \int v f(v) dv \int_{R_3}^{R_D} 2\pi\rho d\rho \sin(2\varphi_p) \tag{35}$$

The cutoff, R_3 [7], is chosen as $2\varphi_p(\rho, v) = 1$. There is no strong collision term for the shift.

3.5. Expressions of $P_{ii'}(\rho, v), P_{ff'}(\rho, v), \varphi_p(\rho, v), \varphi_q(\rho, v)$, Symmetrization and Some Asymptotic Limits

We use CGS units, but we will often use atomic units ($\hbar = 1$, $e = 1$, $m_e = 1$, and thus, $a_0 = 1$, $I_H = 1/2$):
$a_0 = \frac{\hbar^2}{m_e e^2}$ is the first Bohr orbit radius, m_e is the electron mass and e the electron charge. $I_H = \frac{m_e e^4}{2\hbar^2}$ is the ionization energy of hydrogen (1 Rydberg).

3.5.1. Case of Neutral Atoms (Straight Path)

Contribution of the dipolar interaction: expressions of the transition probabilities, of the inelastic cross-sections and of φ_p

$$\frac{1}{2}P_{ii'}(\rho, v) + 2\,\mathrm{i}\varphi_{ii'}(\rho, v) = \frac{a_0^2}{\rho^2} \frac{2\,I_H^2}{E\,\Delta E_{ii'}} f_{ii'} \frac{\mu}{m_e} Z_P^2 \left(A(z) + \mathrm{i}B(z)\right) \tag{36}$$

$z = \frac{\rho \Delta E_{ii'}}{\hbar v}$, $f_{ii'}$ is the oscillator strength of the (ii') transition, and $E = \frac{1}{2}\mu v^2$ is the incident kinetic energy of the perturber.

$$A(z) + \mathrm{i}B(z) = \frac{1}{2} \int_{-\infty}^{+\infty} du \int_{-\infty}^{u} du' \left(\frac{1+\sinh u\,\sinh u'}{\cosh^2 u\,\cosh^2 u'} \exp\left(\mathrm{i}z\left(\sinh u - \sinh u'\right)\right)\right) \tag{37}$$

$A(z)$ and $B(z)$ are connected together by the Hilbert transform:

$$B(z) = \tfrac{1}{\pi}\mathrm{pv} \int_{-\infty}^{+\infty} dz' \tfrac{A(z')}{z-z'}$$

$$A(z) = \tfrac{1}{\pi}\mathrm{pv} \int_{-\infty}^{+\infty} dz' \tfrac{B(z')}{z-z'}$$

where pv denotes the Cauchy principal value.

$A(z)$ can be expressed by means of the modified Bessel functions, K_0 and K_1 [14]:

$$A(z) = z^2 \left(|K_0(z)|^2 + |K_1(z)|^2 \right) \tag{38}$$

and the expression of $B(z)$ was obtained by [21]:

$$B(z) = \pi z^2 \left(|K_0(z)| \, |I_0(z)| - |K_1(z)| \, |I_1(z)| \right). \tag{39}$$

Then, we integrate over the impact parameter for obtaining the inelastic, ii', contribution:

$$\int_{R_1}^{R_D} 2\pi\rho \, d\rho \; P_{ii'}(\rho, v) = \pi a_0^2 \, \frac{8 \, I_H^2}{E \, \Delta E_{ii'}} \, f_{ii'} \, \frac{\mu}{m_e} \, Z_P^2 \, (a(z_D) - a(z_1)) \tag{40}$$

with [14]:

$$a(z) = z \, |K_1(z)| \, K_0(z). $$

Symmetrization

We now introduce the symmetrization of the transition probabilities and cross-sections, in order to satisfy the reciprocity relations [14]. We replace E by E_{sym} and z by z_{sym}:

$$E_{sym} = \frac{1}{2} \left(2E - \Delta E_{ii'} \right) \tag{41}$$

$$z_{sym} = \frac{\rho}{a_0} \sqrt{\frac{\mu}{m_e}} \sqrt{\frac{E}{I_H}} \frac{\Delta E_{ii'}}{2E - \Delta E_{ii'}}. \tag{42}$$

Remark concerning the shift

Note that in the SCP method and computer code, we calculate the shift with Equation (35). Thus, the integration over the impact parameter is not analytical. However, we have also calculated the shift with the more usual formula, where the $b(z)$ function [21] appears:

$$\int_{R_4}^{R_D} 2\pi\rho \, d\rho \, 2\varphi_{ii'}(\rho, v) = \pi a_0^2 \, \frac{4 \, I_H^2}{E \, \Delta E_{ii'}} \, f_{ii'} \, \frac{\mu}{m_e} \, Z_P^2 \, (b(z_D) - b(z_4)) \tag{43}$$

$$b(z) = \frac{\pi}{2} - \pi z \, K_0(z) \, I_1(z) \tag{44}$$

In that case, we have chosen the lower cutoff, R_4, as equal to $\frac{1}{2} (\langle r_i \rangle + \langle r_f \rangle)$ where $\langle r_i \rangle$ is the mean radius of the i level and $\langle r_f \rangle$ the mean radius of the f level. In fact, the obtained results are quite sensitive to the cutoff, and we have preferred our method of calculation given by Equation (35). Therefore, our shift results with the $b(z)$ function appear neither in our publications nor in the STARK-B database.

Asymptotic limits and series expansions of $A(z)$, $a(z)$, $B(z)$

The asymptotic limits of $A(z)$, $B(z)$ and $a(z)$ are recalled hereafter, because they are useful for understanding some systematic trends: At high energies (or very small ΔE):

$A(z) \to 1$, $\frac{\partial A}{\partial z} \to$ zero,

$a(z) \to \ln\left(\frac{2}{\gamma z}\right)$ and

$B(z) \to 0$ and $\frac{\partial B}{\partial z} \to$ zero.

$\gamma = \exp C$ and $C = 0.5772156649$ is the Euler constant.

At low energies (or high ΔE): $A(z) \to \pi z \exp(-2z)$, $B(z) \to \pi/4z + 9\pi/32z^2 + ...$ and $a(z) \to \frac{\pi}{2}\exp(-2z)$.

In fact, at low energies, the Lindholm limit due to a polarization potential of the phase shift, φ_p, is obtained (*cf.* [22]). The limit ($\beta \to 0$) is $\varphi_p = \varphi_i - \varphi_f$, with:

$$\varphi_i = \sum_{i' \neq i} \frac{\pi}{2} Z_P^2 \left(\frac{a_0}{\rho}\right)^3 \sqrt{\frac{I_H}{E}} \sqrt{\frac{\mu}{m_e}} f_{ii'}\left(\frac{I_H}{\Delta E_{ii'}}\right)$$

with an analogous expression for φ_f. *Contribution of the quadrupolar interaction: expression of φ_q*

This only concerns the elastic term of the width (*cf.* above). The contribution of the inelastic transitions between fine structure levels of the initial and final level (if any) are included in the elastic term. The details of the calculations are given in [6], in [8] for the B_i, B_f, B_{if} angular coefficients of complex atoms and in [13] for still more complex atoms. We have [6]:

$$\varphi_q^2 = \left[\left(B_i\langle r_i^2\rangle\right)^2 + \left(B_f\langle r_f^2\rangle\right)^2 - B_{if}\langle r_i^2\rangle\langle r_f^2\rangle\right] Z_P^2 \frac{\mu}{m_e} \frac{a_0^4}{\rho^4} \frac{I_H}{E} \tag{45}$$

In the SCP computer code, the calculations of the Bessel functions use the Fortran library of [23]. The integration over the impact parameter of the elastic part of the width and of the shift use Gauss integration techniques and Gauss–Laguerre integration techniques for the integration over the Maxwell distribution of velocities. The Gauss–Laguerre technique is not suitable for the inelastic part, because at high energies, the inelastic dipolar cross-sections decrease as $(\ln E)/E$. A trapezoidal method with an increasing exponential step is used; it is suitable and accurate.

3.5.2. Case of Ionized Atoms (Hyperbolic Path)

Contribution of the dipolar interaction: expressions of the transition probabilities, of the inelastic cross-sections and of φ_p

We define:

$$\xi = \frac{a\,\Delta E_{ii'}}{\hbar v} = \frac{1}{2} Z_A Z_P \sqrt{\frac{\mu}{m_e}} \frac{\Delta E_{ii'}}{I_H}\left(\frac{I_H}{E}\right)^{\frac{3}{2}}$$

and we obtain:

$$\frac{1}{2}P_{ii'}(\varepsilon, v) + 2\,\mathrm{i}\,\varphi_{ii'}(\varepsilon, v) = \frac{a_0^2}{(a\varepsilon)^2} \frac{2\,I_H^2}{E\,\Delta E_{ii'}} f_{ii'} \frac{\mu}{m_e} Z_P^2 \left(A(\xi, \varepsilon) + \mathrm{i}B(\xi, \varepsilon)\right) \tag{46}$$

$$A(\xi, \varepsilon) = (\xi\varepsilon)^2 \exp(\pm\pi\xi) \left(|K_{\mathrm{i}\xi}(\xi\varepsilon)|^2 + \frac{\varepsilon^2-1}{\varepsilon^2}|K'_{\mathrm{i}\xi}(\xi\varepsilon)|^2\right) \tag{47}$$

where $\pm$ means $+$ for the attractive case and $-$ for the repulsive case.

We recognize the Bessel functions of imaginary order $i\xi$ and real argument $\xi\varepsilon$:

$$
\begin{aligned}
K_{i\xi}(\xi\varepsilon) &= \int_0^\infty e^{-\xi\varepsilon\cosh u}\cos\xi u\ \mathrm{d}u \\
K'_{i\xi}(\xi\varepsilon) &= \int_0^\infty e^{-\xi\varepsilon\cosh u}\cos\xi u\ \cosh u\ \mathrm{d}u
\end{aligned}
\tag{48}
$$

As for the straight path case, $A(\xi,\varepsilon)$ and $B(\xi,\varepsilon)$ are connected together by the Hilbert transform, but now, the variable is ξ:

$$
B(\xi,\,\varepsilon) = \frac{1}{\pi}\mathrm{pv}\int_{-\infty}^{+\infty}\mathrm{d}\xi'\frac{A(\xi',\,\varepsilon)}{\xi - \xi'}
\tag{49}
$$

Contrary to the straight path case, we do not know any other analytical formula for $B(\xi,\varepsilon)$. Therefore, it will be calculated numerically by using asymptotic formulae, which will be given hereafter.

For the attractive case, $B(\xi,\varepsilon)$ has also been calculated by use of the Hilbert transform when asymptotic formulae are not adequate. However, Equation (49) is not suitable for numerical calculations. In the SCP computer code, the dispersion relation has been used:

$$
B(\xi,\,\varepsilon) = \frac{1}{\pi}\int_{0+}^{+\infty}\mathrm{d}\xi'\frac{A(\xi - \xi',\,\varepsilon) - A(\xi + \xi',\,\varepsilon)}{\xi'}
\tag{50}
$$

and also the second imaginary term of the Dyson series:

$$
B(\xi,\,\varepsilon) = \tfrac{1}{2}\int_{-\infty}^{+\infty}\mathrm{d}u\int_{-\infty}^{u}\mathrm{d}u'\,[\sin\left[\xi\left(\varepsilon(\sinh u - \sinh u') - (u - u')\right]\right.
\tag{51}
$$

$$
\left.\times\ \frac{\varepsilon^2 + (\varepsilon^2 - 1)\sinh u\sinh u' + \cosh u\cosh u' - \varepsilon(\cosh u + \cosh u')}{(\varepsilon\cosh u - 1)^2(\varepsilon\cosh u' - 1)^2}\right]
$$

The integration over the impact parameter of the transition probability, $P_{ii'}(\xi,\varepsilon)$, has an analytic solution:

$$
\int_{R_1}^{R_D} 2\pi\rho\,P_{ii'}(\rho,v)\,\mathrm{d}\rho = 8\pi a_0^2\,\frac{I_H^2}{E\,\Delta E_{ii'}}\,f_{ii'}\,\frac{\mu}{m_e}\,Z_P^2\,(a(\xi,\varepsilon_D) - a(\xi,\varepsilon_1))
$$

with:

$$
a(\xi,\varepsilon) = \exp(\pm\pi\xi)\ \xi\varepsilon\,K_{i\xi}(\xi\varepsilon)\,K'_{i\xi}(\xi\varepsilon)
\tag{52}
$$

where $\pm$ means $+$ for the attractive case and $-$ for the repulsive case.

Sample graphs of the $A(\xi,\varepsilon)$, $a(\xi,\varepsilon)$ and $B(\xi,\varepsilon)$ functions are displayed in [5,24,25].

Symmetrization of the transition probabilities and cross-sections

As for the straight path case, the transition probabilities and cross-sections are also symmetrized [7,19,24].

a and ξ become a_{sym} and ξ_{sym}, with $E_i = E$, $E'_i = E_i + \Delta E_{ii'}$:

$a_{sym} = a_0\,I_H\frac{Z_A Z_P}{\sqrt{E_i E_{i'}}}$ and

$\xi_{sym} = Z_A Z_P\sqrt{\frac{\mu}{m_e}}\left(\sqrt{\frac{I_H}{E_i}} - \sqrt{\frac{I_H}{E_{i'}}}\right).$

Asymptotic and series expansions for A, a and B, used in the computer code

(1) First, we recall that:

$$A(0,\ \varepsilon) = 1$$
$$\tfrac{\partial A}{\partial \xi}(0^+,\ \varepsilon) = \pm\pi$$

with $\pm$: $+$ for the attractive case and $-$ for the repulsive case.

$$B(0,\ \varepsilon) = 1$$
$$\tfrac{\partial B}{\partial \xi}(0^+,\ \varepsilon) = \mp\infty$$

with $\mp$: $-$ for the attractive case and $+$ for the repulsive case.

(2) At high energies, E, or small $\Delta E_{ii'}$, $\xi \to$ zero, one obtains:

$$a(\xi,\ \delta) = (1 \pm \pi\xi)\ln\left(\frac{2}{\gamma + \delta}\right) \tag{53}$$

with $\delta = \xi(\varepsilon - 1)$ and with $\gamma = e^C$. C is the Euler constant. We have used this asymptotic expression for calculating $A(\xi,\delta)$ and $a(\xi,\delta)$ for $\xi < 0.1$ and for $\delta < 0.025$.

At very high energies, the Coulomb attraction (or repulsion) becomes weak; the contribution of high impact parameters predominates, $\xi\varepsilon \to \rho$, and the straight path case is recovered.

(3) For high ξ and high $\xi\varepsilon$, the following asymptotic expansion—valid for order and argument both high and nearly equal ([26] p. 245 and [27] p. 88, [28], such as $\delta = \xi\varepsilon - \xi = o(\xi\varepsilon)^{1/3}$)—has been used:

$$K_{i\xi}(\xi\varepsilon) = \tfrac{1}{3}e^{-\frac{\pi\xi}{2}}\sum_{m=0}^{\infty}\sin\left(\frac{(m+1)\pi}{3}\right)\Gamma\left(\frac{m+1}{3}\right)\left(\frac{\xi\varepsilon}{6}\right)^{-\frac{m+1}{3}}A_m$$

$$K'_{i\xi}(\xi\varepsilon) = \tfrac{1}{6}e^{-\frac{\pi\xi}{2}}\sum_{m=0}^{\infty}\sin\left(\frac{(m+1)\pi}{3}\right)\Gamma\left(\frac{m+1}{3}\right)\left(\frac{\xi\varepsilon}{6}\right)^{-\frac{m+1}{3}}A'_m \tag{54}$$
$$+\tfrac{1}{3}e^{-\frac{\pi\xi}{2}}\sum_{m=0}^{\infty}\sin\left(\frac{(m+1)\pi}{3}\right)\Gamma\left(\frac{m+1}{3}\right)\left(\frac{\xi\varepsilon}{6}\right)^{-\frac{m+4}{3}}\left(-\frac{m+1}{18}\right)A_m$$

We have expanded the series up to $m = 4$ [28] in the SCP code. The A_m and A'_m coefficients are given in Table 2.

Table 2. Values of the coefficients A_m and A'_m in Equation (54) [28].

m	A_m	A'_m
0	1	0
1	$-\delta$	-2
2	$\delta^2/2 + 1/20$	2δ
3	$-\delta^3/6 - \delta/15$	$-\delta^2 - 2/15$
4	$\delta^4/24 + \delta^2/24 + 1/280$	$\delta^3/3 + \delta/6$

Note that there is a typo in [28]. Therefore, the correct formulae are:

$$K_{i\xi}(\xi\varepsilon) = \frac{\sqrt{3}}{6}e^{-\frac{\pi\xi}{2}}\left[\Gamma\left(\tfrac{1}{3}\right)\left(\tfrac{6}{\xi\varepsilon}\right)^{\frac{1}{3}} - \Gamma\left(\tfrac{2}{3}\right)\left(\tfrac{6}{\xi\varepsilon}\right)^{\frac{2}{3}}\delta + \Gamma\left(\tfrac{4}{3}\right)\left(\tfrac{6}{\xi\varepsilon}\right)^{\frac{4}{3}}\frac{\delta}{3}\left(0.5\delta^2 + 0.2\right)\right.$$
$$\left. - \Gamma\left(\tfrac{5}{3}\right)\left(\tfrac{6}{\xi\varepsilon}\right)^{\frac{5}{3}}\left(\frac{\delta^4+\delta^2}{3} + \tfrac{1}{35}\right)\right]$$

(55)

$$K'_{i\xi}(\xi\varepsilon) = \frac{\sqrt{3}}{12}e^{-\frac{\pi\xi}{2}}\left(\tfrac{6}{\xi\varepsilon}\right)^{\frac{2}{3}}\left[2\,\Gamma\left(\tfrac{2}{3}\right) + \left(\tfrac{6}{\xi\varepsilon}\right)^{\frac{2}{3}}\left(-\Gamma\left(\tfrac{4}{3}\right)\left(\delta^2 + \tfrac{2}{15}\right)\right)\right.$$
$$\left. + \tfrac{1}{9}\Gamma\left(\tfrac{1}{3}\right) + \left(\tfrac{6}{\xi\varepsilon}\right)\left(\Gamma\left(\tfrac{5}{3}\right)\tfrac{\delta}{6}\left(1 + \delta^2\right) - \tfrac{2}{9}\Gamma\left(\tfrac{2}{3}\right)\delta\right)\right]$$

(4) When $\varepsilon \to \infty$ (or $\delta \to \infty$), we have used the expansion ([26] p. 202):

$$K_{i\xi}(\xi\varepsilon) = \left(\tfrac{\pi}{2\xi\varepsilon}\right)^{\frac{1}{2}}e^{-\xi\varepsilon}\left[\sum_{k=0}^{\infty}\frac{\left(-4\xi^2-1\right)\left(-4\xi^2-9\right)...\left(-4\xi^2-(2k-1)^2\right)}{k!\,(8\xi\varepsilon)^k}\right]$$

$$K'_{i\xi}(\xi\varepsilon) = -\left(\tfrac{\pi}{2\xi\varepsilon}\right)^{\frac{1}{2}}e^{-\xi\varepsilon}\left[\sum_{k=0}^{\infty}\frac{\left(-4\xi^2+16k^2-1\right)...\left(-4\xi^2-(2(2k-1)-1)^2\right)}{(2k)!\,(8\xi\varepsilon)^k}\right.$$
$$\left. + \sum_{k=0}^{\infty}\frac{\left(-4\xi^2+4(2k+1)^2-1\right)...\left(-4\xi^2-(4k-1))^2\right)}{(2k+1)!\,(8\xi\varepsilon)^{2k+1}}\right]$$

(56)

We have used this expansion for calculating $A(\xi,\delta)$ and $a(\xi,\delta)$ for $\xi < 10$ and $\delta > 90$.

(5) Beyond the validity of these above expansions, the $K_{i\xi}(\xi,\delta)$ and $K'_{i\xi}(\xi,\delta)$ functions have been calculated with an unpublished Fortran subroutine developed in [24]. The $A(\xi,\delta)$ and $a(\xi,\delta)$ follow.

(6) Asymptotic expansions for $B(\xi,\varepsilon)$ for the attractive case (collisions with electrons).

We have used two expansions. The first one is valid for high values of ξ, and the first two terms were calculated by [6]. We have added the third term of the expansion in the numerical SCP code. It reads:

$$B(\xi,\varepsilon) = \frac{\varepsilon^2}{2\xi(\varepsilon^2-1)^2}\left[3 + (2+\varepsilon^2)\frac{\pi-\arccos\left(\frac{1}{\varepsilon}\right)}{\sqrt{\varepsilon^2-1}}\right]$$
$$+ \frac{\varepsilon^2}{16\xi^3(\varepsilon^2-1)^5}\left[\frac{124+592\,\varepsilon^2+229\varepsilon^4}{3} + (16+152\varepsilon^2+138\varepsilon^4+9\varepsilon^6)\frac{\pi-\arccos\left(\frac{1}{\varepsilon}\right)}{\sqrt{\varepsilon^2-1}}\right]$$
$$+ \frac{\varepsilon^2}{256\xi^5(\varepsilon^2-1)^8}\left[\frac{4576+144160\,\varepsilon^2+539524\varepsilon^4+392088\varepsilon^6+45777\varepsilon^8}{5}\right.$$
$$\left. + (256+13952\varepsilon^2+82752\varepsilon^4+101600\varepsilon^6+25990\varepsilon^8+675\varepsilon^{10})\frac{\pi-\arccos\left(\frac{1}{\varepsilon}\right)}{\sqrt{\varepsilon^2-1}}\right]$$

(57)

For $\varepsilon < 1.01$, we have used the expansion derived by [29], and we have limited it to the first two terms in the SCP code. It reads:

$$B(\xi,\varepsilon) = \sqrt{3\pi}\,\varepsilon^{\frac{3}{2}}\,\xi^{\frac{2}{3}}\left(-0.940 + 2.228\,(\varepsilon-1)\,\xi^{\frac{2}{3}}\right).$$

(58)

For high values of ξ and high values of ε, we have used the asymptotic form:

$$B\left(\xi,\,\varepsilon\right)=\frac{3\pi\varepsilon^2}{2\xi(\varepsilon^2-1)^{\frac{5}{2}}}=\frac{2\pi a^5}{3\,\xi\,\rho^5}. \tag{59}$$

(7) Beyond the validity of these expansions, $B(\xi,\varepsilon)$ has been calculated by means of a numerical integration of the Hilbert transform (Equation (50)) or by a numerical integration of the second imaginary term of the Dyson series (Equation (52)).

(8) Asymptotic expansions for $B(\xi,\varepsilon)$ for the repulsive case (collisions with positive ions).

Due to the mass effect, ξ is always high in typical conditions for Stark broadening studies. Therefore, an expansion valid for high values of ξ is sufficient and has been introduced in the SCP computer code. It reads [6]:

$$B\left(\xi,\,\varepsilon\right)=\frac{\varepsilon^2}{2\xi\left(\varepsilon^2-1\right)}\left(\frac{2+\varepsilon^2}{\sqrt{\varepsilon^2-1}}\arccos\left(\frac{1}{\varepsilon}\right)-3\right)+(...). \tag{60}$$

If in addition, $\varepsilon\rightarrow1$:

$$B\left(\xi,\,\varepsilon\right)=\frac{\varepsilon^2}{2\xi}\left(\frac{2}{15}-\frac{4}{35}\left(\varepsilon^2-1\right)+\frac{2}{21}\left(\varepsilon^2-1\right)^2-\frac{8}{99}\left(\varepsilon^2-1\right)^3+...\right). \tag{61}$$

Contribution of the quadrupolar interaction: expression of φ_q for radiating ions

After an elementary calculation, one obtains [6]:

$$\varphi_q^2=\left[\left(B_i\left\langle r_i^2\right\rangle\right)^2+\left(B_f\left\langle r_f^2\right\rangle\right)^2-B_{if}\left\langle r_i^2\right\rangle\left\langle r_f^2\right\rangle\right]\frac{\mu}{m_e}\frac{Z_P^2}{(a^2\varepsilon^2)^2}\frac{I_H}{E}$$
$$\times\left[\frac{1}{4}+\frac{3\,\varepsilon^4}{4(\varepsilon^2-1)^2}\left(1+\frac{\mp\pi+2\arcsin\left(\frac{1}{\varepsilon}\right)}{2\sqrt{\varepsilon^2-1}}\right)\right] \tag{62}$$

where $\mp$ means $+$ for the attractive case and $-$ for the repulsive case.

At high eccentricities or high energies (as expected), we recover the straight path limit. At low energies and small eccentricities, by using the series expansion of $\arcsin1/\varepsilon$, we obtain [6] the asymptotic phase shift found in [22] for electron collisions.

4. Results and Discussion

4.1. Validity of the Impact Approximation

The impact approximation is at the basis of the SCP method. Its validity is checked in all of the calculations for every line, every temperature and every density by using Equation (6).

The collision volume is calculated by writing $W=N_P v_{typ}\rho_{typ}^2$, where W is the calculated width, N_P the density of the perturbers and v_{typ} is the mean relative velocity. Thus, ρ_{typ} can be derived. One of the outputs of the code is $N_P V$, where $V=\rho_{typ}^3$ is the collision volume.

109

We give, in the following (Tables 3–6), two examples. Table 3 (electron collisions) and Table 4 (proton collisions) concern the case of Ne VIII $3s - 3p$ at 10^{19} cm^{-3}. The data are taken from the cases studied at the Spectral Line Shapes in Plasmas workshop (April, 2012). Table 5 (electron collisions) and Table 6 (proton collisions) concern the case of Li I $2s - 2p$ at 10^{16} cm^{-3}.

The results for $N_P V$ show that the impact approximation is valid both for electron and proton colliders in these two cases.

In the STARK-B database, the values of the widths and shifts are provided in the tables, except when $N_P V > 0.5$, where the cells are empty and marked with an asterisk preceding the cell. Widths and shifts values for $0.1 < N_P V < 0.5$ are given and marked by an asterisk in the cell preceding the value.

The format of the data of Tables 3–6 is in ASCII. Hence, E + 05 means $\times 10^5$, and so on.

Table 3. Results of the SCP code for Ne VIII $3s - 3p$, electron collisions. Angular frequency units, density $N_P = 10^{19}$ cm^{-3}, temperatures T in Kelvin.

T		0.580E + 05	0.174E + 06	0.580E + 06
$N_P V$		0.579E − 02	0.121E − 02	0.238E − 03
Full width at half maximum		0.105E + 14	0.639E + 13	0.394E + 13
Strong collisions contribution		0.347E + 13	0.202E + 13	0.112E + 13
Inelastic collision contribution from the upper level		0.233E + 13	0.176E + 13	0.125E + 13
Inelastic collision contribution from the lower level		0.687E + 12	0.942E + 12	0.801E + 12
Feshbach resonances contribution from the upper level		0.431E + 12	0.915E + 11	0.156E + 11
Feshbach resonances contribution from the lower level		0.151E + 13	0.409E + 12	0.767E + 11
Elastic collisions contribution (polarization + quadrupole)		0.744E + 13	0.369E + 13	0.189E + 13
Elastic collisions contributions (without quadrupole)		0.435E + 11	0.202E + 11	0.799E + 10

Table 4. Results of the SCP code for Ne VIII $3s - 3p$, proton collisions. Angular frequency units, density $N_P = 10^{19}$ cm^{-3}, temperatures T in Kelvin.

T	0.580E + 05	0.174E + 06	0.580E + 06
$N_P V$	0.205E − 03	0.918E − 03	0.183E − 02
Full width at half maximum	0.269E + 11	0.126E + 12	0.394E + 13
Strong collisions contribution	0.383E + 04	0.253E + 09	0.204E + 11
Inelastic collision contribution from the upper level	0.489E + 05	0.557E + 09	0.310E + 11
Inelastic collision contribution from the lower level	0.511E − 02	0.231E + 06	0.861E + 09
Elastic collisions contribution (polarization + quadrupole)	0.269E + 11	0.126E + 12	0.334E + 12
Elastic collisions contribution (without quadrupole)	0.607E + 09	0.184E + 11	0.138E + 12

Table 5. Results of the SCP code for Li I $2s - 2p$, electron collisions. Angular frequency units, density $N_P = 10^{16}$ cm^{-3}, temperatures T in Kelvin.

T	5,000	10,000	20,000
$N_P V$	0.328E − 04	0.221E − 04	0.186E − 04
Full width at half maximum	0.976E + 10	0.106E + 11	0.134E + 11
Strong collisions contribution	0.583E + 10	0.603E + 10	0.721E + 10
Inelastic collision contribution from the upper level	0.445E + 10	0.461E + 10	0.567E + 10
Inelastic collision contribution from the lower level	0.963E + 08	0.827E + 09	0.267E + 10
Elastic collisions contribution (polarization + quadrupole)	0.522E + 10	0.518E + 10	0.502E + 10
Elastic collisions contribution (without quadrupole)	0.252E + 10	0.189E + 10	0.113E + 10

Table 6. Results of the SCP code for Li I $2s - 2p$, proton collisions. Angular frequency units, density $N_P = 10^{16}$ cm^{-3}, temperatures T in Kelvin.

T	5,000	10,000	20,000
$N_P V$	0.281E − 02	0.167E − 02	0.998E − 03
Full width at half maximum	0.471E + 10	0.472E + 10	0.472E + 10
Strong collisions contribution	0.220E + 10	0.220E + 10	0.221E + 10
Inelastic collision contribution from the upper level	0.206E + 01	0.113E + 01	0.602E + 02
Inelastic collision contribution from the lower level	0.228E − 06	0.897E − 02	0.200E + 02
Elastic collisions contribution (polarization + quadrupole)	0.471E + 10	0.472E + 10	0.473E + 10
Elastic collisions contribution (without quadrupole)	0.789E + 09	0.885E + 09	0.994E + 09

In the wings, $\Delta\omega$ being the detuning in angular frequency units, the validity condition of the generalized impact approximation becomes $\tau \, \Delta\omega \ll 1$. Therefore, if we approach the limit of validity of the impact approximation ($0.1 < N_P V < 0.5$), the impact approximation becomes invalid in the wings. This can be the case of collisions with ions at high densities. However, the contribution of ion collisions is most often weaker than the contribution of electron collisions (about 10%), and a rough accuracy for the contribution of ion collisions is generally sufficient. Of course, exceptions exist, which arise when some perturbing levels are very close to the upper ones: see below and [30], as well as the explanation given in [31]).

Concerning isolated lines perturbed by electron colliders, the impact approximation is quite always valid, due to the high velocity of these light particles. The only exceptions concern radiating ions at very high densities, which can occur in laser plasmas, for instance.

Besides, it is interesting to recall that hydrogen lines arising from low levels (Balmer lines, for instance), which are not isolated, can be treated within the impact approximation at low densities typical of stellar atmospheres and of some laboratory plasmas: in [32], it was shown that the profile of Hα is Lorentzian in the central part of the profile. A good agreement, for collisions with electrons and protons, as well, was obtained between the impact model and the MMM (model microfield method) one at $N_P = 10^{13}$ cm^{-3} and $T = 5,000$ K and $10,000$ K.

4.2. Validity of the Isolated Line Approximation

At high densities or for lines arising from high levels, the electron impact width can be comparable to the separation, $\Delta E(nl, nl \pm 1)$, between the perturbing energy levels and the initial or final level: the corresponding levels become degenerate, and the isolated line approximation is no longer valid. In order to check the validity of this approximation, we have defined a parameter, C, in [10], which is given in the STARK-B database and the corresponding articles.

4.3. Comparisons of the Different Contributions

An extensive discussion concerning comparisons of the different contributions can be found in [31]. Tables 3–6 give the different contributions to the full width for the two cases cited.

4.3.1. Electron Collisions

The contribution of strong collisions is generally less important for radiating ions than for neutrals. This is due to the Coulomb attraction. The contribution of strong collisions decreases when the energy (or temperature) increases, as expected. Generally, inelastic collisions predominate at high energies and elastic ones at low energies.

Concerning inelastic collisions, the contribution of high impact parameters becomes very important when some perturbing levels are close to the upper one (or to the lower one) or when the energy (or temperature) is high. Then, the SCP method is the most accurate. In addition, one can notice that the summation over the incident electron orbital l-quantum numbers for obtaining cross-sections in quantum methods no longer works if l is too high (l about 30 for S VII). Therefore, the summation is generally completed by using the Born, or the Bethe, or the Coulomb–Bethe approximation (cf., for instance, [33]).

For elastic collisions, the contribution of the quadrupolar interaction is never negligible in these two cases. The contribution of small impact parameters is most often predominant. The contribution of elastic collisions can be negligible when some perturbing levels are very close to the upper ones [31].

In addition, a detailed comparison between close-coupling and SCP calculations is given in [34] for the electron impact broadening of the Li I resonance line. In particular, it is shown that the close-coupling and SCP calculations converge at $l = 3$ for the width and $l = 4$ for the shift.

The contribution of the Feshbach resonances, which only concern ionized atoms, is only important at low energies.

4.3.2. Ion Collisions

The two examples cited here (Tables 4 and 6) illustrate the following. For radiating ions, due to the Coulomb repulsion, the contribution of strong collisions is generally small. It increases with the temperature, because the Coulomb repulsion decreases. The contribution of inelastic collisions is very small for the same reason. Of course, there are no Feshbach resonances. The contribution of the quadrupole potential is predominant.

For neutrals, the situation is different, since there is no Coulomb repulsion. However, the contribution of inelastic collisions remains generally weak, and the contribution of the quadrupolar interaction is important, except when high levels are involved [31].

4.4. Accuracy of the SCP Method

The accuracy of a theoretical method is difficult to assess. Therefore, we estimate the accuracy of the SCP method by comparing to the experimental results. This has been made in all our papers that are cited in the STARK-B database. As examples, we will only cite here [9,35–39]. In spite of the fact that the strong collision contribution is never very small (see Tables 3–5), the accuracy is about 20%–30% for the widths of simpler spectra, but is worse for very complex spectra, particularly when configuration mixing is present in the description of energy levels, but not always [39]. If the shifts are of the same order of magnitude as the widths, their accuracy are similar to that of the widths. If they are smaller or much smaller, their accuracy is worse, because of the cancellation effects between the initial and final level. However, such an accuracy is enough for the needs of stellar physics and laboratory physics.

We note that within the semiclassical perturbation method, the weak inelastic collisions are the most reliable. Therefore, especially when their contribution is dominant, the obtained results are of good accuracy. Since the more sophisticated close-coupling method is not suitable for large-scale calculations, semiclassical perturbation data are still the best available data in many cases.

4.5. Ab Initio and Automatic Codes for Obtaining a Great Number of Data in a Same Run and the Influence of the Chosen Atomic Structure

For obtaining widths and shifts of a given line, in addition to the charges Z_A, Z_P, the chosen temperatures and densities, one must also input the energy levels and the $\langle r \rangle$ of the initial, final levels of the line and all the perturbing levels, and the oscillator strengths between the initial and all the perturbing levels. The $\langle r \rangle^2$ and B_i, B_f, B_{if} values of the initial and final levels also have to be input. The results of Stark broadening parameters determination performed by Dimitrijević, Sahal-Bréchot *et al.* using the semiclassical perturbation method are contained in more than 130 publications and have been implemented in the STARK-B database [15]. Thanks to the creation of *ab initio* [40], automatic codes coupling the atomic data and the SCP code, more than several hundred lines (and sometimes about one thousand) can be treated in a same run. The calculations are very fast: only one night is sufficient with a laptop for treating several hundred lines in a same run for several temperatures and densities.

In the older papers, the energies of the levels were taken from measurements and various publications, and the oscillator strengths were obtained using the Coulomb approximation with the quantum defect (Bates and Damgaard approximation, [41], improved for high n values by [42]). As the $\langle r \rangle$, the $\langle r \rangle^2$ were obtained by means of the hydrogenic value with a quantum defect.

Then TOPbase has often been used since the 1990s [43], when the needed sets of energy levels and oscillator strengths became available: see, e.g., [35] and further papers, e.g., [38]. The TOPbase atomic data have been obtained within the close-coupling scattering theory by means of the R-matrix method with innovative asymptotic techniques. Thus they are especially appropriate for low and moderately ionized light atoms, because LS coupling is assumed.

Since the turn of the century, the SUPERSTRUCTURE (SST) code [44] has been used for ionized atoms, e.g., [45] and further papers. SST is well suited for moderately and highly charged ions. The wave functions are determined by the diagonalization of the nonrelativistic Hamiltonian using orbitals calculated in a scaled Thomas–Fermi–Dirac–Amaldi potential. Relativistic corrections are introduced according to the Breit–Pauli approach. Atomic data are obtained in intermediate coupling.

The Cowan code [46] is interesting for complex atoms and has been coupled to the SCP code [39]. The Cowan code, based on a Hartree–Fock–Slater multi-configuration expansion method with statistical exchange, contains relativistic corrections treated by perturbations. Therefore, this method is especially suited to neutral and moderately-ionized heavy atoms. The Cowan code is also useful, because the $\langle r \rangle$ and the $\langle r \rangle^2$ are provided, and then, we can use better values than the hydrogenic ones.

The difference between the use of the different atomic data codes does not exceed 30%, except in exceptional cases, such as in [38], for instance (see below).

Si V and Ne V line widths and shifts data have been calculated with both Bates and Damgaard and SST atomic data [45,47]. The difference does not exceed 30%. C II widths and shifts data have been calculated with both TOPbase and Bates and Damgaard atomic data, [38], and the difference does not exceed a few percent, except when configuration interaction plays an important role by allowing a forbidden transition.

Notice that widths and shifts due to positive ion impacts can be, in certain cases, larger than those due to electron impacts: for Cr I lines studied in [30], there are perturbing levels that are very close to the upper initial ones (4.26 and 14.14 cm^{-1}). This special situation is due to configuration interaction effects.

This gives an idea of the importance of the chosen atomic structure for obtaining Stark broadening data.

4.6. Modifications in Progress and Prospects

Until now, the B_i, B_f, B_{if} coefficients have been "manually" calculated before entering the input data of the code. A new subroutine is in progress, which will automatically calculate these coefficients in the code.

It would be also interesting to study the difference obtained in the SCP results by introducing the large and small ξ asymptotic expansions provided in [48].

To continue on the same train of thought, it would be interesting to include penetrating orbits in the SCP method and code. The difference between the SCP method and the close-coupling ones is considered to be due to close collisions, which should be overestimated by the SCP method. Penetrating orbits were included in semiclassical calculations [49,50], and the discrepancies between quantal and semiclassical calculations became much smaller. However, our SCP method provides results that are in agreement within 20%–30% with the experimental results, whereas the majority of the most accurate (close-coupling) quantum results disagree within a factor of roughly two ([36,37] and references to close-coupling calculations therein). This remains unexplained.

4.7. Regularities and Systematic Trends

Regularities and systematic trends have been observed for many years. A number of them can be understood and, thus, predicted by looking at the formulae provided by the SCP method. This can be easily checked by using the SCP data provided by STARK-B [15], for instance. Some of them are discussed in [31].

In order to interpret the regularities, the widths and shifts data must be expressed in frequency (or angular frequency) units and not in units of wavelength. We will discuss the principal systematic trends in the following.

4.7.1. Behavior with temperature

(1) High temperatures:

At high temperatures (or very small $\Delta E_{ii'}$), the Coulomb attraction or repulsion for ion emitters is small; the behavior is the same for neutrals and ions: $a(z)$ behaves as $ln(E)$, and thus, the cross-sections as $(\mu/m_e)ln(E)/E$. With a rough reasoning, it can be deduced that the widths decrease as $\sqrt{\mu/m_e}\, ln(T)/T$; cf. in particular [51,52]. In addition, due to the mass effect, the contribution of the ion colliders can be greater than that of electrons, e.g., [30].

(2) Low temperatures:

At low temperatures, the behavior is different for neutral and ion emitters. For ions colliding with electrons, the collision strengths tend towards a finite limit; the cross-sections decrease as $1/E$ near the threshold, and the width decreases as $1/\sqrt{T}$. For neutrals colliding with electrons, the width begins to increase with the temperature.

4.7.2. Behavior with the charge of the perturber

As expected by the SCP formulae, the shifts increase linearly with Z_P; cf. [53] for instance.

4.7.3. Behavior with the charge of the radiating ion

The widths are predicted to vary as Z_{eff}^{-2}, with $Z_{eff} = Z_A + 1$. This is shown in [37], for instance; cf. Figure 20 of that paper, which shows a -1.84 slope for the $3s - 3p$ transitions from C IV to P XIII. This is due to the behavior of the line strengths (and, thus, oscillator strengths) with the charge of the radiating ion in the Coulomb approximation.

4.7.4. Behavior of the width of a spectral series of transitions of a given neutral or ionized atom with increasing principal quantum number n

Due to the behavior of the dipolar line strengths in the Coulomb approximation, it is expected that the width of a spectral series $n_1 l_1 - n l$ increases as n^4, when the principal quantum number, n, increases. This has been verified; *cf.* [38,39], for instance.

The dependence of the broadening parameters of spectral lines due to impacts with charged particles *versus* the principal quantum number within a spectral series is important information. If we know the trend of Stark broadening parameters within a spectral series, it is possible to interpolate or extrapolate the eventually missing values within the considered series.

4.7.5. Importance of the fine-structure splitting

Finally, it will be pointed out that the behaviors of the fine structure widths of a multiplet are not very sensitive to the fine structure splitting: for the $3s - 3p$ multiplets of the Li-like series, the ratio of the widths of the two components only attains 1.12 for P XIII [37]. This is quite negligible by looking at the accuracy of the calculations.

5. Conclusions

The SCP code is now extensively used for the needs of spectroscopic diagnostics and modeling, and the results of the published calculations are displayed in the STARK-B database, [15]. Data for 123 neutral and ionized atoms (49 chemical elements) are currently included in STARK-B. In the present paper, we have presented the main approximations leading to the impact semiclassical perturbation method, and we have given the formulae entering the numerical SCP code. This would permit us to better understand the validity conditions of the method and of the results; and also to understand and predict some regularities and systematic trends. If we know the systematic trends of the Stark broadening parameters, it would be possible to interpolate or extrapolate the existing and provided data for obtaining missing values.

This would also allow us to compare the method and its results to those of other methods and codes.

Acknowledgments

The support of IAEA is gratefully acknowledged. The support of the Ministry of Education, Science and Technological Development of Republic of Serbia through projects 176002 and III44022 is also acknowledged. This work has also been supported by the Paris Observatory, the CNRS and the PNPS (Programme National de Physique Stellaire, INSU-CNRS). The cooperation agreements between Tunisia (DGRS) and France (CNRS) (project code 09/R 13.03, No.22637) are also acknowledged. This paper has also been written within the LABEX Plas@par project and received financial state aid managed by the Agence Nationale de la Recherche, as part of the programme "Investissements d'avenir" under the reference, ANR–11–IDEX–0004–02.

Author Contributions

The original method and the associated code was created and developed by S. Sahal-Bréchot during the sixties and seventies [5–9]. M.S. Dimitrijević has contributed to update and operate the code since the eighties [10–12,28] and further papers. In particular, he developed during this decade the automatic code coupled to the Bates and Damgaard's oscillator strengths leading to more than one hundred widths and shifts in a same run. Then N. Ben Nessib joined the cooperation in the second part of the nineties. In particular, he extended in the beginning of this new century the quadrupole part of the width to very complex atoms with one of his Tunisian's PhD's student [13]. Then he developed with the Tunisian team the coupling of the SCP code to TOPbase ([43]), to SST [44] and to the Cowan code [46]. This allowed to obtain *"ab initio"* codes [40] permitting to obtain several hundred widths and shifts data in a same run e.g., [38,39,45]. All the resulting data and publications appear in the database STARK-B [15].

Conflicts of Interest

The authors declare no conflict of interest

References

1. Baranger, M. Simplified Quantum-Mechanical Theory of Pressure Broadening. *Phys. Rev.* **1958**, *111*, 481–493.
2. Baranger, M. Problem of Overlapping Lines in the Theory of Pressure Broadening. *Phys. Rev.* **1958**, *111*, 494–504.
3. Baranger, M. General Impact Theory of Pressure Broadening. *Phys. Rev.* **1958**, *112*, 855–865.
4. Baranger, M. Spectral Line Nroadening in Plasmas. In *Atomic and Molecular Processes*; Bates, D.R., Ed.; Academic Press: New-York, NY, USA; London, UK, 1962; pp. 493–548.
5. Bréchot, S. On electron impact broadening of positive ion lines. *Phys. Lett. A*, **1967**, *24A* 476–477.
6. Sahal-Bréchot, S. Impact theory of the broadening and shift of spectral lines due to electrons and ions in a plasma. *Astron. Astrophys.* **1969**, *1*, 91–123.
7. Sahal-Bréchot, S. Impact theory of the broadening and shift of spectral lines due to electrons and ions in a plasma (continued). *Astron. Astrophys.* **1969**, *2*, 322–354.
8. Sahal-Bréchot, S. Stark nroadening of isolated lines in the impact approximation. *Astron. Astrophys.* **1974**, *35*, 319–321.
9. Fleurier, C.; Sahal-Bréchot, S.; Chapelle, J. Stark profiles of some ion lines of alkaline earth elements. *J. Quant. Spectroscop. Ra.* **1977**, *17*, 595–603.
10. Dimitrijević, M.S.; Sahal-Bréchot, S. Stark broadening of neutral helium lines. *J. Quant. Spectroscop. Ra.* **1984**, *31*, 301–313.
11. Dimitrijević, M.S.; Sahal-Bréchot, S.; Bommier, V. Stark broadening of spectral lines of multicharged ions of astrophysical interest. I-C IV lines. *Astron. Astrophys.* **1991**, *89*, 581–590.

12. Dimitrijević, M.S.; Sahal-Bréchot, S.; Bommier, V. Stark broadening of spectral lines of multicharged ions of astrophysical interest. II- Si IV lines. *Astron. Astrophys.* **1991**, *89*, 591–598.

13. Mahmoudi, W.F.; Ben Nessib, N.; Sahal-Bréchot, S. Stark broadening of isolated lines: Calculation of the diagonal multiplet factor for complex configurations ($n_1 l_1^n \, n_2 l_2^m \, n_3 l_3^p$). *Eur. Phys. J. D* **2008**, *47*, 7–10.

14. Seaton, M.J. The Impact parameter method for electron excitation of optically allowed atomic transitions. *Proc. Phys. Soc.* **1962**, *79*, 1105–1117.

15. Sahal-Bréchot, S.; Dimitrijević, M.S.; Moreau, N. STARK-B Database. LERMA, Observatory of Paris, France and Astronomical Observatory: Belgrade, Serbia, 2014. Available online: http://stark-b.obspm.fr (accessed on 21 April 2014).

16. Mihalas, D. *Stellar Atmospheres*; W. H. Freeman and Company: San Francisco, CA, USA, 1978; pp. 29 and 411–446.

17. Omont, A.; Smith, E.W.; Cooper, J. Redistribution of resonance radiation. I. The effect of collisions. *Astrophys. J.* **1972**, *175*, 185–199.

18. Bommier, V. Master equation theory applied to the redistribution of polarized radiation, in the weak radiation field limit. I. Zero magnetic field case. *Astron. Astrophys.* **1997**, *328*, 706–725.

19. Alder, A.; Bohr, A.; Huus, B.; Mottelson, B.; Winther, A. Study of nuclear structure by electromagnetic excitation with accelerated ions. *Rev. Mod. Phys.* **1956**, *28*, 433–542.

20. Seaton, M.J. Excitation of coronal lines by proton impact. *Mon. Not. R. Astron. Soc.* **1964**, *127*, 191–194.

21. Klarsfeld, S. Exact calculation of electron-broadening shift functions. *Phys. Lett. A* **1970**, *32A*, 26–27.

22. Bréchot, S.; van Regemorter, H. L'élargissement des raie spectrales par chocs. 1.- La contribution adiabatique. *Ann. Astroph.* **1964**, *27*, 432–449.

23. Press, W.H.; Teukolsky, S.A.; Vetterling, W.T.; Flannery, B.F. *Numerical Recipes in Fortran, The Art of Scientific Computing*, 2nd ed.; Cambridge University Press: New York, NY, USA, 1992; pp. 229–233.

24. Feautrier, N. Calcul semi-classique des sections d'excitation par chocs électroniques pour les ions. Application à l'élargissement des raies. *Ann. Astroph.* **1968**, *31*, 305–309.

25. Griem, H.R. *Spectral Line Broadening by Plasmas*; Academic Press: New York, NY, USA; London, UK, 1974.

26. Watson, G.N. *A Treatise on the Theory of Bessel Functions*; Cambridge University Press: London, UK, 1966.

27. Bateman, H. Higher transcendental functions. In *Bateman Manuscript Project*; Erdelyi, A., Ed.; McGraw-Hill: New York, NY, USA, 1953–1955; Volume 2, p. 88.

28. Dimitrijević, M.S.; Sahal-Bréchot, S. Asymptotic behavior of the A and a functions for ionized emitters in semiclassical Stark-broadening theory. *J. Quant. Spectroscop. Ra.* **1992**, *48*, 349–351.

29. Klarsfeld, S. On Stark broadening functions for nonhydrogenic ions. *Phys. Lett. A* **1970**, *33A*, 437–438.

30. Dimitrijević, M.S.; Ryabchikova, T.; Popović, L.Č.; Shulyak, D.; Khan, S. On the influence of Stark broadening on Cr I lines in stellar atmospheres. *Astron. Astrophys.* **2005**, *435*, 1191–1198.

31. Sahal-Bréchot, S.; Dimitrijević, M.S.; Ben Nessib, N. Comparisons and Comments on Electron and Ion Impact Profiles of Spectral Lines. *Balt. Astron.* **2011**, *20*, 523–530.

32. Stehlé, C.; Mazure, A.; Nollez, G.; Feautrier, N. Stark broadening of hydrogen lines: New results for the Balmer lines and astrophysical consequences. *Astron. Astrophys.* **1983**, *127*, 263–266.

33. Elabidi, E.; Sahal-Bréchot, S.; Ben Nessib, N. Fine structure collision strengths for S VII lines. *Phys. Scripta* **2012**, *85*, 065302:1–065302:13.

34. Dimitrijević, M.S.; Feautrier, N.; Sahal-Bréchot, S. Comparison between quantum and semiclassical calculations of the electron impact broadening of the Li I resonance line. *J. Phys. B* **1981**, *14*, 2559–2568.

35. Dimitrijević, M.S.; Sahal-Bréchot, S. Stark broadening of Mg I spectral lines. *Phys. Scripta* **1995**, *52*, 41–51.

36. Elabidi, E.; Ben Nessib, N.; Cornille, M.; Dubau, J.; Sahal-Bréchot, S. Electron impact broadening of spectral lines in Be-like ions: Quantum calculations. *J. Phys. B* **2008**, *41*, 025702:1–025702:11.

37. Elabidi, H.; Sahal-Bréchot, S.; Ben Nessib, N. Quantum Stark broadening of 3s-3p spectral lines in Li-like ions; Z-scaling and comparison with semi-classical perturbation theory. *Eur. Phys. J. D* **2009**, *54*, 51–64.

38. Larbi-Terzi, N.; Sahal-Bréchot S.; Ben Nessib, N.; Dimitrijević, M.S. Stark-broadening calculations of singly ionized carbon spectral lines. *Mon. Not. R. Astron. Soc.* **2012**, *423*, 766–773.

39. Hamdi, R.; Ben Nessib, N.; Dimitrijević, M.S.; Sahal-Bréchot S. Stark broadening of Pb IV spectral lines. *Mon. Not. R. Astron. Soc.* **2013**, *431*, 1039–1047.

40. Ben Nessib, N. Ab initio calculations of Stark broadening parameters. *New Astron. Rev.* **2009**, *53*, 255–258.

41. Bates, D.R.; Damgaard, A. The calculation of the absolute strengths of spectral lines. *Trans. Roy. Soc. Lond.* **1949**, *242*, 101–122.

42. Van Regemorter, H.; Hoang Binh, Dy.; Prudhomme, M. Radial transition integrals involving low or high effective quantum numbers in the Coulomb approximation. *J. Phys. B* **1979**, *12*, 1053–1061.

43. Cunto, W.; Mendoza, C.; Ochsenbein, F.; Zeippen, C.J. TOPbase at the CDS. *Astron. Astrophys.* **1993**, *275*, L5–L8.

44. Eissner, W.; Jones, M.; Nussbaumer, H. Techniques for the calculation of atomic structures and radiative data including relativistic corrections. *Comp. Phys. Com.* **1974**, *8*, 270–306.

45. Ben Nessib, N.; Dimitrijević, M.S.; Sahal-Bréchot, S. Stark broadening of the four times ionized silicon spectral lines. *Astron. Astrophys.* **2004**, *423*, 397–400.

46. Cowan R.D.; Robert, D. Cowan's Atomic Structure Code. *The Theory of Atomic Structure and Spectra*; University of California Press: Berkeley, CA, USA, 1981. Available online: https://www.tcd.ie/Physics/people/Cormac.McGuinness/Cowan/ (accessed on 21 April 2014).

47. Hamdi, R.; Ben Nessib, N.; Dimitrijević, M.S.; Sahal-Bréchot, S. Stark broadening of the spectral lines of Ne V. *Astrophys. J. Suppl. S.* **2007**, *170*, 243–250.

48. Alexiou, S. Calculations of the semiclassical Stark broadening dipole width functions for isolated ion lines. *J. Quant. Spectroscop. Ra.* **1994**, *51*, 849–852.

49. Alexiou, S.; Glenzer, S.; Lee, R.W. Line shape measurement and isolated line width calculations: Quantal versus semiclassical methods. *Phys. Rev. E* **1999**, *60*, 6238–6240.

50. Alexiou, S.; Lee, R.W. Semiclassical calculations of line broadening in plasmas: Comparison with quantal results. *J. Quant. Spectroscop. Ra.* **2006**, *99*, 10–20.

51. Zmerli, B.; Ben Nessib, N.; Dimitrijević, M.S. Temperature dependence of atomic spectral line widths in a plasma. *Eur. Phys. J. D* **2008**, *48*, 389–395.

52. Elabidi, E.; Sahal-Bréchot, S. Checking the dependence on the upper level ionization potential of electron impact widths using quantum calculations. *Eur. Phys. J. D* **2011**, *61*, 285–290.

53. Dimitrijević, M.S. Stark broadening data tables for some analogous spectral lines along Li and Be isoelectronic sequences. *Serb. Astron. J.* **1999**, *159*, 65–72.

Reprinted from *Atoms*. Cite as: Duan, B.; Bari, M.A.; Wu, Z.; Yan, J. Electron-Impact Widths and Shifts of B III 2p-2s Lines. *Atoms* **2014**, *2*, 207-214.

Article

Electron-Impact Widths and Shifts of B III 2p-2s Lines

Bin Duan [1,*], **Muhammad Abbas Bari** [2], **Zeqing Wu** [1] **and Jun Yan** [1]

[1] Institute of Applied Physics and Computation Mathematics, Beijing 100088, China;
   E-Mails: wu_zeqing@iapcm.ac.cn; yan_jun@iapcm.ac.cn
[2] Pakistan Atomic Energy Commission, P. O. Box 1114, Islamabad 44000, Pakistan;
   E-Mail: kmabari@gmail.com

* Author to whom correspondence should be addressed; E-Mail: alexduan1967@hotmail.com;
   Tel.: +86-10-6193-5108; Fax: +86-10-6201-0108.

Received: 17 March 2014; in revised form: 23 April 2014 / Accepted: 5 May 2014 / Published: 15 May 2014

Abstract: In this paper, we present results for the relativistic quantum mechanical calculations of electron-impact line widths and shifts of 2p-2s transitions in doubly ionized boron (B III) ions. We use the Dirac R-matrix methods to solve $(N + 1)$-electron colliding systems for the scattering matrices that are required. The line widths are calculated for an electron density 1.81×10^{18} cm^{-3} and electron temperature 10.6 eV. The obtained results agree well with all the semiempirical calculations and most of the semiclassical calculations, and are closer to the experimental results published by Glenzer and Kunze (Glenzer, S.; Kunze, H.-J. Stark broadening of resonance transitions in B III. *Phys. Rev. A* **1996**, *53*, 2225–2229). Our line widths are almost twice as large as the earlier quantum mechanical calculations for the set of particular plasma conditions.

Keywords: electron-impact widths and shifts; $(N + 1)$-electron colliding systems; line shapes

1. Introduction

The broadening of spectral lines due to collisions with charged particles appears to be important in the studies on the behavior of atomic interactions, and is indispensable for interpreting the spectra of astrophysical and laboratory plasmas. In atmospheres of moderately hot stars (A-type star) to very hot (B- and O-type) stars and in white dwarfs, a large number of isolated ion lines has been observed [1] and Stark broadening is the well proven dominant pressure-broadening mechanism especially for white dwarfs [2]. The knowledge of Stark broadening parameters is crucial for astrophysical applications. The electron-impact broadening data of width and shift for a large number of lines of various elements and their ions are required for radiation transport in stellar plasma. Dimitrijević and Sahal-Bréchot [3] pointed out that line profiles of Boron in various ionization stages play an important role for accurate radiative opacity calculations [4,5]. Moreover, Stark broadening parameters of B III lines are of prime interest not only for the investigation and diagnosis of laboratory and laser-produced plasma, but also for the research of regularities and systematic trends. Consequently, numerous experimental and theoretical studies of Stark widths and shifts of spectral lines have been undertaken since the fundamental and pioneering works by Baranger [6–8]. In fact, Stark broadening parameters have been extensively calculated with both the sophisticated semiclassical [9] and quantum-mechanical methods, although they often require a considerable effort even for the evaluation of a single line width and shift. Apart from the several successful semiclassical frameworks [9–13] and semi-empirical formula [13–15] for performing calculations of electron-impact broadening parameters, it is equally important to perform detailed and systematic quantum mechanical calculations to provide a quantitative check of the different approximations involved in the usual semi-classical methods.

As a continuation of our previous papers [16,17], the aim of present paper is two-fold. First, we want to extend the results of our previous quantum calculations in the impact approximation to investigate the validity and accuracy of our quantum mechanical approach for more spectral lines of the large number of elements. Second, accurate line profile measurements of the resonance doublet transitions $2p^2 P^o_{3/2,1/2} \rightarrow 2s^2 S_{1/2}$ in Li-like boron were performed by Glenzer and Kunze [18]. These experimental results of Stark widths of the investigated transitions are found to disagree substantially with various theoretical calculations [3,11,12,19,20] even for the advanced quantum-mechanical calculations [4,5,21,22]. For instance, the experimental Stark width of the B III 2s-2p resonance doublet was found to be about a factor of two larger than the convergent close coupling (CCC) and Coulomb-Born exchange (CBE) calculations [21]. This disagreement makes it necessary that systematic theoretical calculations of line shifts and widths in B III ions in the particular selected plasma conditions should be carried out using our new quantum mechanical approach. Furthermore, the agreement obtained between our quantum results [16,17] within the framework of impact approximation theory and the available experimental data encourages us to perform calculations for the $2p^2 P^o_{3/2,1/2} \rightarrow 2s^2 S_{1/2}$ resonance transitions in B III ions. In the present work, we use the fully relativistic Dirac R-matrix codes [23] (DARC from hereafter) to calculate electron-impact collision strengths Ω and scattering matrices of B III ions. The target states of B III ions are determined for the low lying levels up to $n = 5$ by employing the General Purpose Relativistic Atomic Structure Package (GRASP2) [24,25] (partly improved by us). The rest of this paper is organized as follows. An outline of the basic procedure for our calculations is presented in section II. In Section III, our results are compared and discussed with available experimental and theoretical data. The conclusions are summarized in section IV.

2. Computational Procedure

We briefly outline here the essential features of the computational procedure where the values of some relevant parameters are presented for the sake of completeness. More details of our computational formalism to address electron-impact broadening parameters have been outlined elsewhere [16,17,26,27], and will not be detailed here. By using DARC, a fully relativistic package, the width and shift of spectral lines due to electron collisions are written as [4,5,28–30]

$$w + id = \alpha N_e \int_0^\infty T_e^{-3/2} exp\,(-\varepsilon/T)\Omega(\varepsilon)d\varepsilon. \tag{1}$$

where $\alpha = 2.8674 \times 10^{-23}$ eV cm^3, T_e is the electron temperature in eV, N_e is the electron density in cm^{-3} and ε is the energy of colliding electron in Rydberg. In above expression, w and d represent the half-width(a half of full width at half-maximum) and shift of spectral line in eV, respectively. The dimensionless collision strength $\Omega(\varepsilon)$ as a function of energy of colliding electrons is evaluated in j-j coupling scheme according to Equation (2) of [17]. The computational details about the collision strength of $(N + 1)$-electron colliding systems (colliding electron+target ion) can be found in [17]. The shift d in Equations (1) and (2) of [16] describes that a negative shift d in photon energy always corresponds to red shift in wavelength, which corresponds to a small reduction in the energy of the transition line.

Table 1. The relative atomic structure data of B III 2p-2s lines are obtained by General Purpose Relativistic Atomic Structure Package (GRASP2) [24,25].

Lines	Energy of Lower State (eV)	Energy of Upper State (eV)	Oscillator Strength $g_i f_{if}$	
			Coulomb Gauge	Babushkin Gauge
$2p^- \to 2s$	−636.1998	− 630.1390	0.2768	0.2503
$2p^+ \to 2s$	−636.1998	−630.1351	0.5543	0.5010

In order to calculate width and shift of spectral lines $2p \to 2s$ in B III ions, we calculate atomic structure data from 13 non-relativistic configurations $1s^2 nl(n \le 5, l \le n - 1$) which give rise to 22 bound states. The relative data of B III 2p-2s lines are shown in Table 1. The lowest 8 bound states are selected as the target states. The relativistic orbitals generated by the GRASP2 code are then used by the DARC package to construct the colliding system that consists of upper or lower state of the desired transition in target ion and a colliding free-electron. In our calculations, the colliding electron is limited by two constraints. First, the quantum number of the orbital angular momentum is $\ell \le 25$, and second, the total number of its continuum basis function for a given $\kappa(\kappa = -\ell - 1,$ if $\kappa < 0,$ else $\kappa = \ell)$ is 30. The required K- matrices and their corresponding symmetry information of the colliding system is obtained by using DARC. Since the dimensionless collision strength $\Omega(\varepsilon)$ defined in Equation (2) [17] is a function of incident electron-impact energies ε and is computed in an increasing energy sequence with an energy increment $\Delta\varepsilon = 0.008$ Rydberg by repeating the same procedure.

Finally, we obtain the width and shift of spectral lines by solving Equation (1) numerically by means of trapezoidal integration rule.

3. Numerical Results and Discussion

The calculated electron-impact broadened full width at half-maximum (FWHM $= 2w$) and shifts in the B III ($2p \rightarrow 2s$) doublet transition lines as a function of electron temperature for an electron density of 1.81×10^{18} cm^{-3} are displayed in Figure 1. Figure 1 clearly demonstrates that widths and shifts of doublet transition lines $2p\ ^2P_{3/2,1/2} \rightarrow 2s\ ^2S_{1/2}$ have very similar behavior with the increase of temperature. One can see that the line broadening parameters decrease with an increase in temperature. We find that electron-impact broadened full-widths ($2w$) are always much larger than the shifts of both the spectral lines for any given plasma condition. An example is provided at an electron temperature $T_e = 10.6$ eV, the calculated FWHM ($2w$) and shift d for the transition line $2p\ ^2P_{3/2} \rightarrow 2s\ ^2S_{1/2}$ are 0.212 and 0.0517 (about 4 times higher), respectively, whereas for the transition line $2p\ ^2P_{1/2} \rightarrow 2s\ ^2S_{1/2}, 2w = 0.216$ and $d = 0.0518$.

Figure 1. The electron-impact widths (FWHM) and shifts of doublet transitions ($2p \rightarrow 2s$) in B III ions as a function of electron temperature for an electron density of 1.81×10^{18} cm^{-3}. (**a**) $2p\ ^2P_{3/2} \rightarrow 2s\ ^2S_{1/2}$; (**b**) $2p\ ^2P_{1/2} \rightarrow 2s\ ^2S_{1/2}$.

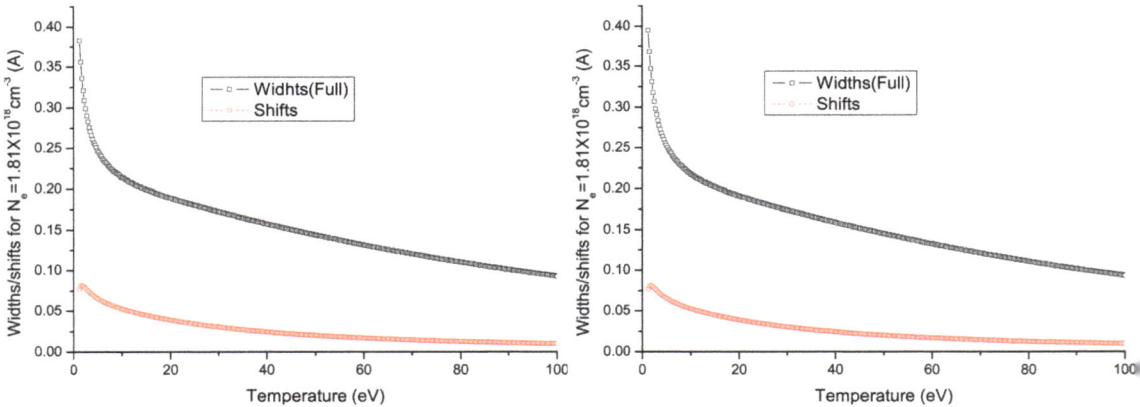

As previously mentioned, experimental Stark linewidths were measured by Glenzer and Kunze [18] by performing an accurate line profile measurements of the $2p \rightarrow 2s$ fine-structure components of the resonance doublet in Li-like boron. The authors used a homogeneous plasma region in a gas-liner pinch discharge, and plasma parameters (local electron density and ion temperature) were determined independently by $90°$ collective Thomson scattering. The Stark linewidths (FWHM) of doublet transitions in B III were measured to be $2w = 0.22$ for an electron density $N_e = 1.81 \times 10^{18}$ cm^{-3} and electron temperature $T_e = 10.6$ eV. Comparison reveals that measured linewidths agree very well with our presented values (within 3.6% accuracy).

In Table 2 our calculated values are compared both with the experimental data [18] and with various theoretical calculations [3–5,9,11,12,19–21] for an electron temperature of $T_e = 10.6$ eV and an electron density of $N_e = 1.81 \times 10^{18}$ cm^{-3}. Table 2 contains theoretical widths w_{th} calculated on the basis of various theoretical approaches discussed ahead. In this table, the last column displays our quantum mechanical values of line widths for the doublet transitions in B III. It should be noted that both the semiclassical calculations [3,20] agree well with the experimental measurements.

Historically, Griem [14] provided a simplified semi-empirical method with the use of an effective Gaunt-factor to calculate halfwidths of an isolated ion line broadened by electron-impact. The ratios of measured Stark widths [18] for the 2p-2s transitions of B III to the semi-empirically [9] calculated ones are 1.19 (see Table 2) and are noticeably different from the semi-classically calculated ones. Later in 1980, Dimitrijević and Konjević modified this well known semi-empirical approach by considering dipole-allowed electron collisions and by selecting the Gaunt factor as a suitable empirical value [15]. Hey and Breger [19] performed calculations based on the quasi-classical Gaunt factor approximation different from Griem [9] to calculate the minimum and maximum impact parameters of the electron collision process.

Table 2. Ratio of experimental electron-impact half-widths w_{exp} [18] of 2p-2s line in B III to different theoretical half-widths w_{th} for an electron temperature of $T_e = 10.6$ eV and an electron density of $N_e = 1.81 \times 10^{18}$ cm^{-3}. The letters from a to h denote theoretical widths w_{th} calculated on the basis of various theoretical approaches; **a**: semiempirical calculations [9]; **b**: Semiempirical calculations [19]; **c**: semiclassical calculations [3]; **d**: semiclassical calculations [20]; **e**: semiclassical calculations (NPSCII) [11,12]; **f**: R-matrix calculations [4,5]; **g** and **h**: calculations of CCC and CBE methods, respectively [21].

w_{exp}/w_{th}	Wavelength (0.1 nm)	a	b	c	d	e	f	g	h	Present
$2p\,{}^2P_{3/2} \rightarrow 2s\,{}^2S_{1/2}$	2065.78	1.19	1.24	1.03	1.09	1.77	1.81	2.1	1.9	1.04
$2p\,{}^2P_{1/2} \rightarrow 2s\,{}^2S_{1/2}$	2067.24	1.19	1.24	1.02	1.08	1.76	1.79	2.1	1.9	1.02

By using the semiclassical perturbation formalism [31], Dimitrijević and Sahal-Bréchot studied and also devoted a series of papers to the calculations of Stark broadening parameters of spectral lines of multicharged ions more than a dozen times. In particular, they studied the line widths and shifts for Be II and B III spectral lines due to interactions with electrons, protons and singly ionized helium ions [3]. The nonperturbative semiclassical methods (NPSC) were introduced and formulated by Alexiou [11,12], which are characterized by choosing and defining a minimum impact parameter $\rho_{min}(v)$ of collisions. These NPSC results show excellent agreement with the highest quality experimental data from the Bochum gas-liner pinch [32,33] for a number of measured isolated lines as exhibited by the results of Dimitrijević and Sahal-Bréchot [3]. Alexiou and Lee [11,12] further improved the NPSC by dropping the long-range approximation and using the full Coulomb interaction, and named it NPSCII. In the case of B III $2p \rightarrow 2s$ lines, NPSCII yielded the values of FWHM which are 10% larger than that of the R-matrix calculations performed by Seaton [4,5] and 5% smaller than that of semiempirical results [21]. The NPSCII results are in excellent agreement with these two quantal calculations.

In contrast to the case of semiclassical calculations, almost all of the quantum-mechanical calculations of doublet transitions ($2p \rightarrow 2s$) of Li-like B are about half of the corresponding experimental data [18], even though they are consistent with each other. The widths obtained with five-state close-coupling quantum mechanical calculations [4,5] are about half of the experimental values (differ by a factor $\approx$ 2). In addition to the CC work, Griem *et al.* [21] performed calculations for Stark broadening parameters of the B III $2p - 2s$ lines with the convergent close coupling (CCC) and the Coulomb-Born exchange (CBE) methods. These two sets of linewidths are in good agreement with each other as well as with previous quantum mechanical R-matrix calculations [4,5].

Finally, the ratio of measured-to-calculated widths tabulated by us (Table 2) for the various theoretical calculations provide guidance on the degree of agreement between the experimental data and theory, and thus provide a valuable indication of the quality of the calculated as well as measured data. It is seen that our calculations along with semiclassical calculations compare well with the experimental data [18], which was also observed in earlier articles. Our calculated widths are about twice as large as the earlier mentioned quantum mechanical calculations, including the NPSCII formalism as well.

4. Conclusions

We have performed fully quantum mechanical calculations of electron-impact widths and shifts of the B III ($2p \rightarrow 2s$) doublet transition lines by employing GRASP2 and DARC atomic packages for plasmas under electron temperature in the range ($0 \sim 100$ eV). We have found that our quantum results agree well with both the semiempirical and semiclassical theory, and are more closer to the experimental ones. Our calculations are generally more consistent with the semiclassical theory and are about twice as large as other quantum mechanical ones. This is not similar to what had already been predicted with quantum mechanical calculations for Stark broadening parameters for B III ions [21]. Unfortunately, we still can not answer these puzzling discrepancies existing among these theoretical calculations. Therefore, comparison of the theory with experiment indicate that further studies and improvements of the theory for more new transition lines of boron ions are required. However, just for the values, our results are much closer to the measured values as compared to the other theoretical calculations. Finally, we believe that our results are useful for new plasma diagnostics.

Acknowledgements

This work is supported by the National Natural Science Foundation of China (NSFC) (Grant No. 11275029 and No. 11204017)) and Foundation for Development of Science and Technology of Chinese Academy of Engineering Physics (Grant No. Grant No. 2013A0102005). One of us is gratefully indebted to the support of IAEA too.

Author Contributions

B. D. contributed to the original theory calculation and analysis. The other authors performed supporting calculation and the data analysis. All authors were involved in the discussion of results and commented on the manuscript.

Conflicts of Interest

The authors declare no conflict of interest.

References

1. Dimitrijević, M.S. Stark broadening in astrophysics (Applications of Belgrade school results and collaboration with Soviet republics). *Astron. Astrophys. Trans.* **2003**, *22*, 389–412.
2. Sahal-Bréchot, S. Case studies on recent Stark broadening calculations and STARK-B database development in the framework of the European project VAMDC (Virtual Atomic and Molecular Data Center). *J. Phys. Conf. Ser.* **2010**, *257*, 012028.
3. Dimitrijević, M.S.; Sahal-Bréchot, S. Stark broadening of specral lines of multicharged ions of astrophysical interest. XIV Be III and B III. *Astrophys. Suppl. Ser.* **1996**, *119*, 369–371.
4. Seaton, M.J. Atomic data for opacity calculations: V. Electron impact broadening of some CIII lines. *J. Phys. B* **1987**, *20*, 6431–6446.
5. Seaton, M. J., Atomic data for opacity calculations: VIII. Line-profile parameters for 42 transitions in Li-like and Be-like ions. *J. Phys. B* **1988**, *21*, 3033–3053.
6. Baranger, M. Simplified quantum-mechanical theory of pressure broadening. *Phys. Rev.* **1958**, *111*, 481–493.
7. Baranger, M. Problem of overlapping lines in theory of pressure broadening. *Phys. Rev.* **1958**, *111*, 494–504.
8. Baranger, M. General impact theory of pressure broadening. *Phys. Rev.* **1958**, *112*, 855–865.
9. Griem, H.R. *Spectral Line Broadening by Plasmas*; Academic Press: New York, NY, USA, 1974.
10. Sahal-Bréchot, S. Impact theory of the broadening and shift of specral lines due to electrons and ions in plasma. *Astron. Astrophys.* **1969**, *1*, 91–123.
11. Alexiou, S.; Lee, R.W. Electron line broadening in plasmas: Resolution of the quantum vs. semiclassical calculations puzzle. In Proceedings of the AIP Conference Proceedings, Berlin, Germany, 10–14 July 2000; Volume 559, pp. 135–143; doi:10.1063/1.1370605.

12. Alexiou, S. Review of Isolated Lines. In Proceedings of the 2nd Spectral Line Shapes in Plasmas Code Comparison Workshop, Vienna, Austria, 5–9 August, 2013.

13. Griem, H.R. *Plasma Spectroscopy*; McGraw-Hill: New York, NY, USA, 1964; pp. 129; 445.

14. Griem, H.R. Semiempirical Formulas for the Electron-Impact Widths and Shifts of Isolated Ion Lines in Plasmas. *Phys. Rev.* **1968**, *165*, 258–266.

15. Dimitrijević, M.S.; Konjević, N. Starkwidths of doubly and triply ionized atom lines. *J. Quant. Spectrosc. Radiat. Transf.* **1980**, *24*, 451–459.

16. Duan, B.; Bari, M.A.; Wu, Z.Q.; Yan, J.; Li, Y.M.; Wang, J.G. Relativistic quantum mechanical calculations of electron-impact broadening for spectral lines in Be-like ions. *Astron. Astrophys.* **2012**, *547*, A4.

17. Duan, B.; Bari, M.A.; Wu, Z.Q.; Yan, J.; Li, Y.M. Widths and shifts of spectral lines in He II ions produced by electron impact. *Phys. Rev. A* **2012**, *86*, 052502:1–052502:6

18. Glenzer, S.; Kunze, H.-J. Stark broadening of resonance transitions in B III. *Phys. Rev. A* **1996**, *53*, 2225–2229.

19. Hey, J.D.; Breger, P. Calculated stark widths of oxygen ion lines. *J. Quant. Spectrosc. Radiat. Transf.* **1980**, *24*, 349–364.

20. Alexiou, S. In Proceedings of the 13th International Conference on Spectral Line Shapes, Firenze, Italy, 16–21 June 1996; Zoppi, M., Ulivi, L., Eds.; AIP Conf. Proc. No. 386; AIP: New York, NY, USA, 1997; pp. 79–98.

21. Griem, H.R.; Ralchenko, Y.V.; Bray, I. Stark broadening of the B III 2s-2p lines. *Phys. Rev. E* **1997**, *56*, 7186–7192.

22. Stambulchik, E.; Maron, Y. A study of ion-dynamics and correlation effects for spectral line broadening in plasma: K-shell lines. *J. Quant. Spectrosc. Radiat. Transf.* **2006**, *99*, 730–749.

23. Norrington, P. DARC manual. Available online: http://www.am.qub.ac.uk /users/p.norrington (DARC-20040123.tar) (accessed on 23 April 2014).

24. Dyall, K.G.; Grant, I.P.; Johnson, C.T.; Parpia, F.A. ; Plummer, E.P. GRASP: A general-purpose relativistic atomic structure program. *Comput. Phys. Commun.* **1989**, *55*, 425–456.

25. Duan, B.; Bari, M.A.; Zhong, J.Y.; Yan, J.; Li, Y.M.; Zhang, J. Energy levels and radiative rates for optically allowed and forbidden transitions of Ni XXV ion. *Astron. Astrophys.* **2008**, *488*, 1155-1157.

26. Duan, B.; Bari, M.A.; Wu, Z.Q.; Yan, J.; Li, Y.M. Electron-impact widths and shifts of Sr II lines in ultracold neutral plasmas. *Phys. Rev. A* **2013**, *87*, 032505:1–032505:5.

27. Duan, B.; Bari, M.A.; Wu, Z.Q.; Yan, J.; Li, Y.M. Electron-impact broadening parameters for Be II, Sr II, and Ba II spectral lines. *Astron. Astrophys.* **2013**, *555*, A144.

28. Bely, O.; Griem, H.R. Quantum-mechanical calculations for the electron-impact braodening of resonance lines of singly ionized magnesium. *Phys. Rev. A* **1970**, *1*, 97–103.

29. Peach, G. Theory of the pressure broadening and shift of spectral lines. *Adv. Phys.* **1981**, *30*, 367–474.

30. Elabidi, H.; Nessib, N.B.; Cornille, M.; Dubau J.; Sahal-Brechot, S. Electron impact broadening of spectral lines in Be-like ions: quantum calculations. *J. Phys. B* **2008**, *41*, 025702.

31. Sahal-Bréchot, S. Impact theory of the broadening and shift of specral lines due to electrons and ions in plasma(Continued). *Astron. Astrophys.* **1969**, *2*, 322–354.

32. Glenzer, S.; Uzelac, N.I.; Kunze, H.J. Stark broadening of spectral lines along the lsoelectronic sequence of Li. *Phys. Rev. A* **1992**, *45*, 8795–8802.

33. Wrubel, T.; Glenzer, S.; Buescher, S.; Kunze, H.-J.; Alexiou, S. Line profile measurements of the 2s3s-2s3p singlet and triplet transitions in Ne VII. *Astron. Astrophys.* **1996**, *307*, 1023-1028.

Reprinted from *Atoms*. Cite as: Koubiti, M.; Goto, M.; Ferri, S.; Hansen, S.B.; Stambulchik, E. Line-Shape Code Comparison through Modeling and Fitting of Experimental Spectra of the C II 723-nm Line Emitted by the Ablation Cloud of a Carbon Pellet. *Atoms* **2014**, *2*, 319-333.

Article

Line-Shape Code Comparison through Modeling and Fitting of Experimental Spectra of the C II 723-nm Line Emitted by the Ablation Cloud of a Carbon Pellet

Mohammed Koubiti [1],*, **Motoshi Goto** [2], **Sandrine Ferri** [1], **Stephanie B. Hansen** [3] and **Evgeny Stambulchik** [4]

[1] Aix-Marseille Université—CNRS, PIIM UMR7345, 13397 Marseille, France;
    E-Mail: sandrine.ferri@univ-amu.fr

[2] National Institute for Fusion Science, Toki, 509-5292, Japan;
    E-Mail: goto@nifs.ac.jp

[3] Sandia National Laboratories, Albuquerque, NM 87185, USA;
    E-Mail: sbhanse@sandia.gov

[4] Faculty of Physics, Weizmann Institute of Science, Rehovot 7610001, Israel;
    E-Mail: evgeny.stambulchik@weizmann.ac.il

*   Author to whom correspondence should be addressed; E-Mail: mohammed.koubiti@univ-amu.fr;
    Tel.: +33-491-282-721.

Received: 9 May 2014; in revised form: 28 May 2014 / Accepted: 25 June 2014 / Published: 14 July 2014

Abstract: Various codes of line-shape modeling are compared to each other through the profile of the C II 723-nm line for typical plasma conditions encountered in the ablation clouds of carbon pellets, injected in magnetic fusion devices. Calculations were performed for a single electron density of 10^{17} cm^{-3} and two plasma temperatures (T = 2 and 4 eV). Ion and electron temperatures were assumed to be equal ($T_e = T_i = T$). The magnetic field, B, was set equal to either to zero or 4 T. Comparisons between the line-shape modeling codes and two experimental spectra of the C II 723-nm line, measured perpendicularly to the B-field in the Large Helical Device (LHD) using linear polarizers, are also discussed.

Keywords: carbon pellet ablation; plasma spectroscopy; LHD; Stark/Zeeman broadening; line-shape codes; atomic physics

Classification PACS: 52.25.Os; 32.60.+i; 32.70.Jz; 52.70.-m; 52.55.Hc

1. Introduction

Comparing synthetic line profiles in plasmas carried out with different codes and simulation methods is certainly an interesting issue, which can help to validate the underlying models. In order to use them for plasma diagnostics, such models have to be reliable. However, checking the reliability of a model requires its comparison with experimental data. This does not exclude model-model comparisons. Therefore, code comparisons become more challenging if they include confrontation with experimental measurements. Such comparisons, between theory and experiment, were not scheduled in the first workshop on Spectral Line Shapes in Plasmas: code comparison [1]. To fill this gap, data from two experiments were introduced in the second workshop (5–9 August, 2013, Vienna). These experimental cases were aimed to allow detailed discussion on the approaches adopted by different research groups to analyze experimental spectra. To have a better understanding of why different approaches might end up with different best-fit plasma parameters, it was recommended to contributors willing to analyze the proposed experimental data cases to calculate the relevant line profiles for a small prescribed grid of parameters. The first experimental case concerned the C II 723-nm $1s^2 2s^2 3p\ ^2P°$-$1s^2 2s^2 3d\ ^2D$ line emitted by the ablation cloud of a carbon pellet injected in the Large Helical Device (LHD) [2]. The data for this case consisted of two spectra, both measured along a line-of-sight that is nearly perpendicular to the magnetic field line but with a polarizer rotated, either nearly parallel "horizontal", or nearly perpendicular "vertical", to the magnetic field (see Figure 1). The second experimental case concerned the Li 460.3-nm $1s^2 2p$-$1s^2 4d$ line and its forbidden components. The experimental setup, data processing, and plasma diagnostic techniques are described in [3]. The present paper deals only with the first experimental case, *i.e.*, the C II 723-nm line.

Figure 1. Geometry of observations for the experimental spectra of the C II 723-nm line measured from the ablation cloud of a carbon pellet injected in the stellarator LHD.

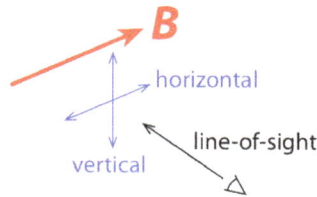

2. Description of the Atomic System and the Line-Shape Modeling Codes

In this section, we introduce all the line-shape codes used for the modeling and/or for the fitting of the previously mentioned experimental spectra of the C II 723-nm line and we briefly describe the atomic physics data necessary for the line profile calculations.

2.1. Description of the Line-Shape Modeling Codes

We present here the five numerical simulation codes and models used by the contributors to model the line-shapes of the C II 723-nm line for four cases sharing the same electron density of $n_e = 10^{17}$ cm^{-3} but for two distinct values of the electron temperature $T_e = 2$ eV and $T_e = 4$ eV, without and with a magnetic field (B = 4 T). Note that the electron and ion temperatures were assumed equal. Even though there are differences in the treatment of the Stark effect by the various codes, they can be separated into two groups, according to the adopted approach to treat the Zeeman effect. Indeed, the line-shape codes used here can be divided into two groups. Those of the first group treat the Zeeman effect within the weak-field approximation [4,5] in which the magnetic field is a perturbation of the emitter fine-structure energy levels, shown in Figure 2a. This approximation is valid when the fine structure splitting exceeds the Zeeman splitting ($\Delta E_{FS} \gg \Delta E_Z$). Three methods belong to this group: SCRAM (Sandia National Laboratory), PPP-B and WEAKZEE (CNRS/ Aix-Marseille Université). The PPP-B code [6] is an extension of the PPP standard Stark line-shape code [7,8], which accounts for ion dynamics. In PPP-B, the Zeeman effect is described in either the weak-field approximation or the opposite one, i.e., the strong-field approximation [4,5]. The latter is valid when the Zeeman splitting is higher than the fine structure one ($\Delta E_Z \gg \Delta E_{FS}$). Note that, when input MCDF atomic data are used, an asterisk is added to the code name PPP-B which becomes PPP-B *. WEAKZEE is a very simple version of PPP-B where only the electron Stark broadening is accounted for, the ion Stark broadening being neglected. In WEAKZEE, the Zeeman components are dressed by a Lorentzian shape with a given width. The later can be obtained from the Stark-B database [9,10]. In [9], one can find Stark broadening parameters (FWHM: Full Width at Half Maximum) by electrons and ions for few values of the electron density and the electron temperature. For all other temperatures, the Stark widths w (in Å units) can be obtained using the following fit formula [11]:

$$Log(w) = a_0 + a_1 Log(T_e) + a_2 \left[Log(T_e)\right]^2, \tag{1}$$

where a_0, a_1, and a_2 are fitting parameters depending on the line, perturbers (ions or electrons) and the electron density. In this relation, the electron temperature is expressed in Kelvin. For the present calculations, the Lorentzian width w used by WEAKZEE was calculated using Equation (1). SCRAM [12,13], which is primarily used for non-LTE diagnostics of emission spectra that cover a wide range of energies and access many charge states, satellites, etc., uses the electron impact approximation for collisional broadening (based on allowed distorted wave transitions among fine structure states), the quasi-static approximation for the ionic Stark broadening, and can interpolate between the weak and strong field limits. Including forbidden collisional transitions increases the widths by about 10%. The second group of codes contains two models: SIMU [14,15] and INTDPH [16] (Weizmann Institute of Science). In these codes, the Zeeman effect is treated non-perturbatively, via a numerical solution of the static (INTDPH) or time-dependent (SIMU) Schrödinger equation. The initial atomic system is that shown on Figure 2a (see also Table 1). More precisely, SIMU is a combination of two codes: a molecular dynamics (MD) simulation of variable complexity and a solver for evolution of an atomic system with the MD field history used as a time-dependent perturbation. INTDPH is another Stark–Zeeman line-shape code using the quasi-static approximation for the ions. In order to account for the electron broadening, the output is convolved with a shifted Lorentzian at the post-processing step. The width and shift of the Lorentzian should be obtained separately from another code or a database. This is a very fast and accurate procedure (when electrons are strictly impact, e.g., for isolated lines). An application of INTDPH to diagnostics of magnetized plasmas can be found in [17]. For the calculations presented here, the impact broadening parameters were inferred from the SIMU line shapes obtained assuming B = 0.

Figure 2. Schematic energy diagrams of the radiator considered for the present study without (**a**) and with (**b**) the fine structure effect. Energy splitting between the $1s^2 2s^2 3p$ $^2P°$ and $1s^2 2s^2 3d$ ^2D doublets is exaggerated (magnified by a factor of 500) for both of levels. Arrows represent radiative dipolar transitions with solid ones representing those transitions considered for the present study.

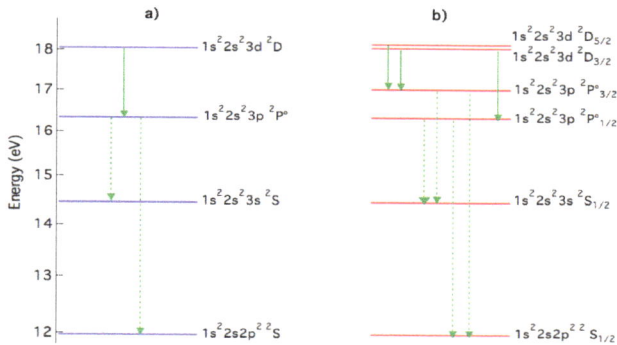

133

Table 1. Summary of the atomic data of the $1s^22s^23s\ ^2S$-$1s^22s^23p\ ^2P^\circ$ and $1s^22s^23p$ $^2P^\circ$-$1s^22s^23d\ ^2D$ transitions used by the different line-shape codes described in Subsection 2.2. For each transition i→k, the wavenumber σ_{ik} (cm^{-1}) is given in the 3rd column while either the line strength S_{ik} or the line oscillator f_{ki}, both in atomic units, are given in columns 4 and 5. The last column shows the corresponding line-shape codes. The sources of the atomic data are indicated in the table: National Institute for Standard and Technology NIST [18], Multiconfiguration Dirac-Fock MCDF [19], and Flexible Atomic Code FAC [20]. Note that PPP-B and PPP-B* codes differ only by the atomic physics: NIST for PPP-B and MCDF for PPP-B *.

Terms	Energies E_i–E_k (cm^{-1})	σ_{ik} (cm^{-1})	S_{ij}(a.u)	f_{ki}(a.u.)	Line-Shape Code(s)
3s ^{2}S-3p ^{2}P°	116,537.65–131 731.80	15,194.15		0.715[18]	INTDPH SIMU
3p ^{2}P°-3d ^{2}D	131,731.80–145 550.13	13,818.33		0.547[18]	INTDPH SIMU
3p ^{2}P°$_{1/2}$-3d ^{2}D$_{3/2}$	131,724.37–145,549.27	13,824.92	26.00[18]	-	PPP-B
			//		WEAKZEE
			11.66[19]	-	PPP-B *
			13.84[20]	-	SCRAM
3p ^{2}P°$_{3/2}$-3d ^{2}D$_{5/2}$	131,735.52–145 550.70	13,815.18	46.90[18]	-	PPP-B
			//		WEAKZEE
			21.00[19]	-	PPP-B *
			33.31[20]	-	SCRAM
3p ^{2}P°$_{3/2}$-3d ^{2}D$_{3/2}$	131,735.52–145 549.27	13,813.75	5.21[18]	-	PPP-B
			//		WEAKZEE
			2.33[19]	-	PPP-B *
			5.56[20]	-	SCRAM

2.2. Brief Description of the Atomic System Representing the Emitter

The calculation of the profile of the C II 723-nm line does not require a complicated atomic data system. However, even though only four energy levels and three dipolar radiative transitions are sufficient, calculations of such atomic physics data is complicated for this weakly charged non-hydrogen-like ion. For our case, different atomic codes give different values of the dipole reduced matrix elements. In the absence of magnetic field, one can use the atomic system shown on Figure 2, where the fine structure effect is shown only for the right part of the figure. In this figure, the fine structure splitting between the $1s^22s^23p\ ^2P^\circ$ and $1s^22s^23d\ ^2D$ doublets have been magnified by a factor of 500. Energies are expressed with respect to the ground level of the C$^+$ ion, *i.e.*, $1s^22s^22p\ ^2P^\circ_{1/2}$.

All line shape codes require some atomic information of the radiator including the energies, labels and quantum numbers of all the radiator energy levels involved in the considered radiative transitions, as well as the reduced matrix elements of the electric dipole or their squares, known as line strengths. Equivalently to line strengths, one can use line oscillators. Different atomic data have been used for the present code comparison. The atomic data including the electric dipolar matrix transitions were taken from NIST ASD [18]. One of the authors has used atomic data calculated by an MCDF code [19], differing only by reduced matrix elements of the dipole transitions about 1.5 times lower than those extracted from NIST ASD; another has used strength data from FAC [20]. The line strengths and/or line oscillators of the radiative transitions considered here are summarized on Table 1.

3. Cross-Comparison of the Line Profiles Computed with the Different Codes

3.1. Magnetic Field-Free Case

Let us start with the modeling of the C II 723-nm line profiles for the magnetic field-free cases with the following plasma parameters: $n_e = 10^{17}$ cm^{-3}, $T_e = T_i = 2$ eV and $T_e = T_i = 4$ eV. Whatever are the positively charged perturbers (D$^+$ or C$^+$ ions), calculations show that the above line is dominated by electron broadening, the ion contribution being close to zero. This is demonstrated in Figure 3, where the electron broadened profile of the C II 723 nm line is compared to the full pure Stark profile of the same line. The profiles shown on this figure, calculated with the PPP-B code using atomic data from NIST, demonstrate the dominance of the electron broadening over the ionic one for these typical conditions. Note that, in this paper, all the computed profiles are plotted against the wavenumber shift (in cm^{-1}) with respect to the line center wavenumber σ_0.

Before comparing the results of the different codes, it is interesting to discuss briefly the electron broadening which is treated in the frame of the impact theory in all the used codes. Different models and formulae exist for the electron collision operator [21,22]. The PPP-B and PPP-B * (as well as PPP) line shape codes use a modified electron broadening operator. This modified electron collision operator, which is based on the semi-classical GBK model due to Griem, Blaha, and Kepple [23], can be written as follows:

$$\Phi(\Delta\omega = 0) = \left(\frac{4\pi}{3}\right)\left(\frac{2m_e}{\pi k T_e}\right)^{1/2} n_e \left(\frac{\hbar}{m_e}\right)^2 \mathbf{R.R}\left[C_n + \frac{1}{2}\int_y^\infty \frac{e^{-x}}{x}dx\right], \quad (2)$$

where:

$$y(\Delta\omega = 0) \approx \left(\frac{hn^2}{2(z_e+1)}\right)^2 \frac{\omega_p^2 + \Delta\omega_{\alpha\alpha''}^2}{E_H \; kT_e} \quad (3)$$

Equation (3) arises from conditions on the limits of the integral over impact parameters ρ_{min} and ρ_{max} appearing in Equation (12) of [23]. More precisely, a strong collision term is added and different cutoffs are included in this Equation.

Figure 3. Comparison of Stark profiles of the C II 723-nm line emitted by a pure carbon plasma (electrons and C^+ ions) computed with the following parameters: $n_e = 10^{17}$ cm^{-3}, $T_e = T_i = 2$ eV and B = 0. Dashed line represents the electron broadening only while the solid one accounts for both ions and electrons, all other broadening mechanisms were ignored. Note the use of a semi-logarithmic scale. Both profiles were calculated using PPP-B with atomic data from NIST and by setting B = 0 T.

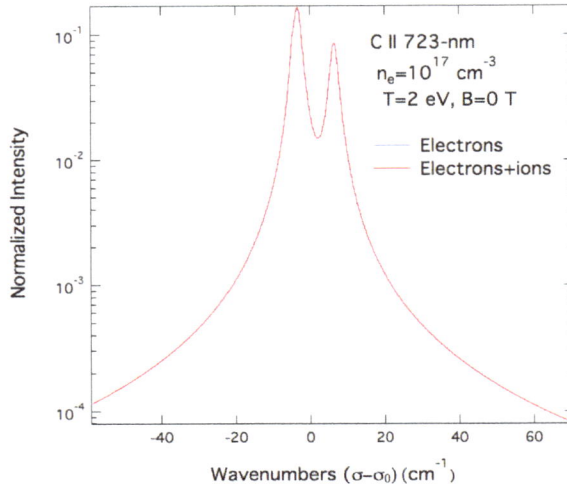

Here, m_e is the electron mass and **R** is the position operator of the radiator. C_n is a strong collision term whose value depends on the principal quantum number n as: $C_2 = 1.5$, $C_3 = 1.0$, $C_4 = 0.75$, $C_5 = 0.5$ and $C_n = 0.4$ for $n > 5$. In Equation (4), E_H and z_e represent respectively the hydrogen ionization potential and the net charge of the emitter while ω_p and $\Delta\omega_{\alpha\alpha''}$ designate the electron plasma frequency and the frequency separation between the state α involved in a given transition and its perturbing states α'' [24]. Moreover, the impact limit has been taken, *i.e.*, $\Delta\omega = 0$. Equation (2) indicates that the electron interactions with the emitter result in a homogeneous broadening represented by a Lorentzian function whose full width at half maximum (FWHM) is proportional to the plasma electron density n_e and inversely proportional to the square root of the electron temperature T_e. It should be noted that the dipole reduced matrix elements are involved through the operator **R.R** present in relation Equation (2). As the line considered here is of the same type as the Li-like (Li I, B III, N V) 2s-2p, *i.e.*, $\Delta n = 0$ transitions, we will not discuss the validity of impact electronic collision operators such the above one. Readers interested by that issue may refer to the discussion found in [25].

In Figure 8 of [2], various experimental and theoretical data representing Stark broadening widths (FWHM) of the C II 723-nm line were fitted with a linear function of the plasma electron density. An electron density $n_e = 10^{17}$ cm^{-3} corresponds to a Stark FWHM of about 0.1 nm or 1.9 cm^{-1} in terms of wavenumbers. A very close value can be obtained from the Stark-B database [9]. The C II 723-nm line profiles calculated by the various codes for $n_e = 10^{17}$ cm^{-3}, $T_e = T_{iv} = 2$ eV and B = 0 are shown on Figure 4. As the profile calculated with WEAKZEE was obtained using Lorentzian functions with

FWHM taken from the Stark-B database, it can be considered as a "reference" profile with a Stark width $\Delta\sigma_0$. Several points can be noted from Figure 4. First, the calculations with SIMU and INTDPH as well as PPP-B* (PPP-B code but with MCDF atomic data) give lower Stark widths, *i.e.*, $\Delta\sigma < \Delta\sigma_0$. Note the agreement between SIMU/INTDPH and PPP-B * despite the differences between the used atomic physics: NIST atomic data for the former and MCDF data for the latter. Second, in term of Stark widths, the profiles obtained with PPP-B slightly higher than ($\Delta\sigma \geq \Delta\sigma_0$) while those obtained with SCRAM are very close to ($\Delta\sigma \approx \Delta\sigma_0$) the "reference" one. This means that the electronic collision operator given by Stark-B and GBK lead to the close results. This is confirmed in Table 2, which presents the ratios of the Stark FWHM $\Delta\sigma$ to the "reference" value $\Delta\sigma_0$ for the different codes. In terms of line-shapes, it can be seen from Figure 4 that PPP-B agrees with WEAKZEE for the line wings. While SCRAM appears to overestimate the line wings, which is due to its inclusion of continuum emission where the other codes included only the line features. The same remarks and conclusions about both Stark widths and shapes of the C II 723-nm line can be drawn for the case with $n_e = 10^{17}$ cm^{-3} and $T_e = T_i = 4$ eV in the absence of the magnetic field (B = 0).

Table 2. Comparison of the Stark Full Width at Half Maximum (FWHM) $\Delta\sigma$ of the C II 723 nm line extracted from the profiles synthetized by the different codes with respect to a reference FWHM $\Delta\sigma_0$ for $n_e = 10^{17}$ cm^{-3}, $T_e = T_i = 2$ eV and B = 0.

Code	PPP-B	PPP-B *	SIMU/INTDPH	SCRAM	WEAKZEE
$\Delta\sigma/\Delta\sigma_0$	1.2	0.6	0.5	0.9	1.0

3.2. Magnetic Field Case

Let us now compare the synthetic profiles in the presence of a magnetic field B = 4 T. The presence of the magnetic field imposes a constraint on the radiation polarization. Photons whose polarizations are parallel or perpendicular to the B-field form respectively the π and σ components of the spectral line profile. These components as calculated by the previously mentioned codes are shown in Figures 5 and 6. On the other hand, assuming a perpendicular observation with respect to **B**, the total spectral line profile is calculated from the σ- and π-polarized ones I_σ and I_π using the following formula:

$$I_{tot}(\sigma) = I_\pi(\sigma) + 2I_\sigma(\sigma) \tag{4}$$

Total profiles are compared on Figure 7.

While the overall agreement between codes in Figures 5–7 is quite good within a factor two in terms of Stark FWHM, there are some significant differences: The profiles provided by SIMU/INTDPH show more structures around the line center than the other results. This can be attributed to the fact the Zeeman effect is fully treated by these two codes but at the same time the Stark broadening is smaller than that of the other codes, as illustrated in the B-field free case. Note that for C II 723-nm line, the fine structure splitting of its lower and upper energy levels are $\Delta E_{3p} = E(3p\ ^2P^\circ_{3/2}) - E(3p\ ^2P^\circ_{1/2}) \approx 11.2$ cm^{-1} and $\Delta E_{3d} = E(3d\ ^2D_{5/2}) - E(3d\ ^2D_{3/2}) \approx 1.4$ cm^{-1} respectively. The energy splitting of the same levels due to Zeeman effect is about $\Delta E_B \approx 0.8$ cm^{-1} for B = 2 T and 1.6 cm^{-1} for B = 4 T. It is clear that for these values of the magnetic field, the Zeeman splitting is comparable to the fine-structure one for the upper energy levels and therefore the use of the weak-field approximation becomes questionable. As for the field-free case, PPP-B * reproduces well the wings computed by both SIMU and INTDPH codes. The profiles computed by WEAKZEE, PPP-B and SCRAM show less features because of the weak-field approximation used to treat the Zeeman effect but the line widths are more correct as compared to those obtained with SIMU/INTDPH. In terms of Stark widths these remarks corroborate those concerning the field-free case. Calculations of profiles of the C II 723-nm line for the same plasma parameters as above with T = 4 eV instead of 2 eV lead to the same conclusions.

Figure 4. Comparison of synthetic Stark profiles of the C II 723-nm line emitted by a plasma with $n_e = 10^{17}$ cm^{-3}, $T_e = T_i = 2$ eV and B = 0 in a semi-logarithmic scale. The dashed line represents the profile obtained with the simulation code SIMU. Note that the profile calculated with INTDPH code is not shown here as it is almost identical to the one obtained with SIMU. The solid thick red line represents the profile calculated with the PPP-B code using the most accurate atomic data provided by NIST while the solid thin red line is the one obtained with the same code PPP-B (PPP-B *) but with atomic data calculated with an MCDF code. Solid blue and green lines represent the profiles obtained respectively with SCRAM (which includes continuum as well as line emission) and WEAKZEE.

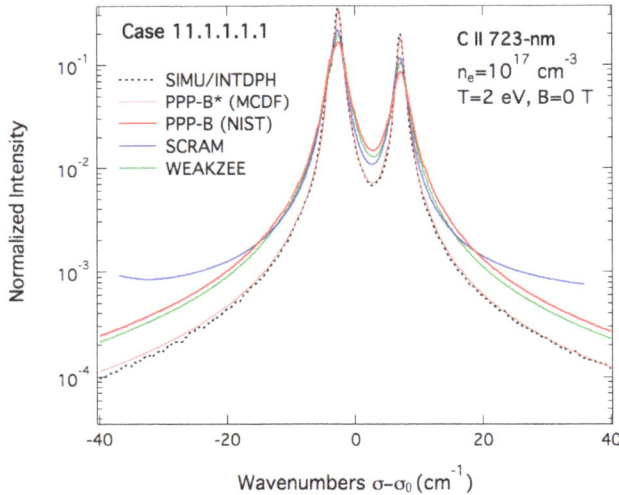

Figure 5. Comparison of the Stark–Zeeman σ-component of the C II 723-nm line as computed by the different line-shape codes for a carbon plasma with the following parameters: $n_e = 10^{17}$ cm^{-3}, $T_e = T_i = 2$ eV and $B = 4$ T.

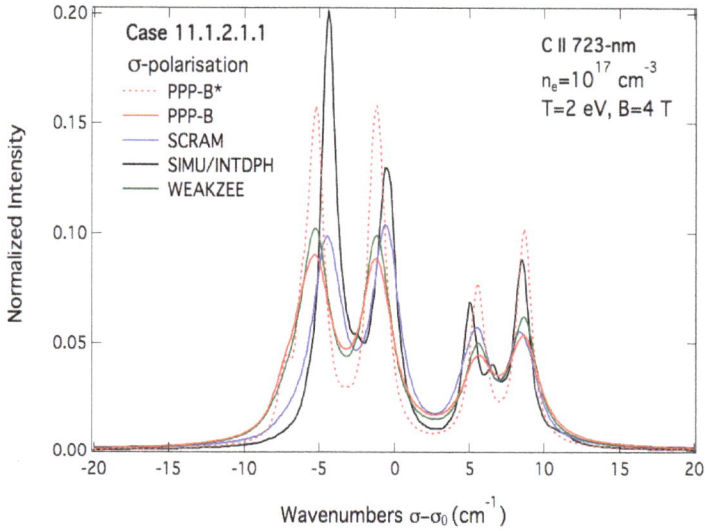

Figure 6. Comparison of the Stark–Zeeman π-component of the C II 723-nm line as computed by the different line-shape codes for the same conditions as in Figure 5.

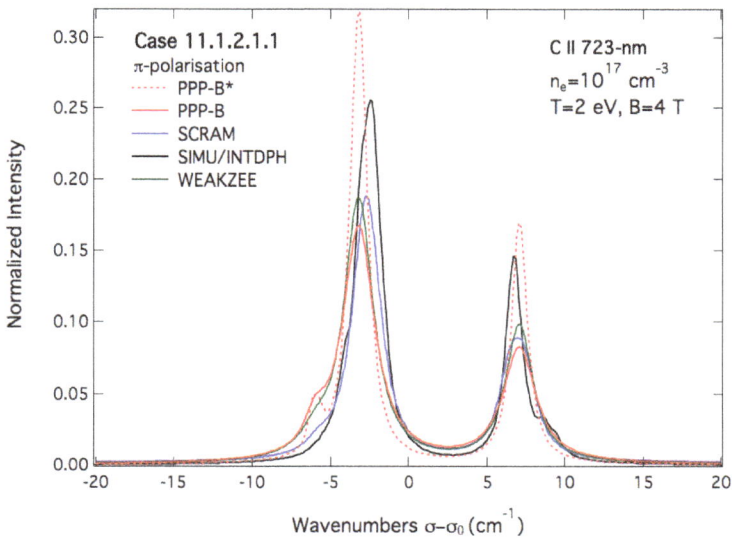

Figure 7. Comparison of the total Stark–Zeeman profiles of the C II 723-nm line as computed by the different line-shape modeling codes for the same conditions as in Figures 5 and 6 with an angle of observation of $90°$ with respect to the magnetic field direction.

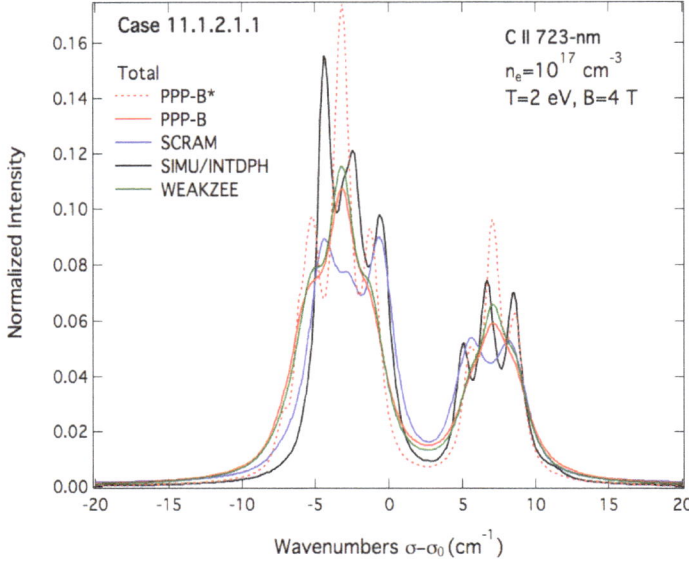

4. Comparison to the Experimental Spectra

As mentioned previously, the experimental data were obtained during the injection of a carbon pellet in the stellarator LHD. As all details concerning the experimental setup, the measurement system and pellet injection in LHD can be found in [2], we give here only the information necessary for the data analysis. The spectra were measured with a very high-resolution visible spectrometer. The instrumental function can be represented by a gaussian function with a FWHM $\Delta\lambda_{1/2} = 0.016$ nm as compared to that of the high-resolution spectrometer used for the data of reference [2] ($\Delta\lambda_{1/2} = 0.075$ nm). The spectra were measured almost perpendicularly to the magnetic field. In addition, thanks to linear polarizers two spectra were obtained: horizontal spectrum (π-polarization) and vertical spectrum (σ-polarization). These spectra are shown in Figure 8. These two spectra were proposed as a challenging case at the 2nd workshop on spectral line shapes in plasmas. Contributors were asked to do their best to fit these spectra in order to find the most reliable set of parameters (magnetic field, angle of observation, electron density and temperature).

Attempts to fit these experimental spectra are shown in Figures 9–11 corresponding, respectively, to the σ- and π-polarized spectra and the built un-polarized total spectrum.

Figure 8. Experimental spectra of the C II 723-nm line as measured in LHD almost perpendicularly to the magnetic field (see Figure 1) using a very high-resolution visible spectrometer and linear polarizers.

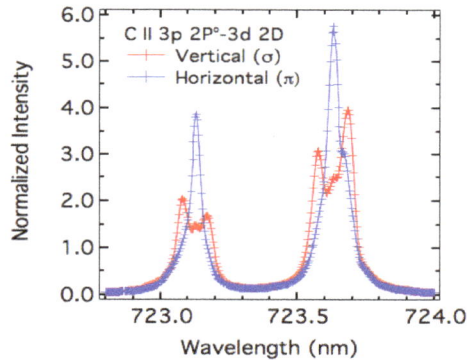

It can be seen from Figures 9–11 that none of the line-shape codes were able to fit perfectly the proposed experimental spectra. However, these attempts were fruitful and can be considered reasonably good. Using the code INTDPH, the best fit of the three spectra was obtained with the following parameters: a carbon plasma with an electron density $n_e = 9 \times 10^{16}$ cm^{-3}, an equal ion and electron temperatures of 1 eV and a magnetic field B = 2.15 T. In the calculations by the INTDPH code, the static profiles were convolved with a shifted Lorentzian whose shift $\delta\sigma$ and FWHM $\Delta\sigma$ were determined using the simulation method SIMU. For the above conditions, the following values were obtained: $\Delta\sigma = 0.9$ cm^{-1} and $\delta\sigma = 0.26$ cm^{-1}. Moreover, a contrast ratio of 4:1 was assumed for the polarizers. This is equivalent to assume that the line-of-sight is not strictly perpendicular to the magnetic field or alternatively to the existence of fluctuations of the field direction in the spectroscopic observation volume observed. It should be noted that a pure LS coupling was assumed and the 3p fine splitting was changed from the NIST value 11.15 to 11.24 cm^{-1} to better fit the data. The second fitting attempt was due to the SCRAM code. Using SCRAM without the ionic Stark broadening, the best-fit calculations were obtained for a magnetic field B = 2 T with the following carbon plasma parameters: an electron density $n_e = 6 \times 10^{16}$ cm^{-3} and an ion/electron temperature T = 2 eV. The angle of observation 0 was set to 90°. The third and last fitting attempt was due to the WEAKZEE method considering only the Stark broadening by electrons. The best fit was obtained for an angle of observation 0 = 70° and the following plasma parameters: electron density $n_e = 4 \times 10^{16}$ cm^{-3}, electron/ion temperature T = 2 eV. The B-field value was set to 2 T. In addition, in an attempt to fit at least partially the π-component, a profile calculated with PPP-B (NIST data) was added to the other results shown on Figure 10. Without the use of any fitting procedure but varying only the electron density, the parameters leading to a good agreement with the right peak of the experimental spectrum (in terms of line-shapes and Stark widths) were the following: $n_e = 4 \times 10^{16}$ cm^{-3}, T = 2 eV, B = 2 T and 0 = 90°. It should be noted that all calculations were done assuming an optical thin plasma emission zone. The spread of the obtained results concerning the plasma electron density was expected since all the parameters, including the angle of observation, were free for the fitting. Introducing constraints on the angle of observation would result in more consistent results providing the same atomic physics is used. Each of the three methods used to fit the experimental spectra has its own advantages and drawbacks. As previously demonstrated, the electron broadening is correctly accounted for by the WEAKZEE and SCRAM codes while underestimated by the couple of codes SIMU/INTDPH. On the other side, Zeeman effect is correctly treated by SIMU/INTDPH while the weak-field approximation was used by both SCRAM and WEAKZEE methods. Therefore, it is clear that a correct plasma electron diagnostics based of the considered line spectra requires a full treatment of the Zeeman effect as well as the use of the appropriate electron line broadening and this should be integrated by these different line-shape codes.

Figure 9. Fitting attempts of the σ-polarized experimental spectrum of the C II 723-nm line as measured in LHD. All spectra are normalized to unity.

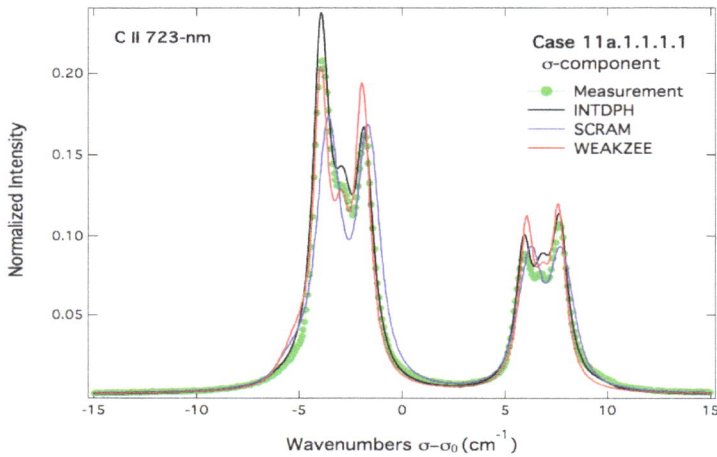

Figure 10. Fitting attempts of the π-polarized experimental spectrum of the C II 723-nm line as measured in LHD. All spectra are normalized to unity.

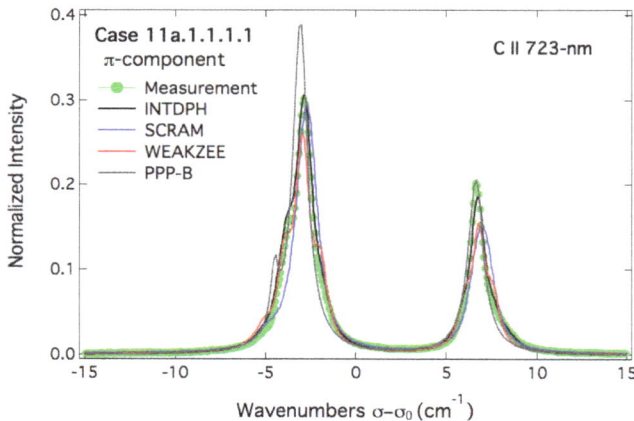

Figure 11. Fitting attempts of the un-polarized "experimental" spectrum of the C II 723-nm line as constructed from σ- and π-polarized spectra of Figures 9 and 10. Superscript [a] indicates that the area of the profile calculated with INTDPH is equal to 0.9 while all other spectra are normalized to unity.

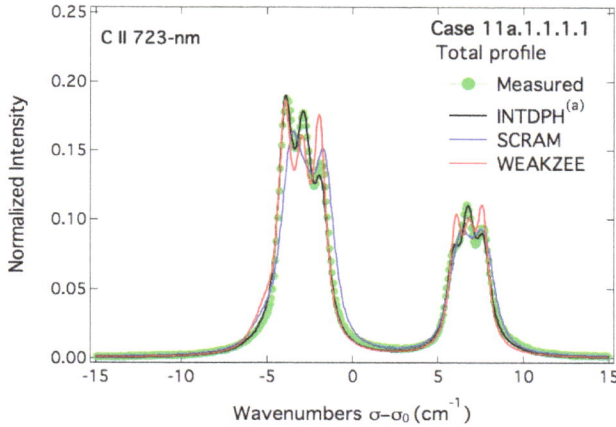

5. Conclusions

Several line-shapes codes were compared through the modeling of the C II 723-nm line profile for four situations corresponding to a fixed electron density of 10^{17} cm^{-3}, two temperatures (T = 2 and 4 eV) with and without the presence of a magnetic field B = 4 T. For these conditions, calculations have shown that the Stark broadening of the above line is mainly due to the plasma electrons. C II 723-nm line profiles calculated by five line-shape codes have been compared. For the magnetic-field free case (B = 0 T), a relatively good overall agreement has been obtained, despite the significant differences in the code results in terms of Stark width (FWHM) of the C II 723-nm line. Indeed, the overall agreement between the five line-shape codes is within a factor of 2. Two factors may contribute to this dispersion of the results: differences in the used atomic physics and/or in the electron collision operator. However, the former factor has been dismissed by partial comparisons between line-shape codes using the same more accurate atomic physics data, *i.e.*, from the NIST ASD database. Therefore, the dispersion in Stark widths is attributed to differences in the electron collision operators used by the various line-shape codes. For the case with B = 4 T, the differences in the treatment of the Zeeman effect by the different codes make more difficult the interpretation of the results. However, it is clear that the weak-field approximation adopted by some of the codes is not valid for the upper level of the C II 723-nm line, *i.e.*, (3d ^{2}D°). Therefore, codes able to deal with intermediate magnetic fields (comparable Zeeman and fine structure splittings of energy levels) are more suitable. In addition to code-code comparisons, σ- and π-polarized experimental spectra of the C II 723-nm line, measured from the ablation cloud of a carbon pellet injected in LHD, were fitted using several line-shape codes. By letting free all the parameters (magnetic field B, electron density n_e, electron/ion temperature T, angle of observation 0), attempts to fit the experimental spectra have shown a large dispersion in the inferred parameters in particular the electron density for which there is a factor of more than 2. However, even though not perfect, these fitting attempts are encouraging and suggest to use the best of each code for the purpose of diagnostics of magnetized plasmas. It is recommended for this case to use a full-treatment of the Zeeman effect and elucidate and reduce the dispersion in the Stark widths due to the electron broadening. This requires more investigations and detailed comparisons with a prescribed atomic system.

Author Contributions

This work is based on the following author contributions: spectral measurements, data—M. Goto; code calculations, comparison with experimental data—all other authors; writing of the manuscript—M. Koubiti with contribution of the other authors.

Conflicts of Interest

The authors declare no conflict of interest.

References

1. Stambulchik, E. Review of the 1st Spectral line shapes in plasmas: Code comparison. *High Energy Density Phys.* **2013**, *9*, 528–534.
2. Goto, M.; Morita, S.; Koubiti, M. Spectroscopic study of a carbon pellet ablation cloud. *J. Phys. B-At. Mol. Opt. Phys.* **2010**, *43*, 144023.
3. Cvejić, M.; Gavrilović, M.R.; Jovićević, S.; Konjević, N. Stark broadening of Mg I and Mg II spectral lines and Debye shielding effect in laser induced plasma. *Spectrochim. Acta Part B: Atomic Spectrosc.* **2013**, *85*, 20–33.
4. Condon, E.U.; Shortley, G.H. One-Electron Spectra. In *The Theory of Atomic Spectra*; Cambridge University Press: London, UK, 1964; pp. 149–157.
5. Weissbluth, M. *Static Fields in Atoms and Molecules*; Student edition, Academic Press: New York, NY, USA, 1978; pp. 346–355.
6. Ferri, S.; Calisti, A.; Mossé, C.; Mouret, L.; Talin, B.; Gigosos, M.A.; Gonzalès, M.A.; Lisitsa, V. Frequency-fluctuation model applied to Stark–Zeeman spectral line shapes in plasmas. *Phys. Rev. E* **2011**, *84*, 026407.
7. Talin, B.; Calisti, A.; Godbert, L.; Stamm, R.; Lee, R.W.; Klein, L. Frequency-fluctuation model for line-shape calculations in plasma spectroscopy. *Phys. Rev. A* **1995**, *51*, 1918.
8. Calisti, A.; Mossé, C.; Ferri, S.; Talin, B.; Rosmej, F.; Bureyeva, L.A.; Lisitsa, V.A. Dynamic Stark broadening as the Dicke narrowing effect. *Phys. Rev. E* **2010**, *81*, 016406.
9. Sahal-Brechot, S.; Dimitrijević, M.S.; Moreau, N. Observatory of Paris, LERMA and Astronomical Observatory of Belgrade Stark-B Database. Available online: http://stark-b.obspm.fr (accessed on 16 April 2014).
10. Mahmoudi, W.F.; ben Nessib, N.; Sahal-Bréchot, S. Semi-classical calculations of Stark broadening impact theory of singly-ionized carbon, nitrogen and oxygen spectral lines. *Phys. Scr.* **2004**, *70*, 142.

11. Sahal-Bréchot, S.; Dimitrijević, M.S.; ben Nessib, N. Comparisons and comments on electron and ion impact profiles of spectral lines. *Balt. Astron.* **2011**, *20*, 523–530.

12. Hansen, S.B. Configuration interaction in statistically complete hybrid-structure atomic models. *Can. J. Phys.* **2011**, *89*, 633–638.

13. Hansen, S.B.; Bauche, J.; Bauche-Arnoult, C.; Gu, M.F. Hybrid atomic models for spectroscopic plasma diagnostics. *High Energy Density Phys.* **2007**, *3*, 109–114.

14. Stambulchik, E.; Maron, Y. A study of ion-dynamics and correlation effects for spectral line broadening in plasma: K-shell lines. *J. Quant. Spectr. Rad. Transf.* **2006**, *99*, 730–749.

15. Stambulchik, E.; Alexiou, S.; Griem, H.R.; Kepple, P.C. Stark broadening of high principal quantum number hydrogen Balmer lines in low-density laboratory plasmas. *Phys. Rev. E* **2007**, *75*, 016401.

16. Stambulchik, E.; Maron, Y. Effect of high-n and continuum eigenstates on the Stark effect of resonance lines of atoms and ions. *Phys. Rev. A* **1997**, *56*, 2713–2719.

17. Tessarin, S.; Mikitchuk, D.; Doron, R.; Stambulchik, E.; Kroupp, E.; Maron, Y.; Hammer, D.A.; Jacobs, V.L.; Seely, J.F.; Oliver, B.V.; *et al.* Beyond Zeeman spectroscopy: Magnetic-field diagnostics with Stark-dominated line shapes. *Phys. Plasmas* **2011**, *18*, 093301.

18. Kramida, A.; Ralchenko, Y.; Reader, J.; NIST ASD Team (**2013**). NIST Atomic Spectra Database (ver. 5.1). Available online: http://physics.nist.gov/asd (accessed on 9 May 2014).

19. Grant, I.P.; McKenzie, B.J.; Norrington, P.H.; Mayers, D.F.; Pyper, N.C. An atomic multiconfigurational Dirac-Fock package. *Comput. Phys. Commun.* **1980**, *21*, 207–231.

20. Gu, M.F. The Flexible Atomic Code. *Can. J. Phys.* **2008**, *86*, 675–689.

21. Calisti, A.; Ferri, S.; Stamm, R.; Talin, B.; Lee, R.W.; Klein, L. Discussion of the validity of binary collision models for electron broadening in plasmas. *J. Quant. Spectrosc. Radiat. Transf.* **2000**, *65*, 109–116.

22. Alexiou, S. Collision operator for isolated ion lines in the standard Stark-broadening theory with applications to the Z scaling of the Li isoelectronic series 3P-3S transition. *Phys. Rev. A* **1994**, *49*, 106–119.

23. Griem, H.R.; Blaha, M.; Kepple, P.C. Stark-profile calculations for Lyman-series lines of one-electron ions in dense plasmas. *Phys. Rev. A* **1979**, *19*, 2421–2432.

24. Calisti, A.; Khelfaoui, F.; Stamm, R.; Talin, B.; Lee, R.W. Model for the line shapes of complex ions in hot and dense plasmas. *Phys. Rev. A* **1990**, *42*, 5433–5440.

25. Alexiou, S.; Dimitrijević, M.S.; Sahal-Brechot, S.; Stambulchik, E.; Duan, B.; Gonzalez-Herrero, D.; Gigosos, M.A. The Second Workshop on Lineshape Code Comparison: Isolated Lines. *Atoms* **2014**, *2*, 157–177.

Reprinted from *Atoms*. Cite as: Dimitrijević, M.S.; Sahal-Bréchot, S. On the Application of Stark Broadening Data Determined with a Semiclassical Perturbation Approach. *Atoms* **2014**, *2*, 357-377.

Article

On the Application of Stark Broadening Data Determined with a Semiclassical Perturbation Approach

Milan S. Dimitrijević [1,2,]* and Sylvie Sahal-Bréchot [2]

[1] Astronomical Observatory, Volgina 7, 11060 Belgrade, Serbia

[2] Laboratoire d'Etude du Rayonnement et de la Matière en Astrophysique, Observatoire de Paris, UMR CNRS 8112, UPMC, 5 Place Jules Janssen, 92195 Meudon Cedex, France;
E-Mail: sylvie.sahal-brechot@obspm.fr

* Author to whom correspondence should be addressed; E-Mail: mdimitrijevic@aob.bg.ac.rs;
Tel.: +381-64-297-8021; Fax: +381-11-2419-553.

Received: 5 May 2014; in revised form: 20 June 2014 / Accepted: 16 July 2014 / Published: 7 August 2014

Abstract: The significance of Stark broadening data for problems in astrophysics, physics, as well as for technological plasmas is discussed and applications of Stark broadening parameters calculated using a semiclassical perturbation method are analyzed.

Keywords: Stark broadening; isolated lines; impact approximation

1. Introduction

Stark broadening parameters of neutral atom and ion lines are of interest for a number of problems in astrophysical, laboratory, laser produced, fusion or technological plasma investigations. Especially the development of space astronomy has enabled the collection of a huge amount of spectroscopic data of all kinds of celestial objects within various spectral ranges. Consequently, the atomic data for trace elements, which had not been of interest in astrophysics before, have become more and more important, and, since we do not know *a priori* the chemical composition of a star, the interest for a very extensive list of such data, as well as for the corresponding databases has been increasing, stimulating the theoretical and experimental work on spectral lineshape research.Such data are particularly needed for interpretation, synthesis and analysis of high resolution spectra with well-resolved line profiles, obtained from space born instruments in space missions like the

Far Ultraviolet Spectroscopy Explorer (FUSE), the Goddard High Resolution Spectrograph (GHRS—the Hubble Space Telescope), the International Ultraviolet Explorer and many others. Space high resolution spectroscopy has demonstrated that ionized manganese, tellurium, gold, indium, tin, chromium, ruthenium, zinc, copper, selenium, rare earths and other trace elements, which prior to the epoch of space born stellar spectroscopy had been completely insignificant for astrophysics, are present in hot stellar atmospheres, where Stark broadening is particularly significant.

In comparison with laboratory plasmas, conditions in astrophysical plasmas, where the Stark broadening mechanism is important, are incomparably more various. This broadening mechanism is of interest for astrophysical plasmas with such extreme conditions like for example the plasma in interstellar molecular clouds, with typical electron temperatures around 30 K or smaller, and typical electron densities of several electrons per cubic centimeter. In plasma of such low density, free electrons may be recombined in a very distant orbit with very large principal quantum number values of several hundreds and deexcited in cascade radiating in the radio domain. Since such distant electrons are weakly bounded with the core, even very weak electric microfields can have a significant influence.

Hydrogen, which is usually the main constituent of stellar atmospheres for temperatures larger or near 10,000 K is ionized in such amount that Stark broadening is the dominant collisional broadening mechanism for spectral lines. This is the case for white dwarfs and hot stars of the O, B and A type. Even for lower temperatures and for cooler stars of the solar type, Stark broadening may be important for spectral lines originating from highly excited atoms, where the distant, weakly bounded, optical electron is significantly influenced by weak electric microfields. This broadening mechanism is also important even for cooler stars for the investigation and modeling of subphotospheric layers.

In the above mentioned cases, when Stark broadening is of interest, the corresponding line broadening parameters (line widths and shifts) are significant e.g., for interpretation, synthesis and analysis of stellar spectral lines, determination of chemical abundances of elements from equivalent widths of absorption lines, estimation of the radiative transfer through the stellar atmospheres and subphotospheric layers, opacity calculations, radiative acceleration considerations, nucleosynthesis research and other astrophysical topics. Stark broadening is of interest for the investigation of neutron stars. The electron densities and temperatures in atmospheres of such stars are orders of magnitude larger than in atmospheres of white dwarfs and are typical for stellar interiors. Temperatures in the extremely thin atmospheric layer where the photospheric emission originates are of the order of 10^6–10^7 K and electron densities of the order of 10^{24} cm^{-3}, which are plasma conditions where Stark broadening dominates.

Stark broadening data are also of interest for laboratory plasma diagnostics, laser produced plasma investigation and modeling, the design of laser devices, inertial fusion plasma and for analysis and modeling of various plasmas in technology, as for example for laser welding and piercing and for plasmas in light sources.

The most sophisticated theoretical method for the calculation of a Stark broadened line profile is the quantum mechanical strong coupling approach, but due to its complexity and numerical difficulties, it is not adequate for large scale determination of Stark broadening parameters, in particular for e.g., complex spectra, heavy elements or transitions between highly excited energy levels. Consequently, in a lot of cases, the semiclassical approach remains the most efficient method for Stark broadening calculations, which has provided the largest set of existing theoretical results.

In order to complete as much as possible the Stark broadening data important for various topics in astrophysics, physics and technology, Stark broadening parameters for a number of spectral lines of various emitters have been determined in a series of papers, using the semiclassical perturbation formalism [1,2]. The corresponding computer code was innovated and optimized several times (see the article Sahal-Bréchot *et al.* in this issue). Up to now, Stark broadening parameters (line widths and shifts) for spectral lines or multiplets of the following atoms and ions have been calculated and published: He I, Li I, Li II, Be I, Be II, Be III, B II, B III, C II, C III, C IV, C V, N I, N II, N III, N IV, N V, O I, O II, O III, O IV, O V, OVI, OVII, F I, F II, F III, F V, F VI, F VII, Ne I, Ne II, Ne III, Ne IV, Ne V, Ne VIII, Na I, Na IX, Na X, Mg I, Mg II, Mg XI, Al I, Al III, Al V, Al XI, Si I, Si IV, Si V, Si VI, Si XI, Si XII, Si XIII, P IV, P V, P VI, S III, S IV, S V, S VI, Cl I, Cl IV, Cl VI, Cl VII, Ar I, Ar II, Ar III, Ar IV, Ar VIII, K I, K VIII, K IX, Ca I, Ca II, Ca V, Ca IX, Ca X, Sc III, Sc X, Sc XI, Ti IV, Ti XI, Ti XII, V V, V XIII, Co I, Cr II, Mn II, Mn III, Fe II, Co III, Ni II, Cu I, Zn I, Ga I, Ga III, Ge I, Ge III, Ge IV, Se I, Br I, Kr I, Kr II, Kr VIII, Rb I, Sr I, Y III, Pd I, Ag I, Cd I, Cd II, Cd III, In II, In III, Te I, I I, Ba I, Ba II, Au I, Hg II, Tl III, Pb IV, and Ra II. In total, Stark broadening parameters have been calculated and published for 123 atomic and ionic species for 49 chemical elements, during a period of more than thirty years.

The obtained Stark broadening data have been used and cited many times for various applications and investigations. The literature where Stark broadening data might be used (as described in more detail previously) has been analyzed many times, but the set of data considered here allows examination of the purposes such data have actually been used for. This is interesting to analyze not only in order to demonstrate the possibilities of their applicability but first of all to see the needs of their principal users in order to adapt the presentation of results and plans for future investigations in accordance with the needs of consumers of such results. We will exclude from this analysis applications and citations concerning the theoretical and experimental research of Stark broadening and consider only applications in other research fields, published in international journals.

2. Applications of Stark Broadening Data Obtained by the Semiclassical Perturbation Method for Astrophysical Research

The analysis of citations of Stark broadening data obtained by semiclassical perturbation method shows that the largest number of citations is for astrophysical applications. For various investigations in astrophysics, our data for He I, Na I, C IV, Si II, Si IV, Li I, N V, Hg II, O VI, S VI, Mg I, Mg II, Ba I, Ba II, Ca I and Ca II have been used.

After hydrogen, helium has the largest cosmic abundance, so it is not surprising that our Stark broadening data for He spectral lines [3–7] have often been used for different investigations in astrophysics [8–83]. For example, they have been used for the following astrophysical problems: non Local Thermodynamical Equilibrium (LTE) model analysis of the interacting binary Beta Lyrrae [8], research of variability of Balmer lines in Ap stars [9], consideration if Delta Orionis C and HD 58260 of peculiar helium-strong stars [10], determination of the chemical composition of two double clusters and of a loose association [11], the critical analysis of the ultraviolet temperature scale of the helium-dominated DB and DBV white dwarfs [12], the effective temperature calibration of MK spectral classes dwarf stars using spectral synthesis [13], investigation of the extreme helium star BD-90-4395 [14], the ionization and excitation of hydrogen and helium in cool giant stars [15], the constitution of the atmospheric layers and the extreme ultraviolet-spectrum of hot hydrogen-rich white dwarfs [16], interpretation of spectral properties of hot hydrogen-rich white dwarfs with stratified H/He model [17], radiative accelerations on iron [18], non-LTE radiative acceleration of helium in the atmospheres of sdOB stars [19], research of hot stars with peculiar helium and noble gas abundances [20,25], a spectroscopic analysis of DAO and hot DA white dwarfs with the investigation of the implications of the presence of helium in the stellar nature [21]. The considered Stark broadening data have been entered into a spectrum synthesis program for binary stars [22] and have been used for the helium surface mapping and spectrum variability considerations of ET Andromedae [23,26], for the investigation of the He $\lambda 10830$ Å spectral line formation mechanism in classical cepheides [24], for the consideration of hot white dwarfs in the Extreme-Ultraviolet Explorer survey [27], for the search for forced oscillations in eclipsing and spectroscopic binaries [28], for investigations of the evolutionary state and helium abundance in He-rich stars [29] and the consideration of how much hydrogen is in white dwarfs [49], for a study of the effect of diffusion and mass-loss on the helium abundance in hot white dwarfs and subdwarfs [30], for the spectral analysis of the low gravity extreme helium stars [51] and a field horizontal-branch B-type star [83], for comparison with theoretical results obtained within the Stark broadening theory of solar Rydberg lines in the far infrared spectrum [58], for the research of dynamic processes in Be star atmospheres based on the example of He I 2P-nD line formation [59], for the investigation of winds of hot stars [60], for a study of the atmospheric variations of a peculiar Be star [61], for the application of a new tool for fitting observations with synthetic spectra [62], for the determination of the abundance of 3He isotope in HgMn star atmospheres [63], and for the investigation of the helium stratification in the atmospheres of magnetic helium peculiar B-type stars [64].

Results for Si II ion lines [84], obtained within the semiclassical perturbation method have been applied in numerous astronomical researches of stellar atmospheres, such as e.g., in [85–131]. The data have been used for silicon abundance analyses for a number of A (mainly) and B type stars [85–88,91–93,97,99,101,103,105,106,109–111,113–118] but also for normal F main sequence stars [89,112]. The discussed Si II Stark broadening data have also been used for an investigation of blue stragglers of M 67 [90], determination of the effective temperature of B-type stars from the Si II lines of the UV multiplet 13.04 at 130.5–130.9 nm [94], analysis of the red spectrum of Ap stars [95], non-LTE analysis of subluminous O type hot subdwarf in the binary system HD 128220 [98], a discussion on the significance of Stark spectral line shifts for element abundance determination with the method of atmospheric model [102], a discussion of the nature of the F STR λ 4077 type stars [104] and have been used for atmosphere research, He surface mapping and spectrum variability considerations of ET Andromedae [107,108].

Results obtained for lithium [132] have been used for a study of the non-LTE formation of Li I lines in cool stars [133] and Stark-broadening parameters of ionized mercury spectral lines [134], have been used for determination of Hg abundances in normal late-B and HgMn stars from co-added spectra from International Ultraviolet Explorer [135].

Our data for Ca II [136,137] have been used for the calcium abundance analysis of the double-lined spectroscopic binary α Andromedae [138], for the investigation of the pressure shifts and abundance gradients in the atmosphere of a DAZ white dwarf [139], for the analysis of VLT/X-shooter observations in order to determine the chemical composition of cool white dwarfs [140], and for abundance analysis of two late A-type stars [141].

Mg I [142,143] Stark broadening parameters have been used for analysis of stellar atmospheric parameters [144], for a non-LTE analysis of Mg I in the solar atmosphere [145], investigation of A-type stars [146,147] and for astrophysical tests of atomic data for the stellar Mg abundance determination [148]. Also, Mg II [149] data have been used for investigation of the pressure shifts and abundance gradients in the atmosphere of a DAZ white dwarf [139].

Na I [150] data were used for non-LTE calculations for neutral Na in late-type stars [151] and, Ba I and Ba II data [152,153] for abundance analysis of late A-type stars [154] and for quantitative spectroscopy of Deneb [155].

Results for multiply charged ions C IV [156], Si IV [157], N V [158], O VI [159] and S VI [160] have been used for non-LTE analysis of a SDO binary [161], analysis of PG 1159 stars [162] with the accent on the influence of gravitational settling and selective radiative forces [163], for high resolution UV spectroscopy of hot (pre-)white dwarfs with the Hubble Space Telescope [164], spectral energy distribution and the atmospheric properties of a helium-rich white-dwarf [165], for an investigation of stellar masses, kinematics, and white dwarf composition for three close DA + dMe binaries [166], for the calculation of C IV, N V, O VI and Si IV resonance lines formed in accretion shocks in T Tauri stars [167], for analysis of UV spectroscopic data for central star of Sh 2-216, obtained by the Far Ultraviolet Spectroscopic Explorer (FUSE) and the Hubble Space Telescope [168], for the spectral analysis of planetary nebulas K 1-27 [169,170] and for their very hot hydrogen-deficient central stars. These data have also been used for the study of the extreme ultraviolet (EUV) spectrum of the unique bare stellar core H1504 + 65 [171], for analysis of the FUSE spectra of a He-poor SDO star [172] and of a hot evolved star [173], GD 605, as well as for the analysis of He I lines in atmospheres of B-type stars [174] and for iron opacity predictions under solar interior conditions [175].

3. Applications of Stark Broadening Data Obtained by the Semiclassical Perturbation Method for Research in Physics and for Plasmas in Technology

If we do not take into account the usage of Stark broadening data for theoretical and experimental research of Stark broadening, the applications of Stark broadening parameters obtained by semiclassical perturbation method are not as numerous as in astrophysics. Semiclassical Stark broadening parameters of He I, Li II, Be II, Na I, Ca I, Ca II, Mg I, Mg II, Sr I, Ba I, Ba II, Zn I, Ag I, Cd I, Cu I, Ar I, Ar VIII, Al III, C IV and S V have been used for various physical problems.

Stark broadening data for helium [3,5] have been used for the analysis of the measurements of the hyperfine structure of the 1s3s3S1 state of Helium 3 [35], for the determination of emission coefficients of low temperature thermal iron-helium plasma [44] and for the analysis of the net emission of Ar–H2–He thermal plasmas at atmospheric pressure [57].

Na I data [174] have been used for the derivation of electron density radial profiles from Stark broadening in a sodium plasma produced by laser resonance saturation [175] and for the study of the mechanisms of resonant laser ionization [176], Be II [177] data for oscillator strength ratio measurements [178], Ca I [179,180] for the determination of differential and integrated cross sections for the electron excitation of the 41Po state of calcium atom [181] and for investigation of charged particle motion in an explosively generated ionizing shock [182], Ca II [137] for c hlorine detection in cement with laser-induced breakdown spectroscopy [183] and for dynamical plasma study during CaCu3Ti4O12 and Ba0.6Sr0.4TiO3 pulsed laser deposition [184], Mg I [185] and Mg II [186] for consideration of plasma plume induced during laser welding of magnesium alloys [187], Sr I [188] for investigation of vapor-phase oxidation during pulsed laser deposition of SrBi2Ta2O9. [189], for the measurement and control of ionization of the depositing flux during thin film growth [190] and for space and time resolved emission spectroscopy of Sr2FeMoO6 laser induced plasma [191], Li II [192] for examination of spatial and temporal variations of electron temperatures and densities from EUV-emitting lithium plasmas [193] and for modeling of continuous absorption of electromagnetic radiation in dense partially ionized plasmas [194], Ba I and Ba II [152,153] for investigation of plasma properties of laser-ablated strontium target [195] and for laser-based optical emission studies of barium plasma [196], Ag I [197] for determination of absolute differential cross sections for electron excitation of silver at small scattering angles [198], Cd I [199] for investigation of cadmium plasma produced by laser ablation, namely for its diagnostics [200], for comparison with zinc plasma [201] and for the research and diagnostics of deposition of wide bandgap semiconductors and nanostructure of deposits [202], Cu I [203] for investigation of characteristics of plume plasma and its effects on ablation depth during ultrashort laser ablation of copper in air [204] and Ar I [205] for spectroscopic investigation of the high-current phase of a pulsed gas metal arc welding (GMAW) process [206], for consideration of characteristics of plasma spray-physical Vapor Deposition (PVD) and impact on coating properties [207], for measurement of the temporal evolution of electron density in a nanosecond pulsed argon microplasma [208] and for the study of metal transfer in CO2 laser + GMAW-P hybrid welding using argon-helium mixtures [209].

Stark broadening parameters of Zn I spectral lines [210] have been used for experimental verification of a radiative model of laser-induced plasma expanding into vacuum [211], analysis of optical emission for the optimization of femtosecond laser processing [212], diagnostics of a laser-induced zinc plasma [213], comparison of zinc and cadmium plasma produced by laser ablation [201], spectroscopic characterization of laser ablation brass plasma [214], stoichiometric investigations of laser-ablated brass plasma [215], investigation of laser ablation and deposition of wide bandgap semiconductors and nanostructure of deposits [202], research of photoluminescence of nanoparticles in vapor phase of colliding plasma [216], consideration of the role of laser pre-pulse wavelength and inter-pulse delay on signal enhancement in collinear double-pulse laser-induced breakdown spectroscopy [217], for comparison of optical emission from nanosecond and femtosecond laser produced plasma in atmosphere and vacuum conditions [218], for investigation of dynamics of laser ablated colliding plumes [219], the investigation of brass plasmoid in external magnetic field [220], and research of emission dynamics of an expanding ultrafast-laser produced Zn plasma [221].

Stark broadening data for Al III [222] spectral lines have been used for examination of a novel plasma source for dense plasma effects [223] and for simulations of spectra from dense aluminum plasmas [224]. Data for C IV [156] have been used for the investigation of long-living plasmoids from an atmospheric discharge [225], data for S V [226] for time-integrated, spatially resolved plasma characterization of steel samples in the vacuum ultraviolet (VUV) [227], and data for Ar VIII [228] for investigation of optical emission spectra of ZnMnO plasma produced by a pulsed laser [229].

4. Conclusions

From the analysis of applications of Stark broadening parameters calculated using semiclassical perturbation method [1,2] one can conclude that principal users of such data are astronomers, using them especially for the investigation of A and B type stars, white dwarfs and hot stars in evolved evolution stages (especially PG1159 type). We note here that in white dwarf and hot pre-white dwarf atmospheres, Stark broadening is the dominant broadening mechanism in comparison with thermal Doppler broadening and for atmospherae modeling, spectra analysis and synthesis, abundance determination, radiative transfer calculation or plasma diagnostics, the knowledge of reliable Stark broadening parameters is essential. For A type stars, Stark broadening is the principal pressure broadening mechanism and often an important correction, the neglect of which may introduce serious errors, especially in abundance determinations. For B-type stars, especially for later types, Stark broadening may also be a non-negligible correction. The most used data are for spectral lines of He I and Si II. Concerning plasmas in physics and technology, the most frequent applications concern laser produced plasma, and the most used data are Stark broadening parameters of Zn I.

In order to make the application and usage of Stark broadening data obtained using the semiclassical perturbation method easier, the here analyzed data are displayed online in the STARK-B database [230], which is part of the Virtual Atomic and Molecular Data Centre—VAMDC [231].

Acknowledgments

This paper is part of the projects 176002 and III44022 of the Ministry of Education, Science and Technological Development of Republic of Serbia. This work was also done within the LABEX Plas@par project and received financial state aid managed by the Agence Nationale de la Recherche, as part of the program "Investissements d'avenir" under the reference ANR-11-IDEX-0004-02.

Conflicts of Interest

The authors declare no conflict of interest.

References

1. Sahal-Bréchot, S. Impact theory of the broadening and shift of spectral lines due to electrons and ions in a plasma. *Astron. Astrophys.* **1969**, 91–123.

2. Sahal-Bréchot, S. Impact theory of the broadening and shift of spectral lines due to electrons and ions in a plasma (continued). *Astron. Astrophys.* **1969**, 322–354.

3. Dimitrijević, M.S.; Sahal-Bréchot, S. Stark broadening of neutral Helium lines. *J. Quant. Spectrosc. Radiat. Transfer* **1984**, *31*, 301–313.

4. Dimitrijević, M.S.; Sahal-Bréchot, S. Stark broadening of neutral helium lines of astrophysical interest: Regularities within spectral series. *Astron. Astrophys.* **1984**, *136*, 289–298.

5. Dimitrijević, M.S.; Sahal-Bréchot, S. Comparison of measured and calculated Stark broadening parameters for neutral-helium lines. *Phys. Rev. A* **1985**, *31*, 316–320.

6. Dimitrijević, M.S.; Sahal-Bréchot, S. Tables for He I lines Stark broadening parameters. *Bull. Obs. Astron. Belgrade* **1989**, *141*, 57–86.

7. Dimitrijević, M.S.; Sahal-Bréchot, S. Stark broadening of He I lines. *Astron. Astrophys. Suppl. Ser.* **1990**, *82*, 519–529.

8. Dimitrov, D.L. The interacting binary Beta Lyr. II. Non-LTE model analysis and evolutionary conclusions. *Bull. Astron. Inst. Czechosl.* **1987**, *38*, 240–252.

9. Musielok, B.; Madej, J. Variability of Balmer lines in Ap stars. *Astron. Astrophys.* **1988**, *202*, 143–152.

10. Bohlender, D.A. Delta Orionis C and HD 58260: Peculiar helium—strong stars? *Astrophys. J.* **1989**, *346*, 459–468.

11. Dufton, P.L.; Brown, P.J.F.; Fitzsimmons, A.; Lennon, D.J. The chemical composition of the northern double cluster H and Chi Persei and the loose association Cepheus OB III. *Astron. Astrophys.* **1990**, *232*, 431–436.

12. Thejll, P.; Vennes, S.; Shipman, H.L. A critical analysis of the ultraviolet temperature scale of the helium-dominated DB and DBV white dwarfs. *Astrophys. J.* **1991**, *370*, 355–369.

13. Gray, R.O.; Corbally, C.J. The calibration of MK spectral classes using spectral synthesis. I. The effective temperature calibration of dwarf stars. *Astron. J.* **1994**, *107*, 742–746.

14. Jeffery, C.S.; Heber, U. The Extreme Helium Star BD 90 4395. *Astron. Astrophys.* **1992**, *260*, 133–150.

15. Luttermoser, D.G.; Johnson, H.R. Ionization and excitation in cool giant stars. 1. Hydrogen and helium. *Astrophys. J.* **1992**, *388*, 579–594.

16. Vennes, S. The constitution of the atmospheric layers and the extreme ultraviolet spectrum of hot hydrogen-rich white dwarfs. *Astrophys. J.* **1992**, *390*, 590–601.

17. Vennes, S.; Fontaine, G. An interpretation of the Spectral Properties of Hot Hydrogen-Rich White Dwarfs with Stratified H/He Model Atmospheres. *Astrophys. J.* **1992**, *401*, 288–310.

18. Alecian, G.; Michaud, G.; Tully, J. Radiative Accelerations on Iron Using Opacity Project Data. *Astrophys. J.* **1993**, *411*, 882–890.

19. Michaud, G.; Bergeron, P.; Heber, U.; Wesemael, F. Studies of hot B subdwarfs. VII Non-LTE radiative acceleration of helium in the atmospheres of sdOB stars. *Astrophys. J.* **1989**, *338*, 417–423.

20. Dufton, P.L.; Conlon, E.S.; Keenan, F.P.; McCausland, R.J.H.; Holmgren, D.E. 3 Stars at High Galactic Latitudes with Peculiar Helium Abundances. *Astron. Astrophys.* **1993**, *269*, 201–208.

21. Bergeron, P.; Wesemael, F.; Beauchamp, A.; Wood, M.A.; Lamontagne, R.; Fontaine, G.; Liebert, J. A spectroscopic analysis of DAO and hot white dwarfs: The implications of the presence of helium and the nature of DAO stars. *Astrophys. J.* **1994**, *432*, 305–325.

22. Linnell, A.P.; Hubeny, I. A spectrum synthesis program for binary stars. *Astrophys. J.* **1994**, *434*, 738–746.

23. Piskunov, N.; Ryabchikova, T.A.; Kuschnig, R.; Weiss, W.W. Spectrum variability of ET Andromedae: Si and He surface mapping. *Astron. Astrophys.* **1994**, *291*, 910–918.

24. Sasselov, D.D.; Lester, J.B. The He I lambda 10830 line in classical cepheides II. Mechanism of formation. *Astrophys. J.* **1994**, *423*, 785–794.

25. Захарова, Л.А. Исследование атмосфер двух HgMn-звезд с предполагаемыми аномалиями благородных газов (Study of Two HgMn-Stars with Suspected Anomalies of Abundances of Noble Gases). *Астрон. Ж.* **1994**, *71*, 588.

26. Kuschnig, R.; Ryabchikova, T.; Piskunov, N.; Weiss, W.W.; LeContel, J.M. The atmosphere of the peculiar binary system Eta Andromedae. *Astron. Astrophys.* **1995**, *294*, 757–762.

27. Vennes, S.; Thejll, P.A.; Wickramasinghe, D.T.; Bessell, M.S. Hot White Dwarfs in the Extreme Ultraviolet Explorer Survey. 1. Properties of a Southern Hemisphere Sample. *Astrophys. J.* **1996**, *467*, 782–793.

28. Harmanec, P.; Hadrava, P.; Yang, S.; Holmgren, D.; North, P.; Koubsky, P.; Kubat, J.; Poretti, E. Search for Forced Oscillations in Binaries. I. The Eclipsing and Spectroscopic Binary V436 Persei = 1 Persei. *Astron. Astrophys.* **1997**, *319*, 867–880.

29. Zboril, M.; North, P.; Glagolevskij, Y.V.; Betrix, F. Properties of He Rich Stars. I. Their Evolutionary State and Helium Abundance. *Astron. Astrophys.* **1997**, *324*, 949–958.

30. Unglaub, K.; Bues, I. The Effect of Diffusion and Mass Loss on the Helium Abundance in Hot White Dwarfs and Subdwarfs. *Astron. Astrophys.* **1998**, *338*, 75–84.

31. Jeffery, C.S.; Woolf, V.M.; Pollacco, D.L. Time-resolved spectral analysis of the pulsating helium star V652 Her. *Astron. Astrophys.* **1998**, *376*, 497–517.

32. Labrosse, N.; Gouttebroze, P. Formation of helium spectrum in solar quiescent prominences. *Astron. Astrophys.* **2001**, *380*, 323–340.

33. Pandey, G.; Kameswara Rao, N.; Lambert, D.L.; Jeffery, C.S.; Asplund, M. Abundance analyses of cool extreme helium stars. *Mon. Not. R. Astron. Soc.* **2001**, *324*, 937–959.

34. Smith, G.R. Enhancement of the helium resonance lines in the solar atmosphere by suprathermal electron excitation – II. Non-Maxwellian electron distributions. *Mon. Not. R. Astron. Soc.* **2003**, *341*, 143–163.

35. Andersson, M.; Pendrill, L.R. Improved measurements of the hyperfine structure of the 1s3s3S1 state of Helium 3. *Phys. Scr.* **1984**, *30*, 403–406.

36. Sakhibullin, N.A.; Schabert, W.J. Role of blending in the formation of helium singlet lines in the atmospheres of Bp stars. *Sov. Astron. Lett.* **1990**, *16*, 231–233.

37. Zboril, M.; Žižnovsky, J.; Zverko, J.; Budaj, J. Elemental abundance analysis of Phi Herculis and Omicron Pegasi with coadded spectra. *Contrib. Astron. Obs. Skaln. Pleso.* **1992**, *22*, 9–24.

38. Catanzaro, G.; Leone, F.; Dall, T.H. Balmer lines as Teff and log g indicators for non-solar composition atmospheres. An application to the extremely helium-weak star HR 6000. *Astron. Astrophys.* **2004**, *425*, 641–648.

39. Ding, M.D.; Li, H.; Fang, C. On the formation of the He I 10830 A line in a flaring atmosphere. *Astron. Astrophys.* **2005**, *432*, 699–704.

40. Mortimore, A.N.; Lynas-Gray, A.E. Helium, Carbon and Silicon abundances in the HW Vir eclipsing binary subdwarf-B primary. *Balt. Astron.* **2006**, *15*, 207.

41. Eisenstein, D.J.; Liebert, J.; Koester, D.; Kleinmann, S.J.; Nitta, A.; Smith, P.S.; Barentine, J.C.; Brewington, H.J.; Brinkmann, J.; Harvanek, M.; *et al.* Hot DB white dwarfs from the Sloan digital sky survey. *Astron. J.* **2006**, *132*, 676–691.

42. Cidale, L.S.; Arias, M.L.; Torres, A.F.; Zorec, J.; Frémat, Y.; Cruzado, A. Fundamental parameters of He-weak and He-strong stars. *Astron. Astrophys.* **2007**, *468*, 263–272.

43. Labrosse, N.; Goutebroze, P.; Vial, J.-C. Effect of motions in prominences on the helium resonance lines in the extreme ultraviolet. *Astron. Astrophys.* **2007**, *463*, 1171–1179.

44. Moscicki, T.; Hoffman, J.; Szymanski, Z. Net emission coefficients of low temperature thermal iron-helium plasma. *Opt. Appl.* **2008**, *38*, 365–373.

45. Latour, M.; Fontaine, G.; Brassard, P.; Green, E.M.; Chayer, P.; Randal, S.K. An analysis of the pulsating star SDSS J160043.6 + 074802.9 using new non-LTE model atmospheres and spectra for hot O subdwarfs. *Astrophys. J.* **2011**, *733*, 100 (1–15).

46. Linnell, A.P.; de Stefano, P.; Hubeny, I. BINSYN: A Publicly Available Program for Simulating Spectra and Light Curves of Binary Systems with or without Accretion Disks. *Publ. Astron. Soc. Pac.* **2012**, *124*, 885–894.

47. Dodin, A.V.; Lamzin, S.A.; Sitnova, T.M. Non-LTE modeling of narrow emission components of He and Ca lines in optical spectra of classical T Tauri stars. *Astron. Lett.* **2013**, *39*, 315–335.

48. Kopylov, I.M.; Leushin, V.V.; Topil'Skaya, G.P.; Tsymbal, V.V.; Gvozd', Yu.A. Investigation of spectral classification and temperature scale criteria of spectral classes. II. Analysis of spectral criteria. *Astrofiz. Issled. Izv. Spets. Astrofiz. Obs.* **1989**, *28*, 72–87.

49. Mac Donald J.; Vennes, S. How much hydrogen is there in a white dwarf? *Astrophys. J.* **1991**, *371*, 719–738.

50. Ryabchikova, T.A.; Stateva, I. Helium lines in the He-weak star 36 Lyncis. In *Model Atmospheres and Spectrum Synthesis*; Adelman, S.J., Kupka, F., Weiss, W.W., Eds.; *ASP Conference Series*, **1996**, *108*, 265–269.

51. Jeffery, C.S.; Hamill, P.J.; Harrison, P.M.; Jeffers, S.V. Spectral Analysis of the Low Gravity Extreme Helium Stars LSS 4357, LS II + 33o5 and LSS 99. *Astron. Astrophys.* **1998**, *340*, 476–482.

52. Eisenstein, D.J.; Liebert, J.; Koester, D.; Kleinmann, S.J.; Nitta, A.; Smith, P.S.; Barentine, J.C.; Brewington, H.J.; Brinkmann, J.; Harvanek, M.; *et al.* Hot DB white dwarfs from the Sloan digital sky survey. *Astron. J.* **2006**, *132*, 676–691.

53. Bouret, J.-C.; Lanz, T.; Martins, F.; Marcolino, W.L.F.; Hillier, D.J.; Depagne, E.; Hubeny, I. Massive stars at low metallicity. Evolution and surface abundances of O dwarfs in the SMC. *Astron. Astrophys.* **2013**, *555*, A1.

54. Leushin, V.V.; Glagolevskij, Yu.V.; North, P. Helium abundance in atmospheres of He-rich stars. In *Magnetic Fields of Chemically Peculiar and Related Stars*, Glagolevskij, Yu. V., Romanyuk, I.I., Eds.; Russian Academy of Sciences, Special Astrophysical Observatory: Moscow, Russia, 2000; pp. 173–179.

55. Glagolevskij, Yu.V.; Leushin, V.V.; Chuntonov, G.A.; Shulyak, D. The Atmospheres of Helium-Deficient Bp Stars. *Astron. Lett.* **2006**, *32*, 54–68.

56. Glagolevskij, Yu.V.; Leushin, V.V.; Chountonov, G.A. Chemical composition of the He-w stars HD 37058, 212454, and 224926. *Astrophys. Bull.* **2007**, *62*, 319–330.

57. Cressault, Y.; Rouffet, M. E.; Gleizes, A. Meillot. E. Net emission of Ar–H2–He thermal plasmas at atmospheric pressure. *J. Phys. D* **2010**, *43*, 335204.

58. Van Regemorter, H.; Hoang-Binh, D. Stark broadening theory of solar Rydberg lines in the far infrared spectrum. *Astron. Astrophys.* **1993**, *277*, 623–634.

59. Smith, M.A.; Hubeny, I.; Lanz, T.; Meylan, T. Dynamic processes in Be star atmospheres II. He I 2P-nD line formation in lambda Eridani (outburst). *Astrophys. J.* **1994**, *432*, 392–402.

60. Butler, K. Atmospheres and winds of hot stars: The impact of new opacity calculations and continuing needs. In *Astrophysical Applications of Powerful New Databases*; Adelman, S.J., Wiese, W.L., Eds.; ASP Conf. Series **1995**, *78*, 509–525.

61. Israelian, G.; Friedjung, M.; Graham, J.; Muratorio, G.; Rossi, C.; de Winter, D. The atmospheric variations of the peculiar B[e] star HD 45677 (FS Canis Majoris). *Astron. Astrophys.* **1996**, *311*, 643–650.

62. Valenti, J.A.; Piskunov, N. Spectroscopy Made Easy A New Tool for Fitting Observations with Synthetic Spectra. *Astron. Astrophys. Suppl.* **1996**, *118*, 595–603.

63. Zakharova, L.A.; Ryabchikova, T.A. The 3He isotope in the atmospheres of HgMn stars. *Astron. Lett.* **1996**, *22*, 152–156.

64. Leone, F.; Lanzafame, A.C. Behavior of the HeI 587.6, 667.8, 706.5 and 728.1 nm Lines in B-Type Stars On the Helium Stratification in the Atmosphere of Magnetic Helium Peculiar Stars. *Astron. Astrophys.* **1997**, *320*, 893–898.

65. Leushin, V.V.; Glagolevskij, Yu.V.; North, P. Helium abundance in atmospheres of He-rich stars. In *Magnetic Fields of Chemically Peculiar and Related Stars*, Glagolevskij, Yu. V., Romanyuk, I.I., Eds.; Russian Academy of Sciences, Special Astrophysical Observatory: Moscow, Russia, 2000; pp. 173–179.

66. Piskunov, N.; Kupka, F. Model atmospheres with individualized abundances. *Astrophys. J.* **2001**, *547*, 1040–1056.

67. Smith, G.R. Enhancement of the helium resonance lines in the solar atmosphere by suprathermal electron excitation – II. Non-Maxwellian electron distributions. *Mon. Not. R. Astron. Soc.* **2003**, *341*, 143–163.

68. Domiciano de Souza, A.; Zorec, J.; Jankov, S.; Vakili, F.; Abe, L.; Janot-Pacheco, E. Stellar differential rotation and inclination angle from spectro-interferometry. *Astron. Astrophys.* **2004**, *418*, 781–794.

69. Lyubimkov, L.S.; Rostopchin, S.I.; Lambert, D.L. Surface abundance of light elements for a large sample of early B-type stars – III. An analysis of helium lines in spectra of 102 stars. *Mon. Not. R. Astron. Soc.* **2004**, *351*, 745–767.

70. Castelli, F.; Hubrig, S. A spectroscopic atlas of the HgMn star HD 175640 (B9 V) λλ 3040 – 10 000 A. *Astron. Astrophys.* **2004**, *425*, 263–270.

71. Przybilla, N.; Butler, K.; Heber, U.; Jeffrey, C.S. Extreme helium stars: Non-LTE matters; Helium and hydrogen spectra of the unique objectsV652 Her and HD 144941. *Astron. Astrophys.* **2005**, *443*, L25–L28.

72. Eisenstein, D.J.; Liebert, J.; Koester, D.; Kleinmann, S.J.; Nitta, A.; Smith, P.S.; Barentine, J.C.; Brewington, H.J.; Brinkmann, J.; Harvanek, M.; *et al.* Hot DB white dwarfs from the Sloan digital sky survey. *Astron. J.* **2006**, *132*, 676–691.

73. Glagolevskij, Yu.V.; Leushin, V.V.; Chuntonov, G.A.; Shulyak, D. The Atmospheres of Helium-Deficient Bp Stars. *Astron. Lett.* **2006**, *32*, 54–68.

74. Mortimore, A.N.; Lynas-Gray, A.E. Helium, Carbon and Silicon abundances in the HW Vir eclipsing binary subdwarf-B primary. *Balt. Astron.* **2006**, *15*, 207–210.

75. Przybilla, N.; Butler, K.; Heber, U.; Jeffrey, C.S. Improved helium line formation for extreme helium stars. *Balt. Astron.* **2006**, *15*, 127–130.

76. Cidale, L.S.; Arias, M.L.; Torres, A.F.; Zorec, J.; Frémat, Y.; Cruzado, A. Fundamental parameters of He-weak and He-strong stars. *Astron. Astrophys.* **2007**, *468*, 263–272. Glagolevskij, Yu.V.; Leushin, V.V.; Chountonov, G.A. Chemical composition of the He-w stars HD 37058, 212454, and 224926. *Astrophys. Bull.* **2007**, *62*, 319–330.

77. Nieva, M.F.; Przybilla, N. Hydrogen and helium line formation in OB dwarfs and giants; A hybrid non-LTE approach. *Astron. Astrophys.* **2007**, *467*, 295–309.

78. Schiller, F.; Przybilla, N. Quantitative spectroscopy of Deneb. *Astron. Astrophys.* **2008**, *479*, 849–858.

79. Bohlender, D. A.; Rice, J. B.; Hechler, P. Doppler imaging of the helium-variable star a Centauri. *Astron. Astrophys.* **2010**, 520, A44.

80. Koester, D. White dwarf spectra and atmosphere modelsoppler imaging of the helium-variable star a Centauri. *Mem. Soc. Astron. Ital.* **2010**, *81*, 921–931.

81. Falcon, R.E.; Winget, D.E.; Montgomery, M.H.; Williams, K.A. A Gravitational Redshift Determination of the Mean Mass of White Dwarfs: DBA and DB Stars. *Astrophys. J.* **2012**, *757*, 116.

82. Ferrero, G.; Gamen, R.; Benvenuto, O.; Fernández-Lajús, E. Apsidal motion in massive close binary systems – I. HD 165052, an extreme case? *Mon. Not. R. Astron. Soc.* **2013**, *433*, 1300–1311.

83. Bonifacio, P.; Castelli, F.; Hack, M. The field horizontal-branch B-type star Feige 86. *Astron. Astrophys. Supp.* **1995**, *110*, 441–468.

84. Lanz, T.; Dimitrijević, M.S.; Artru, M.-C. Stark broadening of visible Si II lines in stellar atmospheres. *Astron. Astrophys.* **1988**, *192*, 249–254.

85. Adelman, S.J. Elemental abundance analyses with co-added DAO spectrograms – IV. Revision of Previous Analyses. *Mon. Not. R. Astron. Soc.* **1988**, *235*, 749–762.

86. Adelman, S.J. Elemental abundance analyses with co-added DAO spectrograms – V. The Mercury-Manganese Stars Phi-Herculis, 28-Herculis and Hr-7664. *Mon. Not. R. Astron. Soc.* **1988**, *235*, 763–785.

87. Adelman, S.J. Elemental abundance analyses with co-added DAO spectrograms – VI. The Mercury-Manganese Stars Nu-Cancri, Iota Coronae Borealis and Hr-8349. *Mon. Not. R. Astron. Soc.* **1989**, *239*, 487–511.

88. Adelman, S.J. Elemental abundance analyses with DAO spectrograms – VII. The late normal B stars Pi Ceti, 134 Tauri, 21 Aquilae and Nu Capricorni and the use of Reticon spectra. *Mon. Not. R. Astr. Soc.* **1991**, *252*, 116–131.

89. Adelman, S.J.; Bolcal, C.; Hill, G.; Kocer, D. Elemental abundance analyses with DAO spectrograms – VIII. The normal F main sequence stars Theta Cygni and Iota Piscium. *Mon. Not. R. Astr. Soc.* **1991**, *252*, 329–333.

90. Mathys, G. The blue stragglers of M 67. *Astron. Astrophys.* **1991**, *245*, 467–484.

91. Adelman, S.J. Elemental abundance analyses with DAO spectrograms – X. The mercury—manganese stars Pi 1 Bootis, v Herculis and HR 7361. *Mon. Not. R. Astron. Soc.* **1992**, *258*, 167–176.

92. Adelman, S.J.; Philippe, A.G.D. Elemental abundances of the B-star and A-star Gamma Geminorum, 7-Sextantis, Hr-4817, and Hr-5780. *Publ. Astron. Soc. Pac.* **1992**, *104*, 316–321.

93. Bolcal, C.; Kocer, D.; Adelman, S.J. Elemental abundance analyses with DAO spectrograms. IX. The metallic-lined stars 15 - Vulpeculae and 32 – Aquarii. *Mon. Not. R. Astron. Soc.* **1992**, *258*, 270–276.

94. Singh, J.; Castelli, F. Effective temperature of B-type stars from the Si II lines of the UV multiplet 13.04 at 130.5–130.9 nm. *Astron. Astrophys.* **1992**, *253*, 431–446.

95. Lanz, T.; Artru, M.C.; Didelon, P.; Mathys, G. The Ga-II lines in the red spectrum of Ap stars. *Astron. Astrophys.* **1993**, *272*, 465–476.

96. Lopez-Garcia, Z.; Adelman, S.J. An abundance analysis of the silicon CP star HD 43819, in Peculiar *versus* normal phenomena in A-type and related stars. *Astron. Soc. Pac. Conf. Series* **1993**, *44*, 149–153.

97. Pintaldo, O.I.; Adelman, S.J. Elemental abundance analyses with DAO spectrograms. XI. B stars Gamma Pegasi and Iota Herculis. *Mon. Not. R. Astron. Soc.* **1993**, *264*, 63–70.

98. Rauch, T. NLTE Analysis of subluminous O stars: The hot subdwarf in the binary system HD 128220. *Astron. Astrophys.* **1993**, *276*, 171–183.

99. Adelman, S.J. Elemental abundance analyses with DAO spectrograms—XII. The mercury—manganese stars HR 4072A and 7775 and the metallic-lined star HR 4072B. *Mont. Not. R. Astron. Soc.* **1994**, *266*, 97–113.

100. Adelman, S.J. Elemental abundance analyses with DAO spectrograms. XIII. The superficially normal early A-type stars 68 Tauri, 21 Lyncis and Alpha Draconis. *Mont. Not. Roy. Soc.* **1994**, *271*, 355–371.

101. Adelman, S.J.; Davis Philip, A.G. Elemental abundances of the B and A stars. II. Gamma Geminorum, HD 60825, 7 Sextantis, HR 4817, and HR 5780. *Publ. Astron. Soc. Pacific* **1994**, *106*, 1239–1247.

102. Khokhlova, V.L. On the significance of Stark line shifts for element abundance determinations by the model atmosphere method. *Astron. Lett.* **1994**, *20*, 89–90.

103. Lopez-Garcia, Z.; Adelman, S.J. Elemental abundance studies of CP stars: The silicon star HD 43819 and the CP star HD 147550. *Astron. Astrophys. Supp.* **1994**, *107*, 353–363.

104. North, P.; Berthet, S.; Lanz, T. The nature of the F STR Lambda 4077 stars V. Spectroscopic data. *Astron. Astrophys. Supp.* **1994**,103, 321–347.

105. Wahlgren, G.M.; Adelman, S.J.; Robinson, R.D. An optical region elemental abundance analysis of the chemically peculiar HgMn star Chi Lupi. *Astrophys. J.* **1994**, *434*, 349–362.

106. Zverko, J.; Zboril, M.; Žižnovsky, J. Abundance determination in the CP star 21 Canum Venaticorum by means of spectrum synthesis. *Astron. Astrophys.* **1994**, *283*, 932–936.

107. Kuschnig, R.; Ryabchikova, T.; Piskunov, N.; Weiss, W.W.; LeContel, J.M. The atmosphere of the peculiar binary system ET Andromedae. *Astron. Astrophys.* **1995**, *294*, 757–762.

108. Piskunov, N.; Ryabchikova, T.A.; Kuschnig, R.; Weiss, W.W. Spectrum variability of ET Andromedae: Si and He surface mapping. *Astron. Astrophys.* **1994**, *291*, 910–918.

109. Adelman, S.J. Elemental Abundance Analyses with DAO Spectrograms. 15. The Superficially Normal Late B-Type and Early A-Type Stars Merak, Pi Draconis and Kappa Cephei. *Mon. Not. R. Astron. Soc.* **1996**, *280*, 130–142.

110. Adelman, S.J.; Philip, A.G.D. Elemental Abundances of the BStar and AStar – 3. Gamma Geminorum, Hr 1397, Hr 2154, HD 60825 and 7 Sextantis. *Mon. Not. R. Astron. Soc.* **1996**, *282*, 1181–1190.

111. Pintaldo, O.I.; Adelman, S.J. Elemental abundance analyses with Complejo Astronomico El Leoncito REOSC echelle spectrograms. I. Kappa Cancri, HR 7245, and Ksi Octantis. *Astron. Astrophys. Supp.* **1996**, *118*, 283–291.

112. Adelman, S.J.; Caliskan, H.; Kocer.; D.; Bolcal, C. Elemental Abundance Analyses with DAO Spectrograms – XVI. The Normal F Main Sequence Stars Sigma Bootis, Theta Cygni and Iota Piscum, and the Am Stars 15 Vulpeculae and 32 Aquarii. *Mon. Not. R. Astron. Soc.* **1997**, *288*, 470–500.

113. Caliskan, H.; Adelman, S.J. Elemental Abundance Analyses with DAO Spectrograms .17. The Superficially Normal Early A Stars 2 Lyncis, Omega Ursa Majoris and Phi Aquilae. *Mon. Not. R. Astron. Soc.* **1997**, *288*, 501–511.

114. Adelman, S.J. Elemental Abundance Analyses with DAO Spectrograms XIX The Superficially Normal B Stars Zeta Draconis, Eta Lyrae, 8 Cygni and 22 Cygni. *Mon. Not. R. Astron. Soc.* **1998**, *296*, 856–862.

115. Adelman, S.J.; Albayrak, B. Elemental Abundance Analyses with DAO Spectrograms 20 The Early A Stars Epsilon Serpentis, 29Vulpeculae and Sigma Aquarii. *Mon. Not. R. Astron. Soc.* **1998**, *300*, 359–372.

116. Pintado, O.I.; Adelman, S.J.; Gulliver, A.F. Elemental Abundance Analyses with Complejo Astronomico El Leoncito REOSC Echelle Spectrograms III Hr 4487, 14 Hydrae, and 3 Centauri A. *Astron. Astrophys. Supp.* **1998**,129, 563–567.

117. Adelman, S.J.; Caliskan, H.; Cay, T.; Kocer.; D.; Tektanali, H.G. Elemental Abundance Analyses with DAO Spectrograms – XXI. The hot metallic-lined stars 60 Leonis and 6 Lyrae. *Mon. Not. R. Astron. Soc.* **1998**, *305*, 591–601.

118. Lopez-Garcia, Z.; Adelman, S.J. Elemental Abundance Studies of CP Stars – II. The Silicon Stars H 133029 and HD 192913. *Astron. Astrophys. Supp.* **1999**, *137*, 227–232.

119. Lopez-Garcia, Z.; Adelman, S.J.; Pintado, O.I. Elemental abundance studies of CP stars III. The magnetic CP stars alpha Scl and HD 170973. *Astron. Astrophys.* **2001**, *367*, 859–864.

120. Albacete-Colombo, J.F.; Lopez-Garcia, Z.; Levato, H.; Malaroda, S.M.; Grosso, M. Elemental abundance study of the CP star HD 206653. *Astron. Astrophys.* **2002**, *392*, 613–617.

121. Alonso, M.S.; Lopez-Garcia, Z.; Malaroda, S.; Leone, F. Elemental abundance studies of CP stars. The helium-weak stars HD 19400, HD 34797 and HD 35456. *Astron. Astrophys.* **2003**, *402*, 331–334.

122. Zboril, M.; Žižnovsky, J.; Zverko, J.; Budaj, J. Elemental abundance analysis of Phi Herculis and Omicron Pegasi with coadded spectra. *Contrib. Astron. Obs. Skalnate. Pleso.* **1992**, *22*, 9–24.

123. Castelli, F.; Hubrig, S. A spectroscopic atlas of the HgMn star HD 175640 (B9 V) λλ 3040 – 10 000 A. *Astron. Astrophys.* **2004**, *425*, 263–270.

124. Saffe, C.; Levato, H.; Lopez-Garcia, Z. Elemental abundance studies of CP stars. The silicon stars HD 87240 and HD 96729. *Revista. Mexicana de Astronomia. y Astropfisica.* **2005**, *41*, 415–421.

125. Lehmann, H.; Tsymbal, V.; Mkrtichian, D.E.; Fraga, L. The helium-weak silicon star HR 7224. I. Radial velocity and line profile variations. *Astron. Astrophys.* **2006**, *457*, 1033–1041.

126. Schiller, F.; Przybilla, N. Quantitative spectroscopy of Deneb. *Astron. Astrophys.* **2008**, *479*, 849–858.

127. Collado, A.; López-García, Z. Chemical Abundances of the magnetic CP star HD 168733. *Rev. Mex. Astron. Astrofís.* **2009**, *45*, 95–105.

128. Fossati, L.; Ryabchikova, T.; Bagnulo, S.; Alecian, E.; Grunhut, J.; Kochukhov, O.; Wade, G. The chemical abundance analysis of normal early A- and late B-type stars. *Astron. Astrophys.* **2009**, *503*, 945–962.

129. Saffe, C.; Levato, H. Elemental abundance studies of CP stars. The silicon stars HD 87405 and HD 146555. *Rev. Mex. Astron. Astrofís.* **2009**, *45*, 171–178.

130. Saffe, C.; Nunez, N.; Levato, H. Upper Main Sequence Stars with Anomalous Abundances. The HgMn stars HR 3273, HR 8118 HR 8567 and HR 8937. *Rev. Mex. Astron. Astrofís.* **2011**, *47*, 219–234.

131. Vennes, S.; Kawka, A.; Németh, P. Pressure shifts and abundance gradients in the atmosphere of the DAZ white dwarf GALEX J193156.8 + 011745. *Mon. Not. R. Astron. Soc.* **2011**, *413*, 2545–2553.

132. Dimitrijević, M.S.; Sahal-Bréchot, S. Stark broadening of Li (I) lines. *J. Quant. Spectrosc. Radiat. Transfer* **1991**, *46*, 41–53.

133. Carlsson, M.; Rutten, R.J., Bruls, J.H.M.J.; Schukina, N.G. The non-LTE formation of Li I lines in cool stars. *Astron. Astrophys.* **1994**, *288*, 860–882.

134. Dimitrijević, M.S.; Sahal-Bréchot, S. Stark-broadening parameters of ionized mercury spectral lines of astrophysical interest. *J. Quant. Spectrosc. Radiat. Transf.* **1992**, *47*, 315–318.

135. Smith, K.C. Elemental Abundances in Normal Late-B and HgMn Stars from Co-Added IUE Spectra. V. Mercury. *Astron. Astrophys.* **1997**, *319*, 928–947.

136. Dimitrijević, M.S.; Sahal-Bréchot, S. Stark broadening of Ca II spectral lines, *J. Quant. Spectrosc. Radiat. Transf.* **1993**, *49*, 157–164.

137. Dimitrijević, M.S.; Sahal-Bréchot, S. Stark broadening parameter tables for Ca II lines of astrophysical interest. *Bull. Astron. Belgrade* **1992**, *145*, 81–99.

138. Ryabchikova, T.A.; Malanushenko, V.P.; Adelman, S.J. Orbital elements and abundance analyses of the double-lined spectroscopic binary alpha Andromedae. *Astron. Astrophys.* **1999**, *351*, 963–972.

139. Vennes, S.; Kawka, A.; Németh, P. Pressure shifts and abundance gradients in the atmosphere of the DAZ white dwarf GALEX J193156.8 + 011745. *Mon. Not. R. Astron. Soc.* **2011**, *413*, 2545–2553.

140. Kawka, A.; Vennes, S. VLT/X-shooter observations and the chemical composition of cool white dwarfs. *Astron. Astrophys.* **2012**, 538, A13.

141. Bikmaev, I.F.; Ryabchikova, T.A.; Bruntt, H.; Musaev, F.A.; Mashonkina, L.I.; Belyakova, E.V.; Shimansky, V.V.; Barklem, P.S.; Galazutdinov, G. Abundance analysis of two late A-type stars HD 32115 and HD 37594. *Astron. Astrophys.* **2002**, *389*, 537–546.

142. Dimitrijević, M.S.; Sahal-Bréchot, S. Stark broadening parameter tables for Mg I lines of interest for solar and stellar spectra research. I. *Bull. Astron. Belgrade* **1994**, *149*, 31–84 (Erratum in *Bull. Astron. Belgrade* **1994**, *150*, 121).

143. Dimitrijević, M.S.; Sahal-Bréchot, S. Stark broadening of solar Mg I lines. *Astron. Astrophys. Supp.* **1996**, *117*, 127–129.

144. Fossati, L.; Ryabchikova, T.; Shulyak, D.V.; Haswell, C.A.; Elmasli, A.; Pandey, C.P.; Barnes, T.G.; Zwintz, K. The accuracy of stellar atmospheric parameter determinations: A case study with HD 32115 and HD 37594. *Mon. Not. R. Astron. Soc.* **2011**, *417*, 495–507.

145. Zhao, G.; Butler, K.; Gehren, T. NonLTE Analysis of Neutral Magnesium in the Solar Atmosphere. *Astron. Astrophys.* **1998**, *333*, 219–230.

146. Przybilla, N.; Butler, K.; Becker, S.R.; Kudritzki, R.P. Non-LTE line formation for Mg I/II: Abundances and stellar parameters; Model atom and first results on A-type stars. *Astron. Astrophys.* **2001**, *369*, 1009–1026.

147. Ryde, N.; Korn, A.J.; Richter, M.J.; Ryde, F. The Zeeman-sensitive emission lines of Mg I at 12 microns in Procyon. *Astrophys. J.* **2004**, *617*, 551–558.

148. Mashonkina, L. Astrophysical tests of atomic data important for the stellar Mg abundance determinations. *Astron. Astrophys.* **2013**, 550, A28.

149. Dimitrijević, M.S.; Sahal-Bréchot, S. Stark broadening parameter tables for Mg I lines of interest for solar and stellar spectra research. II. *Bull. Astron. Belgrade* **1995**, *151*, 101–114.

150. Dimitrijević, M.S.; Sahal-Bréchot, S. Stark broadening of Na (I) lines with principal quantum number of the upper state between 6 and 10. *J. Quant. Spectrosc. Radiat. Transfer* **1990**, 44, 421–431.

151. Lind, K.; Asplund, M. Barklem, P.S.; Belyaev, A.K. Non-LTE calculations for neutral Na in late-type stars using improved atomic data. *Astron. Astrophys.* **2011**, 528, A103.

152. Dimitrijević, M.S.; Sahal-Bréchot, S. On the Stark broadening of Ba II spectral lines. *XVIII Symp. Phys. Ioniz. Gases (Kotor.)* **1996**, 548–551.

153. Dimitrijević, M.S.; Sahal-Bréchot, S. Stark broadening of Ba I and Ba II spectral lines. *Astron. Astrophys. Supp.* **1997**, *122*, 163–166.

154. Bikmaev, I.F.; Ryabchikova, T.A.; Bruntt, H.; Musaev, F.A.; Mashonkina, L.I.; Belyakova, E.V.; Shimansky, V.V.; Barklem, P.S.; Galazutdinov, G. Abundance analysis of two late A-type stars HD 32115 and HD 37594. *Astron. Astrophys.* **2002**, *389*, 537–546.

155. Schiller, F.; Przybilla, N. Quantitative spectroscopy of Deneb. *Astron. Astrophys.* **2008**, *479*, 849–858.

156. Dimitrijević, M.S. ; Sahal-Bréchot, S. ; Bommier, V. Stark broadening of spectral lines of multicharged ions of astrophysical interest – I. C IV lines, *Astron. Astrophys. Supp.* **1991**, *89*, 581–590.

157. Dimitrijević, M.S.; Sahal-Bréchot, S.; Bommier, V. Stark broadening of spectral lines of multicharged ions of astrophysical interest – II. Si IV lines. *Astron. Astrophys. Supp.* **1991**, *89*, 591–598.

158. Dimitrijević, M.S.; Sahal-Bréchot, S. Stark broadening of spectral lines of multicharged ions of astrophysical interest – IV. N V lines. *Astron. Astrophys. Supp.* **1992**, *95*, 109–120.

159. Dimitrijević, M.S.; Sahal-Bréchot, S. Stark broadening of spectral lines of multicharged ions of astrophysical interest – III. O VI lines. *Astron. Astrophys. Supp.* **1992**, *93*, 359–371.

160. Dimitrijević, M.S.; Sahal-Bréchot, S. Stark broadening of spectral lines of multicharged ions of astrophysical interest – VIII. S VI lines. *Astron. Astrophys. Supp.* **1993**, *100*, 91–101.

161. Rauch, T. NLTE Analysis of a SDO binary: HD128220. *Lect. Not. Phys.* **1992**, *401*, 267–269.

162. Werner, K. Analysis of PG 1159 stars. *Lect. Not. Phys.* **1992**, *401*, 273–287.

163. Unglaub, K.; Bues, I. The influence of gravitational settling and selective radiative forces in PG 1159 stars. *Astron. Astrophys.* **1996**, *306*, 843–859.

164. Werner, K.; Dreizler, S.; Heber, U.; Rauch, T.; Fleming, T.A.; Sion, E.M.; Vauclair, G. High resolution UV spectroscopy of two hot (pre-)white dwarfs with the Hubble Space Telescope. KPD 0005 + 5106 and RXJ 2117 + 3412. *Astron. Astrophys.* **1996**, *307*, 860–868.

165. Vennes, S.; Dupuis, J.; Chayer, P.; Polomski, E.F.; Dixon, W.V.; Hurwitz, M. The Complete Spectral Energy Distribution and the Atmospheric Properties of the Helium Rich White Dwarf MCT 05012858. *Astrophys. J.* **1998**, 500, L41–L44.

166. Vennes, S.; Thorstensen, J.R.; Polomski, E.F. Stellar masses, kinematics, and white dwarf composition for three close DA + dMe binaries. *Astrophys. J.* **1999**, *523*, 386–398.

167. Lamzin, S.A. Calculation of profiles of CIV, NV, OVI, and SiIV resonance lines formed in accretion shocks in T Tauri stars: A plane layer. *Astron. Rep.* **2003**, *47*, 498–510.

168. Rauch, T.; Ziegler, M.; Werner, K.; Kruk, J.W.; Oliveira, C.M.; Vande Putte, D.; Mignani, R.P.; Kerber, F. High-resolution FUSE and HST ultraviolet spectroscopy of central star of Sh 2–216. *Astron. Astrophys.* **2007**, *470*, 317–329.

169. Rauch, T.; Koeppen, J.; Werner, K. Spectral analysis of the planetary nebula K 1–27 and its very hot hydrogen-deficient central star. *Astron. Astrophys.* **1994**, *286*, 543–554.

170. Rauch, T.; Koeppen, J.; Werner, K. Spectral analysis of the multiple-shell planetary nebula LoTr4 and its very hot hydrogen-deficient central star. *Astron. Astrophys.* **1996**, *310*, 613–628.

171. Werner, K.; Wolf, B. The EUV spectrum of the unique bare stellar core H1504 + 65. *Astron. Astrophys.* **1999**, 347, L9-L13.

172. Fontaine, M.; Chayer, P.; Wesemael, F.; Fontaine, G.; Lamontagne, R. Analysis of the FUSE spectra of the He-poor SDO star MCT 0019–2441. *Balt. Astron.* **2006**, *15*, 99–102.

173. Fontaine, M.; Chayer, P.; Oliveira, C.M.; Wesemael, F.; Fontaine, G. Analysis of the FUSE spectrum of the hot, evolved star GD 605. *Astrophys. J.* **2008**, *678*, 394–407.

174. Dimitrijević, M.S.; Sahal-Bréchot, S. Broadening of neutral sodium lines. *J. Quant. Spectrosc. Radiat. Transf.* **1985**, *34*, 149–161.

175. Cappelli, M.A.; Measures, R.M. Electron density radial profiles derived from Stark broadening in a sodium plasma produced by laser resonance saturation. *Appl. Optics* **1987**, *26*, 1058–1067.

176. Leonov, A.G.; Chekhov, D.I.; Starostin, A.N. Mechanisms of Resonant Laser Ionization. *J. Exper. Theor. Phys.* **1997**, *84*, 703–715.

177. Dimitrijević, M.S.; Sahal-Bréchot, S. Stark broadening of Be II spectral lines. *J. Quant. Spectrosc. Radiat. Transf.* **1992**, *48*, 397–403.

178. Villoresi, P.; Bidoli, P.; Nicolosi, P. Absorption Spectra and Oscillator Strength Ratio Measurements for $\Delta n = 1$ Transitions from Excited Levels of Be I and Be II. *J. Quant. Spectrosc. Radiat. Transf.* **1997**, *57*, 847–857.

179. Dimitrijević, M.S.; Sahal-Bréchot, S. Stark broadening of neutral calcium spectral lines. *Astron. Astrophys. Supp.* **1999**, *140*, 191–192.

180. Dimitrijević, M.S.; Sahal-Bréchot, S. Stark broadening parameter tables for neutral calcium spectral lines II. *Serb. Astron. J.* **2000**, *161*, 39–88.

181. Milisavljević, S.; Šević, D.; Pejčev, V.; Filipović, D.M.; Marinković, B.P. Differential and integrated cross sections for the electron excitation of the 4 $^1P^o$ state of calcium atom. *J. Phys. B* **2004**, *37*, 3571–3581

182. .Boswell, C.J.; O'Connor, P.D. Charged particle motion in an explosively generated ionizing shock. In *Shock Compression of Condensed Matter*; Elert, M.L.; Buttler, W.T.; Furnish, M.D.; Anderson, W.W.; Proud, W.G., Eds.; *Am. Inst. Phys. Conf. Ser.* **2009**, *1195*, 400–403.

183. Gehlen, C.D.; Wiens, E.; Noll, R.; Wilsch, G.; Reichling, K. Chlorine detection in cement with laser-induced breakdown spectroscopy in the infrared and ultraviolet spectral range. *Spectrochim. Acta B* **2009**, *64*, 1135–1140.

184. Lagrange, J.F.; Hermann, J.; Wolfman, J.; Motret, O. Dynamical plasma study during CaCu3Ti4O12 and Ba0.6Sr0.4TiO3 pulsed laser deposition by local thermodynamic equilibrium modeling. *J. Phys. D Appl. Phys.* **2010**, *43*, 285.

185. Dimitrijević, M.S.; Sahal-Bréchot, S. Stark broadening of Mg I spectral lines. *Phys. Scr.* **1995**, *52*, 41–51.

186. Dimitrijević, M.S.; Sahal-Bréchot, S. Electron-impact broadening of Mg II spectral lines for astrophysical and laboratory plasma research. *Phys. Scr.* **1998**, *58*, 61–71.

187. Hoffman, J.; Szymanski, Z.; Azharonok, V. Plasma plume induced during laser welding of magnesium alloys. In *International Conference on Research and Applications of Plasmas (PLASMA 2005)*, Opole, Poland, 6–9 September 2005; Sadowski, M.J.; Dudeck, M.; Hartfuss, H.J.; Pawelec, E., Eds.; *AIP Conf. Proc.* **2006**, *812*, 469–472.

188. Dimitrijević, M.S.; Sahal-Bréchot, S. Stark broadening of Sr I spectral lines. *Astron. Astrophys. Supp.* **1996**, *119*, 529–530.

189. Christou, C.; Garg, A.; Barber, Z.H. Vapor-phase oxidation during pulsed laser deposition of SrBi2Ta2O9. *J. Vac. Sci. Tech. A* **2001**, *19*, 2061–2068.

190. Barber, Z.H.; Christou, C.; Chiu, K.-F.; Garg, A. The measurement and control of ionization of the depositing flux during thin film growth. *Vac.* **2003**, *69*, 53–62.

191. Santagata, A.; di Trolio, A.; Parisi, G.P.; Larciprete, R. Space and time resolved emission spectroscopy of Sr $_2$FeMoO $_6$ laser induced plasma. *Appl. Surf. Sci.* **2005**, *248*, 19–23.

192. Dimitrijević, M.S.; Sahal-Bréchot, S. Stark broadening of Li II spectral lines. *Phys. Scr.* **1996**, *54*, 50–55.

193. Coons, R.W.; Harilal, S.S.; Polek, M.; Hassanein, A. Spatial and temporal variations of electron temperatures and densities from EUV-emitting lithium plasmas. *Anal. Bioanal. Chem.* **2011**, *400*, 3239–3246.

194. Mihajlov, A.A.; Sakan, N.M.; Srećković, V.A.; Vitel, Y. Modeling of continuous absorption of electromagnetic radiation in dense partially ionized plasmas. *J. Phys. A* **2011**, *44*, 095502.

195. Hafeez, S.; Shaikh, N.M.; Rashid, B.; Baig, M.A. Plasma properties of laser-ablated strontium target. *J. Appl. Phys.* **2008**, *103*, 083117.

196. Hanif, M.; Salik, M.; Shaikh, Nek M.; Baig, M.A. Laser-based optical emission studies of barium plasma. *Appl. Phys. B* **2013**, *110*, 563–571.

197. Dimitrijević, M.S.; Sahal-Bréchot, S. Stark broadening of Ag I spectral lines. *Atom. Data Nucl. Data* **2003**, *85*, 269–290.

198. Tošić, S.D.; Pejčev, V.; Šević, D.; McEachran, R.P.; Stauffer, A.D.; Marinković, B.P. Absolute differential cross sections for electron excitation of silver at small scattering angles. *Nucl. Instr. Method Phys. Res. B* **2012**, *279*, 53–57.

199. Simić, Z.; Dimitrijević, M.S.; Milovanović, N.; Sahal-Bréchot, S. Stark broadening of Cd I spectral lines. *Astron. Astrophys.* **2005**, *441*, 391–393.

200. Shaikh, N.M.; Rashid, B.; Hafeez, S.; Mahmood, S.; Saleem, M.; Baig, M.A. Diagnostics of cadmium plasma produced by laser ablation. *J. Appl. Phys.* **2006**, *100*, 073102.

201. Shaikh, N.M.; Hafeez, S.; Baig, M.A. Comparison of zinc and cadmium plasma produced by laser ablation. *Spectrochim. Acta B* **2007**, *62*, 1311–1320.

202. Sanz, M.; Lopez-Arias, M.; Rebollar, E.; de Nalda, R.; Castillejo, M. Laser ablation and deposition of wide bandgap semiconductors: Plasma and nanostructure of deposits diagnosis. *J. Nanoparticle Res.* **2011**, *13*, 6621–6631.

203. Zmerli, B.; Ben Nessib, N.; Dimitrijević, M.S.; Sahal-Bréchot, S. Stark broadening calculations of neutral copper spectral lines and temperature dependence. *Phys. Scr.* **2010**, *82*, 055301.

204. Hu, Wenqian; Shin, Yung C.; King, Galen. Characteristics of plume plasma and its effects on ablation depth during ultrashort laser ablation of copper in air. *J. Phys. D Appl. Phys.* **2012**, *45*, 355204.

205. Dimitrijević, M.S.; Christova, M.; Sahal-Bréchot, S. Stark broadening of visible Ar I spectral lines. *Phys. Script.* **2007**, *75*, 809–819.

206. Rouffet, M.E.; Wendt, M.; Goett, G.; Kozakov, R.; Schoepp, H.; Weltmann, K.D.; Uhrlandt, D. Spectroscopic investigation of the high-current phase of a pulsed GMAW process. *J. Phys. D* **2010**, *43*, 434003.

207. Mauer, G.; Vaßen, R. Plasma Spray-PVD: Plasma Characteristics and Impact on Coating Properties. *J. Phys. Conf. Series* **2012**, *406*, 012005.

208. Zhu Xi-Ming; Walsh, J.L.; Chen Wen-Cong; Pu Yi-Kang Measurement of the temporal evolution of electron density in a nanosecond pulsed argon microplasma: Using both Stark broadening and an OES line-ratio method. *J. Phys. D* **2012**, *45*, 295201.

209. Zhang, W.; Hua, X.; Liao, W.; Li, F.; Wang, M. Study of metal transfer in CO_2 laser + GMAW-P hybrid welding using argon-helium mixtures. *Opt. Laser Technol.* **2014**, *56*, 158–166.

210. Dimitrijević, M.S.; Sahal-Bréchot, S. Stark broadening of neutral zinc spectral lines. *Astron. Astrophys. Supp.* **1999**, *140*, 193–196.

211. Gornushkin, I.B.; Kazakov, A.Ya.; Omenetto, N.; Smith, B.W.; Winefordner, J.D. Experimental verification of a radiative model of laser-induced plasma expanding into vacuum. *Spectrochim. Acta B* **2005**, *60*, 215–230.

212. Deng, Y.Z.; Zheng, H.Y.; Murukeshan, V.M.; Zhou, W. Analysis of Optical Emission towards Optimisation of Femtosecond Laser Processing. *J. Laser Micro Nanoengineering.* **2006**, *1*, 136–141.

213. Shaikh, N.M.; Rashid, B.; Hafeez, S.; Jamil, Y.; Baig, M.A. Measurement of electron density and temperature of a laser-induced zinc plasma. *J. Physics D Appl. Phys.* **2006**, *39*, 1384–1391.

214. Shaikh, N.M.; Hafeez, S.; Kalyar, M.A.; Ali, R.; Baig, M.A. Spectroscopic characterization of laser ablation brass plasma. *J. Appl. Phys. B* **2008**, *104*, 103108.

215. Patel, D.N.; Pandey, P.K.; Thareja, R.K. Stoichiometric investigations of laser-ablated brass plasma. *Appl. Opt.* **2012**, 51, B192-B200.

216. Gupta, Shyam L.; Thareja, Raj K. Photoluminescence of nanoparticles in vapor phase of colliding plasma. *J. Appl. Phys.* **2013**, *113*, 143308.

217. Diwakar, P.K.; Harilal, S.S.; Freeman, J.R.; Hassanein, A. Role of laser pre-pulse wavelength and inter-pulse delay on signal enhancement in collinear double-pulse laser-induced breakdown spectroscopy. *Spectrochim. Acta B Phys. Plasmas* **2013**, *87*, 65–73.

218. Freeman, J.R.; Harilal, S.S.; Diwakar, P.K.; Verhoff, B.; Hassanein, A. Comparison of optical emission from nanosecond and femtosecond laser produced plasma in atmosphere and vacuum conditions. *Spectrochim. Acta B Phys. Plasmas* **2013**, *87*, 43–50.

219. Gupta, Shyam L.; Pandey, P.K.; Thareja, Raj K. Dynamics of laser ablated colliding plumes. *Phys. Plasmas* **2013**, *20*, 013511.

220. Patel, D.N.; Pandey, Pramod K.; Thareja, Raj K. Brass plasmoid in external magnetic field at different air pressures. *Phys. Plasmas* **2013**, *20*, 103503.

221. Smijesh, N.; Philip, Reji Emission dynamics of an expanding ultrafast-laser produced Zn plasma under different ambient pressures. *J. Appl. Phys.* **2013**, *114*, 093301.

222. Dimitrijević, M.S.; Sahal-Bréchot, S. Stark broadening of spectral lines of multicharged ions of astrophysical interest. VII. Al III lines. *Astron. Astrophys. Supp.* **1993**, *99*, 585–589.

223. Heading, D.J.; Benett, G.R.; Wark, J.S.; Lee, R.W. Novel plasma source for dense plasma effects. *Phys. Rev. Lett.* **1995**, *74*, 3616–3619.

224. Heading, D.J.; Wark, J.S.; Bennett, G.R.; Lee, R.W. Simulations of spectra from dense aluminium plasmas. *J. Quant. Spectrosc. Radiat. Transf.* **1995**, *54*, 167–180.

225. Versteegh, A.; Behringer, K.; Fantz, U.; Fussmann, G.; Juttner, B.; Noack, S. Long-living plasmoids from an atmospheric discharge. *Plasma Sources Sci. T.* **2008**, *17*, 024014.

226. Dimitrijević, M.S.; Sahal-Bréchot, S. Stark broadening of spectral lines of multicharged ions of astrophysical interest. XVI. S V spectral lines. *Astron. Astrophys. Supp.* **1998**, *127*, 543–544.

227. Bengoechea, J.; Kennedy, E.T. Time-integrated, spatially resolved plasma characterization of steel samples in the VUV. *J. Anal. Atom. Spectrom.* **2004**, *19*, 468–473.

228. Dimitrijević, M.S.; Sahal-Bréchot, S. Stark broadening parameter tables for Ar VIII. *Serb. Astron. J.* **1999**, *160*, 15–20.

229. Uzuriaga, J.; Chamorro, J.C.; Marín, R.A.; Riascos, H. Optical emission spectra of ZnMnO plasma produced by a pulsed laser. *J. Phys. Conf. Ser.* **2012**, *370*, 012056.

230. Sahal-Bréchot, S.; Dimitrijević, M.S.; Moreau, N. STARK-B Database. LERMA, Observatory of Paris, France and Astronomical Observatory, Belgrade, Serbia, 2014. Available online: http://stark-b.obspm.fr (accessed on 30 April 2014).

231. Dubernet, M. L.; Boudon, V.; Culhane, J. L.; Dimitrijevic, M. S.; Fazliev, A. Z.; Joblin, C.; Kupka, F.; Leto, G.; *et al.* Virtuel Atomic and Molecular Data Centre, *JQSRT* **2010**, *111*, 2151–2159.

Reprinted from *Atoms*. Cite as: Omar,B.; González, M.Á.; Gigosos, M.A.; Ramazanov, T.S.; Jelbuldina, M.C.; Dzhumagulova, K.N.; Zammit, M.C.; Fursa, D.V.; Bray, I. Spectral Line Shapes of He I Line 3889 Å. *Atoms* **2014**, *2*, 277-298.

Article

Spectral Line Shapes of He I Line 3889 Å

Banaz Omar [1,*], **Manuel Á. González** [2], **Marco A. Gigosos** [3], **Tlekkabul S. Ramazanov** [4], **Madina C. Jelbuldina** [4], **Karlygash N. Dzhumagulova** [4], **Mark C. Zammit** [5], **Dmitry V. Fursa** [5] and **Igor Bray** [5]

[1] Institute of Physics, University of Rostock, Rostock 18051, Germany

[2] Departamento de Física Aplicada, Universidad de Valladolid, Valladolid 47071, Spain;
    E-Mail: manuelgd@termo.uva.es

[3] Departamento de Óptica, Universidad de Valladolid, Valladolid 47071, Spain;
    E-Mail: gigosos@coyanza.opt.cie.uva.es

[4] Scientific Research Institute for Experimental and Theoretical Physics of Al-Farabi
    Kazakh National University, Almaty, Kazakhstan; E-Mails: ramazan@physics.kz (T.S.R.),
    zholboldy@mail.ru (M.C.J.), dzhumagulova.karlygash@gmail.com (K.N.D.)

[5] Institute of Theoretical Physics, Curtin University, Perth, WA 6845, Australia;
    E-Mails: mark.zammit@postgrad.curtin.edu.au (M.C.Z.), d.fursa@curtin.edu.au (D.V.F.),
    I.Bray@curtin.edu.au (I.B.)

* Author to whom correspondence should be addressed; E-Mail: b.omar@dr.com;
  Tel.: +46 722 758100.

Received: 31 March 2014; in revised form: 5 June 2014 / Accepted: 10 June 2014 / Published: 23 June 2014

Abstract: Spectral line shapes of neutral helium 3889 Å (2^3S–3^3P) transition line are calculated by using several theoretical methods. The electronic contribution to the line broadening is calculated from quantum statistical many-particle theory by using thermodynamic Green's function, including dynamic screening of the electron-atom interaction. The ionic contribution is taken into account in a quasistatic approximation, where a static microfield distribution function is presented. Strong electron collisions are consistently considered with an effective two-particle T-matrix approach, where Convergent Close Coupling method gives scattering amplitudes including Debye screening for neutral helium. Then the static profiles converted to dynamic profiles by using the Frequency Fluctuation Model. Furthermore, Molecular Dynamics simulations for interacting and independent particles are used where the dynamic sequence of microfield is taken into account. Plasma parameters are diagnosed and good agreements are shown by comparing our theoretical results with the recent experimental result of Jovićević et al. (*J. Phys. B: At. Mol. Opt. Phys.* 2005, *38*, 1249). Additionally, comparison with various experimental data in a wide range of electron density $n_e \approx (10^{22} - 10^{24})\,\mathrm{m}^{-3}$ and temperature $T \approx (2 - 6) \times 10^4$ K are presented.

Keywords: spectral line shapes; Green's function; T-matrix; molecular dynamics simulations; microfield distribution function; plasma diagnostics

1. Introduction

Calculation of spectral line shapes is a most powerful tool for plasma diagnostic in both the star atmosphere and in laboratory plasmas. Perturbation of the radiating atom or ion by the surrounding particles leads to spectral line broadening (Stark broadening), while the coherent emission process is interrupted by collisions and influenced by plasma microfield. Several theoretical approaches have been applied to calculate Stark broadening, such as the well known semiclassical approximation the standard theory (ST) by Griem [1], or the quantum statistical approach of many-particle theory [2], where the motion of ion perturber is neglected during the inverse halfwidth of the line. Furthermore, the model microfield method (MMM) [3–6], the frequency fluctuation method (FFM) [7] or computer simulations [8–11] are used for calculating the line broadening including ion-dynamics effects, which lead to further broadening of the line shapes.

In this work, we describe several theoretical approaches to study the He I 3889Å (2^3S-3^3P) transition line. Plasma parameters are determined by comparing our theoretical results with the measurement of Jovićević et al. [12]. In our quantum statistical approach, thermodynamic Green's function is used for calculating emitted or absorbed radiation from bound–bound transition of charge particles by using two-particle polarization function [13–16], which is related to the Fourier transformation of the imaginary part of the dipole–dipole correlation function. In principle, this approach is able to describe the dynamic screening and strong collision by electrons in a systematic way. The quantum statistical approach can adequately treat electron collisions. To obtain full profiles, it has been supplemented by a calculation of ion broadening done by other methods and thereby has been successfully applied to calculate the line profile of hydrogen and H-like ions [17]. The strong collisions of an electron-emitter pair are treated within a T-matrix approach by solving close-coupling equations to produce forward scattering amplitudes. Ions are treated in a quasistatic approximation via the microfield distribution function. Then the latest formulation of the FFM is applied to account for ion dynamics [18]. On the other hand, computer simulations accounting simultaneously for the electronic and ionic fields on the same footing are used for comparison. Recently, the shift and width of this line was computed by Lorenzen et al. [19], comparison was made with various theoretical and measurement data in a wide range of electron density and temperature. In Section 2 we describe briefly the experimental setup and result of Jovićević *et al.* [12]. Our theoretical approaches for calculating the line shapes are reviewed in Section 3, followed by results and discussions Section 4.

2. Experiment

In this paper, we analyze theoretically the He I 3889 Å Stark broadening, measured by Jovićević *et al.* [12] from a pulsed low-pressure capillary discharge. The plasma was a mixture of neon, helium and hydrogen, with the predominant H^+ perturber ions. Originally the experiment was set up for the measurement of Ne I spectral line profiles, while He was added to the gas mixture for plasma diagnostic purpose. The presence of hydrogen increases the electron density to the required value until a constant value of the Stark width was recorded. The optimum gas mixture was determined in a series of measurements starting from pure neon and then diluting until optically thin Ne I lines were recorded. In the experiment, a quartz discharge tube with 3 mm inner diameter and length of 7.2 cm was used, and aluminum electrodes with 3 mm diameter holes were located 200 mm apart. The measurements were performed in a premixed gas mixture of 2.4% Ne, 5.6% He and 92% H_2 by volume, at an initial pressure of 4 mbar. Spectroscopic plasma observations of Ne I and He I lines were performed, measuring the line shapes at the same time of plasma decay. The dilution of neon and helium with hydrogen was done in order to increase the electron density, since H has much lower ionization potential than Ne and He, and to generate a plasma with H^+ ions dominating He^+ and Ne^+. Special care was taken to keep the line profiles optically thin during measurements, in order to minimize self-absorption. This was achieved by using a continuous gas flow and diluting neon with helium and hydrogen in the operated arc. Moreover, plasma radiation was observed from the axial narrow discharge tube as well as from the radial expanded part simultaneously. However, the experimental design and procedure may have resulted in plasma inhomogeneity which may cause line distortion and line asymmetry. Electron density was determined from the measured He I line profiles, which were assumed to be optically thin under the same plasma conditions. The experimental profiles were fitted to a semiclassical calculation using two parameters, the electron impact width and the ion-broadening parameter A of the quasistatic approximation [1], which allow the authors to determine the electron density. The mean value of electron density n_e was estimated to be $4.8 \times 10^{22}\,\mathrm{m}^{-3}$ with an estimated uncertainty of $\pm 10\%$. By assuming local thermal equilibrium, the electron temperature $T_e = T_i = 3.3 \times 10^4$ K was determined from the Boltzmann plot of O II impurity lines with an uncertainty of $\pm 12\%$, as the Stark broadening parameters weakly depend upon T_e. More details about the experimental set up and measurement are given in [12].

3. Theoretical Approaches

In this section we present the different kinds of theoretical approaches that we used for calculating He I line profiles.

3.1. Quantum Statistical Approach

In quantum statistical many-particle theory, the medium effects are taken into account by using the thermodynamic Green's function technique to describe the Stark broadening. The perturbing electrons and ions can be treated independently due to variety of mass between them ($m_e << m_i$). The dynamically screened Born approximation applied for electron-emitter interaction, while almost stationary heavy ions are treated in a quasistatic ion approximation due to static microfield. Spectral line broadening is given by the imaginary part of the second order two-particle polarization function, which is a bound–bound transition. The polarization function is calculated in a systematic way from thermodynamic Green's functions and is proportional to the Fourier transform of the dipole–dipole autocorrelation function (see Appendix A). The perturber–radiator interaction leads to the pressure broadening, including both the electronic and the ionic contributions in the quasistatic limit by averaging the static microfield distribution function $W(\beta)$ [14–16], and we get

$$I^{\mathrm{pr}}(\omega) \sim \sum_{i',i'',f',f''} \langle i'|\mathbf{r}|f'\rangle\langle f''|\mathbf{r}|i''\rangle \; \frac{\omega^4}{8\pi^3 c^3} \; \mathrm{e}^{-\frac{\hbar\omega}{k_{\mathrm{B}}T}} \int_0^\infty W(\beta)\,d\beta$$
$$\times \mathrm{Im}\,\langle i'|\langle f'| \left[\hbar\omega - \hbar\omega_{if} - \Sigma_{if}(\omega,\beta) + \imath\Gamma^{\mathrm{v}}_{if}\right]^{-1} |f''\rangle|i''\rangle \qquad (1)$$

where $\langle i|\mathbf{r}|f\rangle$ is identified as a dipole matrix-element for the transition probability between initial (n_i, l_i, m_i) and final (n_f, l_f, m_f) states. The ionic microfield strength $\beta = E/E_0$ is normalized to the Holtsmark field E_0, and $\hbar\omega_{if} = E_i - E_f$ is the unperturbed transition energy. The interference correction Γ^{v}_{if} for the overlapping line is related to coupling between the initial and final states, the interference contribution vanishes for this transition because the lower state is a s-state and thus not collisionally connected to itself.

The interference term is less important for neutral helium, and is ignored in the dipole approximation, this is not the case for quadrupole approximation of transition [1,20]. However, more accurate calculations are needed for lines involving transitions with high excited states, which are almost hydrogenic states of the helium atom. If the contribution of the lower levels is neglected, then the interference term is zero [21]. Generally the interference contribution is less important for non-hydrogenic line transitions [20]. The function $\Sigma_{if}(\omega,\beta)$ is determined by the self-energy correction $\Sigma_n(\omega,\beta)$ of the initial and the final states

$$\Sigma_{if}(\omega,\beta) = \mathrm{Re}\left[\Sigma_i(\omega,\beta) - \Sigma_f(\omega,\beta)\right] + \imath\,\mathrm{Im}\left[\Sigma_i(\omega,\beta) + \Sigma_f(\omega,\beta)\right] \qquad (2)$$

where the real part represents the shift Δ_n^{SE} and the imaginary part gives the width Γ_n^{SE} of each state. The electronic and ionic contributions occur in the self-energy. Performing Born approximation with respect to the dynamically screened perturber–radiator potential $V(q)$, the diagonal elements of electronic self-energy operator is obtained as (see Appendix B)

$$\Delta_n^{\mathrm{SE}} + \imath \Gamma_n^{\mathrm{SE}} \;=\; <n|\Sigma^{\mathrm{el}}(E_n,\beta)|n> \;=\; -\frac{1}{e^2}\int\frac{\mathrm{d}^3 q}{(2\pi)^3}V(q)\sum_\alpha |M_{n\alpha}(\mathbf{q})|^2$$

$$\times \int_{-\infty}^{\infty}\frac{\mathrm{d}\omega}{\pi}[1+n_{\mathrm{B}}(\omega)]\frac{\operatorname{Im}\varepsilon^{-1}(\mathbf{q},\omega+\imath 0)}{E_n - E_\alpha(\beta) - \hbar(\omega+\imath 0)}. \qquad (3)$$

Here, the level splitting $(E_\alpha(\beta) \approx E_\alpha)$ due to the microfield has been neglected, $n_{\mathrm{B}}(\omega) = \left[\exp(\hbar\omega/k_{\mathrm{B}}T) - 1\right]^{-1}$ is the Bose distribution function, approximated by the Boltzmann distribution function for non-degenerated plasma. The sum over principal quantum number α runs from $n-2$ to $n+2$ discrete bound states for the virtual transitions. Dynamic screening is obtained from the imaginary part of the inverse dielectric function $\varepsilon^{-1}(\mathbf{q},\omega)$, for which the random phase approximation (RPA) is used. The transition matrix-element $M_{n\alpha}(\mathbf{q})$ describes the interaction of the atom with the Coulomb potential through the vertex function, where the Coulomb interaction with the electron-electron-ion triplet depends on the momentum transfer $\hbar\mathbf{q}$ (see Appendix C). The electron wavefunctions for helium are obtained by applying the Coulomb approximation method [22]. In the Born approximation, the electronic contribution is overestimated. Hence, in order to avoid this, we apply the cut-off procedure and add the strong collisions term [1,16,23] instead of the partial summation via the T-matrix. For He I we adopt the cut-off parameter and close electron-radiator collisions term, evaluated according to the semiclassical estimation, see [16,23]. For non-hydrogenic radiator the ionic contribution to the self-energy is related to the quadratic Stark effect and quadrupole interaction [24], further details are given in [15,16].

The above procedure is used to describe the perturbing effect of free electrons on the radiator from the electronic self energy and the interference correction, which gives exactly the same expression as the semiclassical impact approximation of Griem [1], if the term with q^2 is neglected in the argument of delta function as in the classical limit, see Equation (4.5) in [13]. The comparison between the Green's function approach and impact approximation is proved in detail by Kraeft *et al.* [25].

3.1.1. T-Matrix Approach

The electronic contribution can be evaluated again from the effective T-matrix approach which is based on the partial summation. To avoid the cut-off procedure for strong electron-atom collisions, the electronic contribution to shift and width of spectral lines can be treated via partial summation of the three-particle T-matrix. A three-particle propagator (T_3) in the channel of the two-particle bound state and perturbing electron was adopted by Günter [2,26] for calculating the hydrogen L_α line. It can describe weak and strong collisions within static Debye screening. For a non-degenerate plasma, the electronic contribution can be evaluated from scattering amplitudes, characterized by the self energy [19]

$$\Sigma_n^{\mathrm{el}} = -\frac{2}{\pi}n_e\Lambda_{\mathrm{th}}^3\int_0^\infty dk\,k^2\,e^{-k^2/(k_{\mathrm{B}}T)}\,f_n(0,k). \qquad (4)$$

Here, $\Lambda_{\mathrm{th}} = \sqrt{4\pi/(k_{\mathrm{B}}T)}$ is thermal wave length, and $f_n(\theta = 0, k)$ are forward scattering amplitudes of elastic electron scattering at the Debye screened He I for both state initial $(n = i)$ and final $(n = f)$ states, respectively. The singlet $S = 0$ and triplet $S = 1$ channels for scattering are considered for both initial i and final f states. This expression was also found by Baranger [27]. The single electron-atom and -ion scattering problem is studied to provide reliable results for all transitions of interest including elastic, excitation, ionization, and total cross sections [28–30]. Recently, the electron-helium scattering in weakly coupled hot-dense plasmas has been investigated by using the Convergent Close Coupling method (CCC), where the Laguerre basis expansions are applied to obtain fully convergent scattering amplitudes in the close-coupling equations. The screened Coulomb potential or Debye-Hückel potential [31,32] has been employed. This Yukawa-type potential has been used to describe the plasma Coulomb screening effect by formulating a set of close-coupling equations for the T-matrix approach by providing a complete description of the scattering process. The CCC method solves the close-coupling equations in momentum space and uses the T-matrix to determine the cross sections and forward scattering amplitudes for dipole and spin-allowed transitions on the Debye screened neutral helium atom, The CCC method uses the analytical Born subtraction technique to speed up the convergence of the partial-wave expansion, thus the first Born approximation matrix elements for inelastic scattering is modified. The quantum system involves electron-helium scattering from the ground-state state up to 4f with the correct energies using 153 pseudo-states to describe the scattering, more details are given in [33–35]. Our method gives different results for singlet and triplet scattering channels as well as for scattering at the emitter with initial and final states. Here, we consider only the electron-He forward elastic scattering amplitudes $f(0, k)$ from initial state (n_i, l_i, m_i) to final state (n_f, l_f, m_f) for the self-energies, where n, l, m are the well known principal quantum numbers. Here, the transitions $2S \rightarrow 2S$ and $3P \rightarrow 3P$ are considered for double channel transitions. For a detailed description of the T-matrix calculation see [19].

3.1.2. Microfield Distribution Function

The ionic microfield is considered to be constant during the time of interest of the order of the inverse HWHM of the line, so that a static description of the microfield can be used. In this study, the plasma microfield distribution function is calculated by taking into account the screening and quantum effects of diffraction, the evaluated distribution function is used for calculating line broadening for analyzing the experimental data of Jovićević et al. [12]. Processes such as ionization, dissociation, and excitation of bound states occur continuously in plasma and affect plasma properties and its compositions [36]. The distribution of charged particles in the system within the Debye radius r_D is not homogeneous at distances $r < r_D$, although the plasma is electrically neutral as a whole, but at sufficiently small distances the fluctuation of the local electric field affects the plasma kinetic coefficients, optical and thermodynamic properties, etc. [37]. According to Ecker and Müller [38] the effective Debye screened field is defined as

$$\vec{E} = \frac{e}{r^3}\vec{r}(1 + r/r_D)e^{-r/r_D} \tag{5}$$

The problem of microfield distribution is formally similar to the problem of determining the chemical potential. This allowed Iglesias [39] to reduce the microfield problem to finding the radial distribution functions (RDF) of some fictitious system with a complex potential energy of interaction. This made the basis for the development of the integral equations method to study the problem of the plasma microfield distribution.

In this paper, we use a method for calculating the distribution function of ionic microfield component $P(E)$ proposed by Iglesias [40]. The advantage of this method is that the distribution function is exactly expressed in terms of a two-body function and does not require knowledge of many-body functions. We use a strongly coupled one component plasmas (OCP) model, *i.e.*, in a fully ionized plasma, in which the electron component forms a homogeneous neutralizing background with N positively charged particles. The microfield distribution function $W(\vec{E})$ is defined as the probability of finding an electric field $\vec{E}$ at a charged point located at $\vec{r}_0$. It is generally expressed in terms of the probability density $P_{N+1}(\vec{r}_0, \vec{r}_1, ..., \vec{r}_N)$ of finding a particular configuration of $(N + 1)$ particles:

$$W(\vec{E}) = \int ... \int \delta(\vec{E} - \sum_{j=1}^{N} \vec{E}_j) P_{N+1}(\vec{r}_0, \vec{r}_1, ..., \vec{r}_N) d\vec{r}_0, d\vec{r}_1, ..., d\vec{r}_N \qquad (6)$$

where $\vec{E}_j$ is the electric field, created by $j-$th particle on radiating atoms or ions at $\vec{r}_0$. Assuming the system is isotropic we may introduce Fourier transformation of the function $W(\vec{E})$

$$P(E) = \frac{2E}{\pi} \int_0^\infty l\, T(l) \sin(El)\, dl \qquad (7)$$

where $T(l)$ is the characteristic function with a total pair correlation function $h(\vec{r}; \lambda)$

$$\ln T(l) = n \int_0^1 d\lambda \int_0^\infty d\vec{r}\, \frac{\imath q l \hat{r}}{r^3} h(\vec{r}; \lambda) \qquad (8)$$

with $\hat{l}$ is a unit vector in the direction of $\vec{l}$, and the introduced parameter λ is a magnitude of the vector $\vec{l}$, while q is the charge of ions, immersed in a neutralizing electron background. The next step is a definition of $h(\vec{r}; \lambda)$. The simplest approximation suitable for a Coulomb system is the Debye–Hückel theory. In this approximation we have

$$h(\vec{r}; \lambda) \approx \exp\{-\bar{\beta}[1 - \frac{\imath \lambda \vec{l}}{q\bar{\beta}} \vec{\nabla}_0] \Phi(r)\} - 1. \qquad (9)$$

In [41] with the help of linear response method an effective potential was obtained, which is also used in this paper, by taking in to account both the diffraction effects at short distances and the screening effect at large distances

$$\Phi(r) = -\frac{Z_\alpha e^2}{\sqrt{1 - 4\lambda_{\alpha e}^2/r_D^2}} \left(\frac{e^{-Br}}{r} - \frac{e^{-Ar}}{r}\right) \qquad (10)$$

where

$$A^2 = \frac{1}{2\lambda}\left(1 + \sqrt{1 - 4\lambda_{\alpha e}^2/r_D^2}\right)$$

$$B^2 = \frac{1}{2\lambda}\left(1 - \sqrt{1 - 4\lambda_{\alpha e}^2/r_D^2}\right).$$

The coupling parameter is defined by $\Gamma = \bar{\beta}e^2/r_0$, and

$$r_D = (4\pi n e^2 \bar{\beta})^{1/2}, \quad \bar{\beta} = (k_B T)^{-1}, \quad r_0 = (4\pi n/3)^{-1/3} \tag{11}$$

where r_0 is the mean particle distance. Here the reduced form of pseudopotential is used

$$-\bar{\beta}\Phi(r) = -\frac{\Gamma}{R}\frac{1}{\sqrt{1 - 24\Gamma^2/(\pi r_s)}}\left(e^{-BR} - e^{-AR}\right) \tag{12}$$

with

$$\begin{aligned}
A^2 &= \frac{\pi r_s}{4\Gamma}\left(1 + \sqrt{1 - 24\Gamma^2/(\pi r_s)}\right) \\
B^2 &= \frac{\pi r_s}{4\Gamma}\left(1 + \sqrt{1 - 24\Gamma^2/(\pi r_s)}\right).
\end{aligned}$$

In order to solve Equation (6) numerically, it is better to use the dimensionless field units and to introduce $l^* = E_0 l$. Thus we obtain a formula to calculate the distribution of electric field E^*:

$$\mathcal{P}(E^*) = \frac{2E^*}{\pi}\int_0^\infty dl^* \, l \, T(l^*) \sin(E^* l^*). \tag{13}$$

By substitution of Equation (9) into Equation (8) yields

$$\begin{aligned}
\ln T(l) &= 3\int_0^\infty dR \cdot \exp\left\{\frac{\Gamma}{R}\frac{1}{\sqrt{1 - \chi}}(e^{-AR} - e^{-BR})\right\} \\
&\times \frac{R^2}{e^{-BR}(1 + BR) - e^{AR}(1 + AR)}
\end{aligned} \tag{14}$$

where $\chi = 6\Gamma^2/(\pi r_s)$, $R = r/r_0$ and

$$lb(R) = \frac{lq}{r_0}\frac{e^{-BR}(1 + BR) - e^{-AR}(1 + AR)}{R^2}$$

The one-dimensional integration over R involves the sin transformation in Equation (7), in which $T(l)$ is defined by Equation (8) which is solved numerically.

The microfield distribution function illustrated here is applied for calculating the line profiles to determine the plasma parameters of the measured data by Jovićević *et al.* [12].

3.1.3. Frequency Fluctuation Model

Recently, Calisti *et al.* [18] presented a very fast method to account for ion dynamic effects on atomic spectra in plasmas. This method is based on reformulation for the frequency fluctuation model (FFM), which provides an expression of the dynamic line shapes based on a mixing frequency and the quasistatic Stark profiles. This method allows a very fast and accurate calculation of Stark broadening $I(\omega)$ by taking into account ion-dynamics in this expression

$$I(\omega) \sim \mathrm{Re}\frac{\int \frac{\mathcal{A}(\omega')\,d\omega'}{\gamma + i(\omega - \omega')}}{1 - \gamma\int \frac{\mathcal{A}(\omega')\,d\omega'}{\gamma + i(\omega - \omega')}}. \tag{15}$$

Here, $\mathcal{A}(\omega)$ is the area-normalized line profile in quasistatic limit, evaluated from Equation (1), the inverse state lifetime $\gamma = v_{\text{th}}/d$ is defined by ion thermal velocity of perturbers $v_{\text{th}} = \sqrt{8k_B T/(\pi m_i)}$ and the mean distance between ions $d = (4\pi n_i/3)^{-1/3}$, where m_i and n_i are the mass and density of perturbing ions, respectively.

3.2. Classical Molecular Dynamics Simulations

Spectral line shapes $I(\Delta\omega)$ are also calculated using molecular dynamics (MD) simulations. Using these simulations, the dynamics of the particles in the plasma is reproduced numerically in the computer and the electric field sequences suffered by the emitters are obtained. These field sequences are taken to the differential equations that permit to calculate the emitter evolution, and the final profiles are obtained as the Fourier transform of an average of the emitter dipole-moment autocorrelation function [42].

$$I(\Delta\omega) = \mathrm{Re}\frac{1}{\pi}\int_0^\infty dt\,\{C(t)\}e^{i\Delta\omega t} \tag{16}$$

$$C(t) = \mathrm{tr}[\mathbf{D}(t)\cdot\mathbf{D}(0)\,\rho] \tag{17}$$

$$D(t) = U^+(t)\,\mathbf{D}(0)\,U(t) \tag{18}$$

where ρ is the equilibrium density-matrix operator, D is the dipole operator for the radiating system, and the symbols {} mean an average over a large number of simulated individual processes obtained in order to have a representative sampling of the electric fields in the plasma. $U(t)$ is the emitter time evolution operator that obeys Schrödinger equation

$$i\hbar\frac{d}{dt}U(t) = H(t)U(t) = \left(H_0 + q\mathbf{E}(t)\cdot\mathbf{R}\right)U(t) \tag{19}$$

where the Hamiltonian $H(t)$ includes the structure of the emitter states without perturbation H_0, and the action of the perturbers through the dipole interaction with the electric field $\mathbf{E}(t)$ at the emitter location [43]. For a specific line transition, two groups of states are considered, the upper and lower group of states of the transition. If the difference in energy of those groups is large enough in comparison with the average energy of the collisions in the plasma, then we can assume that both groups of states evolve independently though undergoing the same sequence of perturbing electric field, which is known as the no-quenching approximation. The simulations used in this work consider that the perturbers are classical point independent particles that move along straight line trajectories with constant speeds according to the Maxwell distribution corresponding to their equilibrium temperature. Correlation effects between perturbers are taken into account considering Debye screened electric fields. The relative movement of the heavy perturbers and the emitter is taken into account using the reduced mass of the emitter-perturber pair according to the μ-ion model [44]. This simulation technique is designed to guarantee that the density of particles within the simulation sphere is kept constant along the simulation and that both the spatial distribution of particles and the velocities distribution are kept stationary [9]. On the other hand, this technique allows us to consider non-equilibrium plasmas [45], and can also be applied to non-homogeneous plasmas [46,47]. However, for strongly interacting plasmas (large plasma coupling parameter) or charged emitters, MD simulations accounting for the interactions between all the particles in the plasma (*i.e.*, including the emitter) are required [48,49], though for the calculations in this work the independent particles technique was accurate enough. More details of the calculation can be seen in [11].

4. Results and Discussions

In this study, we present several theoretical approaches for calculating He I line shapes. The plasma parameters are inferred by comparing our results with the experimental measurement of Jovićević et al. [12]. The measured profile is compared with the theoretical profiles at $T = 3.3 \times 10^4$ K for two values of electron densities $n_e = 5.2 \times 10^{22}$ m^{-3} and $n_e = 4.58 \times 10^{22}$ m^{-3}, respectively. In [12] the experimental He I 3889 Å line broadening was fitted by using the $j_{AR}(x)$ line profiles generated on the assumption of electron impact broadening combined with broadening by quasistatic ion field using the ST of Griem [1]. The electron density of $n_e = 4.8 \times 10^{22}$ m^{-3} with an estimated uncertainty of $\pm 10\%$ was determined from the line broadening, while the electron temperature of $T = 3.3 \times 10^4$ K with an uncertainty of $\pm 12\%$ was estimated by using the relative intensities of O II impurity lines.

In our quantum statistical approach electrons and ions are treated separately. Green's function is applied for calculating the electronic contribution from Equation (3), while the dominant ionic contribution arises from the quadratic Stark effect and quadrupole interaction [15,16]. First, the ionic contribution is determined in the quasistatic approximation by using the calculated microfield distribution function in Section 3.1.2, then the new FFM formulation [18] is adopted to account for the fluctuation of the microfield on the spectral line shapes. The dynamic effects of the ionic motion are taken into account by substituting the area-normalized static profile into Equation (15). The result is displayed in Figure 1, which compares measured and theoretical profiles. Since the authors of the original experimental paper [12] did not mention any measurement of absolute line shift, we displaced the calculated line profiles to the unperturbed wavelength $\lambda_0 = 3888.65$ Å in Figures 1–3 to determine the free electron density n_e from the line width.

The electronic contribution is recalculated via the T-matrix by applying the CCC method in order to describe the strong electron-helium collisions within the Debye static screening model [34]. Both real and imaginary part of forward scattering amplitudes are substituted in Equation (4) to provide electronic shift and width [19]. In contrast to the Born approximation, the contribution of each state is different for different values of the magnetic quantum number $|m| = 0, 1$. Furthermore, an average over different spin-scattering channels $S = 0$ and $S = 1$ has been done. The Stark profiles are calculated in the quasistatic approximation from Equation (1) by using the T-matrix approach for electron-emitter collisions, The dynamic profiles are evaluated from Equation (15) and illustrated in Figure 2. It may be seen that the electron density $n_e = 5.2 \times 10^{22}$ m^{-3} gives better agreement with the experimental profile in Figures 1 and 2.

Results from simulations are presented in Figure 3. Dynamic profiles show very good agreement with the measured profile for the inferred plasma density $n_e = 4.58 \times 10^{22}$ m^{-3} and temperature $T = 3.3 \times 10^4$ K. In Table 1, numerical results of the shift and FWHM from our theories are given. We compare the broadening parameters in the dynamically screened Born approximation, the T-matrix approach, and MD simulations for interacting particles.

Figure 1. The calculated He I (2^3S–3^3P) line profiles in dynamically screened Born approximation by including ion dynamics in the FFM . Comparison is made with the measured data by Jovićević *et al.* [12] at $T = 3.3 \times 10^4$ K. The profiles are plotted at the unperturbed transition wavelength $\lambda_0 = 3888.65$ Å, and the shift of profiles are shown.

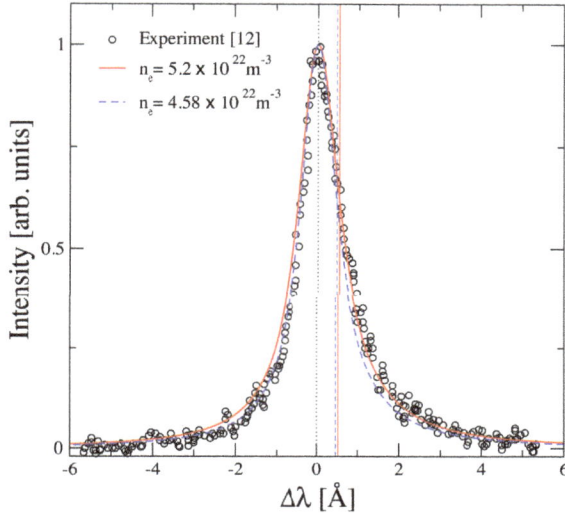

Figure 2. Comparison between the measured [12] and calculated line shapes of He I (2^3S–3^3P) using the T-matrix approach with ion dynamics in the FFM at $T = 3.3 \times 10^4$ K. The profiles are given at the unperturbed transition wavelength $\lambda_0 = 3888.65$ Å, and the shift of profiles are shown.

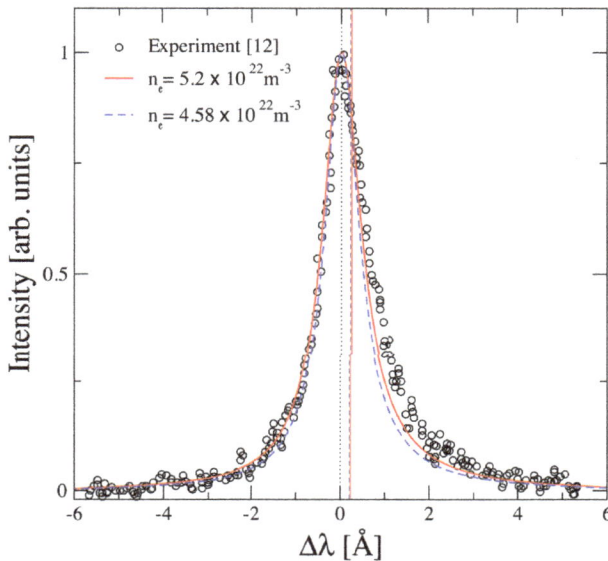

Figure 3. The calculated He I (2^3S-3^3P), line profiles from MD simulations compared with the experimental result of Jovićević *et al.* [12] at $T = 3.3 \times 10^4$ K. The profiles are given at the unperturbed transition wavelength $\lambda_0 = 3888.65$ Å, and the shift of profiles are shown.

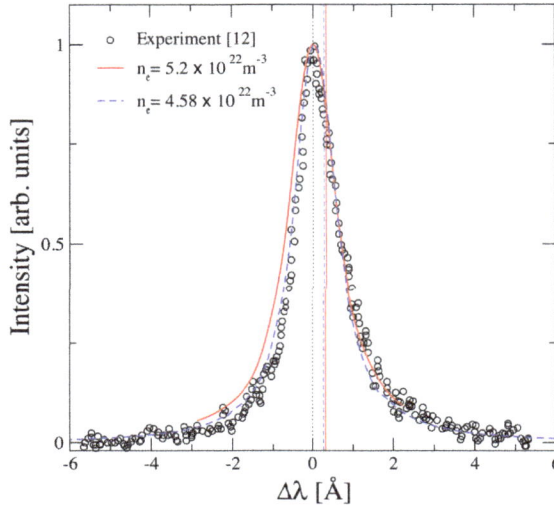

Figure 4. The shift of He I 3889 Å *vs.* electron density. Comparison between our calculated data and the experimental results are shown. The electron temperature is given for the corresponding T-matrix data points.

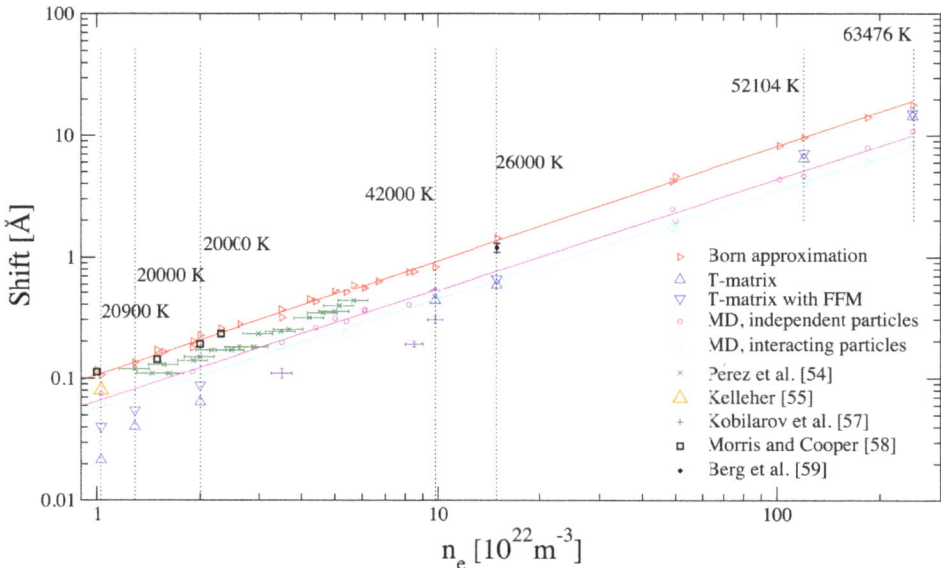

Ion dynamics (FFM) are included in the Born approximation and the T-matrix approach, the dynamic effects lead to further shift. The shift is significantly affected by including strong collisions via the T-matrix and ion-dynamics, while the dynamic profiles appear slightly broadened compared to the static profile. The calculated line shift for different values of n_e is illustrated in Figures 1–3 as well. Note, that the shift of this line is red, towards larger wavelengths with respect to the unperturbed transition line at λ_0. An example of the line measurement can be seen also in [50].

Table 1. Theoretical calculations of shift and FWHM of the He I 3889 Å line are illustrated in [Å], to analyze the experimental result of Jovićević *et al.* [12]. The calculated values correspond to dynamically screened Born approximation (quasistatic/FFM), T-matrix approach (quasistatic/FFM) and MD simulations at $T = 3.3 \times 10^4$ K.

$n_e\ 10^{22}\,\mathrm{m}^{-3}$	Shift [Å]			FWHM [Å]		
	Born	**T-Matrix**	**MD**	**Born**	**T-Matrix**	**MD**
5.2	0.483/0.523	0.178/0.228	0.32	1.245/1.278	1.038/1.087	1.402
4.58	0.427/0.462	0.147/0.197	0.269	1.094/1.123	0.914/0.965	1.245

The illustrated approaches can provide the shift and full width at half maximum (FWHM) in a wide range of n_e and T. The comparison with different measurements for shift and FWHM as a function of electron density $n_e \approx (10^{22} - 10^{24})\mathrm{m}^{-3}$ in temperature range $T \approx (2 - 6) \times 10^4$ K are shown in Figures 4 and 5, respectively. The microfield distribution function of Hooper [51] is used in both the Born approximation and the T-matrix method for evaluating the Stark parameters in Figures 4 and 5 [19]. However, outside the validity range of Hooper's approach, we used the fit formula of Potekhin *et al.* [52] in our previous calculations in [19,53], which is based on Monte Carlo simulations and is appropriate for strongly coupled plasmas as well. The calculated FWHM in the T-matrix approach shows good agreement with the MD simulations data, especially at high densities. The ion dynamics slightly affects the width. However, the shift is rapidly reduced at low electron densities, and the discrepancy can be seen in contrast to MD simulations, but still the contribution of FFM is more pronounced in this region. The shift of this line is overestimated in the Born approximation compared to both the T-matrix method and MD simulations. The correlation between perturbers tends to decrease the line shift and width at high electron densities for interacting particles. In Figures 4 and 5 the comparison is made with the following experiments: The measurement by Pérez *et al.* [54] was done in a low-pressure pulsed arc, in the plasma density range of $n_e = (1 - 6) \times 10^{22}\,\mathrm{m}^{-3}$ and temperature interval of $T = (0.8 - 3) \times 10^4$ K with a mean value of 2×10^4 K . The error bar of n_e was $\pm 10\%$, and uncertainty in the temperature evaluation was about 20%. The experimental result of Kelleher [55] was obtained in a helium plasma generated in a wall-stabilized arc, with $n_e = 1.03 \times 10^{22}\,\mathrm{m}^{-3}$ and $T_e = 2.09 \times 10^4$ K. Recently, the FWHM of the same transition line was measured by Gao *et al.* [56] from a helium arc for the density range $n_e = (0.5 - 4) \times 10^{22}\,\mathrm{m}^{-3}$ at $T = 2.3 \times 10^4$ K. The reported result by Kobilarov *et al.* [57] from a pulsed low-pressure arc at $n_e = (2 - 10) \times 10^{22}\,\mathrm{m}^{-3}$ and $T = (3.1 - 4.2) \times 10^4$ K are included.

Furthermore, values for the shift measured by Morris and Cooper [58] within the density range $n_e = (0.6 - 2.3) \times 10^{22} \, \text{m}^{-3}$ and $T = (1 - 1.6) \times 10^4$ K are shown. The Stark parameters of this line were also measured by Berg *et al.* [59] at $n_e = 1.5 \times 10^{22} \, \text{m}^{-3}$ and $T_e = 2.6 \times 10^4$ K. The measured widths by Milosavljević and Djeniže [60] are included as well, by using a linear low-pressure pulsed arc. The measured electron density and temperature were in the ranges of $(4.4 - 8.2) \times 10^{22} \, \text{m}^{-3}$ and $(1.8 - 3.3) \times 10^4$ K with error bars $\pm 9\%$ and $\pm 10\%$, respectively. Furthermore, a comparison is made with the measurement of Soltwisch and Kusch [61], a wall-stabilized quasistationary pulsed discharge was used as a homogeneous plasma source. The spectra were recorded in a single shot, and the electron density range of $(0.7 - 1.2) \times 10^{23} \, \text{m}^{-3}$ was determined from the plasma reflectivity at two different wavelengths. The estimation of the temperature was about 2×10^4 K.

Figure 5. The FWHM of He I $3889 \, \text{Å}$ as a function of electron density. Our theoretical approaches are compared with the experiments results.

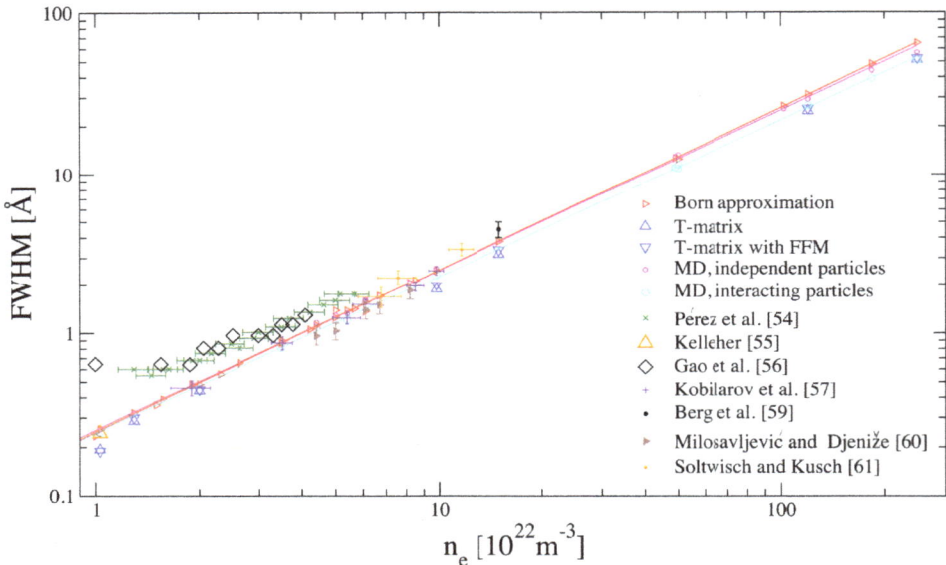

As shown in Figure 5, the T-matrix approach always gives a smaller FWHM than the Born approximation. This trend leads to a better agreement with the MD simulations results for interacting particles. For the data of Pérez *et al.* [54], the calculated width in the T-matrix approach agrees very well with the result of the Born approximation. However, the measurement may be influenced by self-absorption, as mentioned in [54], and then the measured widths may be overestimated. At $n_e = 9.8 \times 10^{22} \, \text{m}^3$, the width in the Born approximation shows a very good agreement with both measurement and MD simulations for independent particle data. However, the results of T-matrix approach and MD simulations for interacting particles give slightly lower values than the experiment data. At the highest measured density, the FWHM of theories is below that of Berg *et al.* [59]. Good agreement is found between the MD simulations result and T-matrix approach. However, the shift of this line is overestimated in the Born approximation, where, the strong collisions with large momentum transfer are not treated appropriately, while the perturbation theory breaks down at small distances and the perturbative expansion has to be avoided [20].

5. Conclusions

Spectral line shapes for He I are investigated theoretically in dense plasmas, the comparison is made with the measurements of Jovićević *et al.* [12] for diagnostic purposes. Free electron-emitter collisions are considered within both the Born approximation, and the T-matrix approach while ions are treated by the quasistatic approximation. The electronic contribution to the shift and width is computed by thermodynamic Green's function by using Born approximation, which is the main contribution to the line broadening. Dynamic screening of the electron-atom interaction is included, which is a collective, many-particle effect. In contrast to ST, this effect modifies the broadening parameters with increasing free electron density, causing a non-linear behavior, where the plasma oscillations become relevant, see Figures 4 and 5. A cut-off procedure for strong collisions is used according to Griem [1], while the second-order Born approximation overestimated the electronic contribution and the strong collisions term are added [23]. The Coulomb approximation is employed to evaluate the wavefunctions of helium. The accuracy of this approximation was approved for He-like ion by comparing our result with the Hartree-Fock wavefunctions, thus the radial part of transition matrix-element from various approaches were compared in [62].

Then the electron-emitter interaction is investigated again by elastic scattering amplitudes in Debye plasmas from a two-particle T-matrix approach by using the close-coupling equations. The Debye–Hückel potential significantly affects the bound states and scattering processes of the metastable states. The result is improved by treating strong electron-helium collisions consistently within T-matrix approach and better agreement can be seen with the measurement than the Born approximation, especially for the line shift. The contribution of perturbing ions is taken into account in a quasistatic approximation, with both quadratic Stark effect and quadrupole interaction. The perturbing ionic microfield is considered as a static field during the radiation. The calculated spectral line shapes in the static limit are modified by applying the FFM for both the Born approximation and T-matrix approach, which provides the dynamic line shapes and leads to further broadening [18]. In addition, MD simulations are used for comparison with the experiments and the analytically obtained profiles.

The shift of this line is over estimated in Born approximation even with a cut-off procedure and better agreement can be seen with the screened T-matrix approach and MD simulations. The FWHM of all theoretical approaches are in good agreement with the experimental result.

Acknowledgments

The authors thank the anonymous referees for their very helpful comments and suggestions. This work was supported by the German Research Foundation DFG within SFB 652. The support of the Australian Research Council, the Australian National Computational Infrastructure Facility, and its Western Australian node iVEC are gratefully acknowledged.

Author Contributions

B.O. did the quantum statistical calculations and analysis, M.Á.G and M.A.G did the computer simulations, T.S.R., M.C.J., and K.N.D. calculated the microfield distribution function, M.C.Z., D.V.F., and I.B. calculated the scattering amplitudes.

Appendix A: Spectral Line Profile

The Green's function methods provide a systematic approach to correlation functions in many-body system. The theory includes correlations and quantum effects of the many-body system. The line broadening is proportional to the imaginary part of the inverse dielectric function of the plasma. The two-particle Green's function is used to calculate the line shapes. Recently, further improvements of this approach have been made by Günter *et al.* [2], Hitzschke *et al.* [13], and Röpke *et al.* [63]. Optical properties of many-particle systems are described by the dielectric function $\varepsilon(\mathbf{q}, \omega)$ based on Green's function theory, which is the response of the medium to an external electromagnetic field. The dielectric tensor can be divided into transversal t and longitudinal l parts, depending on wavenumber $\mathbf{q}$ and frequency ω. $\varepsilon_l(\mathbf{q}, \omega)$ is related to the polarization function $\Pi(\mathbf{q}, \omega)$,

$$\varepsilon_l(\mathbf{q}, \omega) = 1 - V(q)\,\Pi(\mathbf{q}, \omega) \tag{20}$$

where $V(q) = e^2/(\epsilon_0 q^2)$ is the Fourier transformed Coulomb potential. Starting from the relation of the absorption coefficient $\alpha(\omega)$ and the index of refraction $n(\omega)$, the transverse dielectric function $\varepsilon_t(\mathbf{q}, \omega)$ in the long wavelength limit $q^{-1} = c/\omega \to 0$ reads

$$\lim_{\mathbf{q}\to 0} \varepsilon_t(\mathbf{q}, \omega) = [n(\omega) + \frac{\imath c}{2\omega}\alpha(\omega)]^2. \tag{21}$$

In the visible region where the wavelength λ is large compared with the atomic dimension a_B, the transverse and the longitudinal part of dielectric function coincide

$$\lim_{\mathbf{q}\to 0} \varepsilon_t(\mathbf{q}, \omega) = \lim_{\mathbf{q}\to 0} \varepsilon_l(\mathbf{q}, \omega) = \lim_{\mathbf{q}\to 0} \varepsilon(\mathbf{q}, \omega). \tag{22}$$

The absorption coefficient is proportional to the imaginary part of the dielectric function

$$\alpha(\omega) = \frac{\omega}{c\,n(\omega)} \lim_{\mathbf{q}\to 0} \operatorname{Im}\varepsilon(\mathbf{q}, \omega) \tag{23}$$

$$n(\omega) = \frac{1}{\sqrt{2}} \lim_{\mathbf{q}\to 0} \left\{ \operatorname{Re}\varepsilon(\mathbf{q}, \omega) + \left[(\operatorname{Re}\varepsilon(\mathbf{q}, \omega))^2 + (\operatorname{Im}\varepsilon(\mathbf{q}, \omega))^2 \right]^{1/2} \right\}^{1/2}.$$

In the case of thermal equilibrium, the absorption coefficient $\alpha(\omega)[\mathrm{cm}^{-1}]$ is related to the emission coefficient $j(\omega)[\mathrm{erg\,cm^{-3}\,s^{-1}\,Hz^{-1}\,ster^{-1}}]$ by Kirchhoff's law [64]. The emission coefficient is defined as the number of photons with frequency between ν and $(\nu + d\nu)$ emitted per unit time per unit volume per unit frequency interval

$$j(\omega) = \alpha(\omega)B(\omega, T) \tag{24}$$

$$B(\omega, T) = \frac{\hbar\omega^3}{4\pi^3 c^2} \left[\exp(\frac{\hbar\omega}{k_B T}) - 1 \right]^{-1} \tag{25}$$

where $B(\omega, T)$ is the black body radiation, defined in intensity units per steradian $[\mathrm{erg\,cm^{-2}\,s^{-1}\,Hz^{-1}\,ster^{-1}}]$. In an optically thin plasma, $j(\omega)$ can be approximated to the specific intensity (surface brightness) $I(\omega)$. The dielectric function itself can be determined from the polarization function according to Equation (20). The medium modifications of spectral line shapes can be described in a systematic way from the bound–bound two-particle polarization function $\Pi_2(\mathbf{q}, \omega)$ [2,13,25,65]. From Equations (20), (23) and (24) the line emission is given by

$$j(\omega) = \frac{\omega^4}{4\pi^3 c^3 n(\omega)} \left[\exp(\frac{\hbar\omega}{k_{\mathrm{B}}T}) - 1 \right]^{-1} \lim_{q\to 0} \mathrm{Im}\, \varepsilon(\mathbf{q}, \omega). \tag{26}$$

Assuming $n(\omega) = 1$ for $\omega \gg \omega_{\mathrm{pl}}$, and $q \to 0$

$$j(\omega) \approx \frac{\omega^4}{4\pi^3 c^3} \exp(-\frac{\hbar\omega}{k_{\mathrm{B}}T}) \lim_{q\to 0} \mathrm{Im}\, \Pi(\mathbf{q}, \omega). \tag{27}$$

The polarization function $\Pi_2(\mathbf{q}, \omega)$ is related to the dipole–dipole autocorrelation function, which describes the pressure broadening (Stark broadening) [66]. A systematic treatment of the two-particle polarization function can be performed via Green's function

$$\Pi_2(\mathbf{q}, \omega) = \sum_{if} |M_{if}^0(\mathbf{q})|^2 \frac{[g(\omega_i) - g(\omega_f)]}{\hbar\omega - \hbar\omega_{if} - \Delta_{if} - i\Gamma_{if} + i\Gamma_{if}^{\mathrm{v}}} \tag{28}$$

thus the imaginary part of this expression is substituted in Equation (26), which describes the spectral line shapes, see Equation (1). The two-particle Bose distribution function can be approximated by the Boltzmann distribution function for non-degenerated plasmas

$$g_{ei}(\omega) = \frac{1}{e^{\bar{\beta}(\omega - \mu_e - \mu_i)} - 1} \approx \frac{1}{4} n_{ei} \Lambda_{ei}^3 e^{-\bar{\beta}\omega} \tag{29}$$

where $\Lambda_{ei} = \sqrt{2\pi\bar{\beta}/m_i}$ is the thermal wavelength, m_i is the mass of radiator, and $\bar{\beta} = 1/(k_{\mathrm{B}}T)$. The pressure line broadening is achieved by substituting Equation (28) into Equation (27) and averaging over the microfiled distribution function, see Eq. (1). The Rydberg units ($\hbar = 2m_e = e^2/2 = 1$) are used throughout this paper.

Appendix B: Electronic Self-energy

The electronic self-energy Σ_2 has been calculated from the quantum-statistical many-particle approach, by using the two-particle Green's function G_2^0 for the radiating atom in the momentum $\mathbf{P} + \mathbf{q}$ and frequency representation $\Omega_\lambda - \omega_\mu$ [15], where Ω_λ and ω_μ are the bosonic the Matsubara frequency. The statistical information on temperature and chemical potential contains in the definition of the Matsubara frequency. The correction self-energy is proportional to the free electron density, given in dynamically screened Born approximation [67,68]. The self-energy is evaluated be solving this diagram

$$\Sigma_2(n, n', \mathbf{P}, \Omega_\lambda) = \frac{1}{-\imath\beta} \sum_{q,\mu,\alpha} M_{n\alpha}^0(\mathbf{q}) \, V^s(q, \omega_\mu) \, G_2(\alpha, \mathbf{P} + \mathbf{q}, \Omega_\lambda - \omega_\mu) M_{\alpha n'}^0(-\mathbf{q}) \qquad (30)$$

where $M_{n\alpha}^0(\mathbf{q})$ is the matrix-element for virtual transitions from the considered state to the excited state, with principal quantum number n, α, see Appendix C. The dynamically screened potential $V^s(q, \omega_\mu)$ for the radiator-electron interaction is represented by the spectral function of the dielectric function $\varepsilon(\mathbf{q}, \omega)$

$$V^s(q, \omega_\mu) = V(q) + V(q) \int_{-\infty}^{\infty} \frac{d\omega}{\pi\hbar} \frac{\mathrm{Im}\,\varepsilon^{-1}(\mathbf{q}, \omega + \imath 0)}{\omega - \omega_\mu}. \qquad (31)$$

The first term in $V^s(q, \omega_\mu)$ is neglected, which leads to the exchange Fock self-energy, whereas the second term is the correction part. The Matsubara summation $\sum_\mu$ can be represented as a contour integral in the complex plane [67]. After performing the Matsubara frequency summation, the real and imaginary part of the diagonal electronic self-energy can be obtained [15,63,69].

Appendix C: Matrix-Element

The transition matrix-element $M_{n\alpha}(\mathbf{q})$ describes the interaction of the atom with the Coulomb potential through the vertex function. In lowest order, they are determined by the atomic eigenfunctions ψ_n. For helium, the Coulomb interaction with an electron-electron-ion triplet depends on the momentum transfer $\hbar\mathbf{q}$, this can be represented for helium and hydrogen by these diagrams [2,25,68,70].

The matrix-element for hydrogen can be written as

$$M_{n\alpha}(\mathbf{q}) = \imath e \sum_{\mathbf{p}_e, \mathbf{p}_i} \psi_n^*(\mathbf{p}_e, \mathbf{p}_i) \left[Z_n \, \psi_\alpha(\mathbf{p}_e, \mathbf{p}_i + \mathbf{q}) - \psi_\alpha(\mathbf{p}_e + \mathbf{q}, \mathbf{p}_i) \right] \qquad (32)$$

However, the helium matrix-element can be approximated by hydrogen, while the outer electron is screened by the inner electron. Here, the states closely adjacent to n are identified by α, and Z_n is the ion charge. In an adiabatic approximation, the ion is much heavier than the electron $m_i \gg m_e$, therefore the relative momentum is approximated by the momentum of electron

$$\mathbf{p} = \frac{m_i \mathbf{p}_e - m_e \mathbf{p}_i}{m_e + m_i} \approx \mathbf{p}_e$$

$$M_{n\alpha}(\mathbf{q}) = \imath e \sum_{\mathbf{p}} \psi_n^*(\mathbf{p}) \left[Z_n \, e \, \psi_\alpha(\mathbf{p}) - e \, \psi_\alpha(\mathbf{p} + \mathbf{q}) \right]. \tag{33}$$

After Fourier transformation of the above relation, and expanding the exponential term into spherical harmonics,

$$M_{n\alpha}(\mathbf{q}) = \imath e \left[Z_n \, \delta_{n\alpha} - \int \mathrm{d}^3 r \, \psi_n^*(\mathbf{r}) \, \mathrm{e}^{\imath \mathbf{q} \cdot \mathbf{r}} \, \psi_\alpha(\mathbf{r}) \right] \tag{34}$$

$$\mathrm{e}^{\imath \mathbf{q} \cdot \mathbf{r}} = 4\pi \sum_{l=0}^{\infty} \sum_{m=-l}^{l} \imath^l j_l(qr) Y_{lm}^*(\Omega_q) Y_{lm}(\Omega_r)$$

where $j_l(qr)$ is the spherical Bessel function [71], a multipole expansion can be derived, e.g., $l = 0, 1, 2$ give the monopole, dipole and quadrupole contribution of the radiator-electron interaction, respectively. In the long wavelength limit $q \to 0$ for bound–bound transitions, the dipole approximation is valid, where the q-integral is dominated by the small q region [20].

Conflicts of Interest

The authors declare no conflict of interest.

References

1. Griem, H.R. *Spectral Line Broadening by Plasmas*; Academic Press: New York, NY, USA, 1974.
2. Günter, S.; Hitzschke, L.; Röpke, G. Hydrogen spectral lines with the inclusion of dense-plasma effects. *Phys. Rev. A* **1991**, *44*, 6834–6844.
3. Frisch, U.; Brissaud, A. Theory of Stark broadening soluble scalar model as a test. *J. Quant. Spectrosc. Radiat. Transf.* **1971**, *11*, 1753–1766.
4. Seidel, J. Hydrogen Stark Broadening by Model Electronic Microfields. *Z. Naturforschung A* **1977**, *32a*, 1195–1206.
5. Seidel, J. Effects of Ion Motion on Hydrogen Stark Profiles. *Z. Naturforschung A* **1977**, *32a*, 1207–1214.
6. Stehlé, C.; Hutcheon, R. Extensive tabulations of Stark broadened hydrogen line profiles. *Astron. Astrophys. Suppl. Ser.* **1999**, *140*, 93–97.
7. Talin, B.; Calisti, A.; Godbert, L.; Stamm, R.; Lee, R.W.; Klein, L. Frequency-fluctuation model for line-shape calculations in plasma spectroscopy. *Phys. Rev. A* **1995**, *51*, 1918–1928.
8. Stamm, R.; Smith, E.W.; Talin, B. Study of hydrogen Stark profiles by means of computer simulation. *Phys. Rev. A* **1984**, *30*, 2039–2046.

9. Gigosos, M.A.; Cardeñoso, V. New plasma diagnosis tables of hydrogen Stark broadening including ion-dynamics. *J. Phys. B Atom. Mol. Opt. Phys.* **1996**, *29*, 4795–4838.

10. Gigosos, M.A.; González, M.A. Stark broadening tables for the helium I 447.1 line. Application to weakly coupled plasmas diagnostics. *Astron. Astrophys.* **2009**, *503*, 293–299.

11. Gigosos, M.A.; Djurovic, S.; Savic, I.; GonzÃ̧lez-Herrero, D.; Mijatovic, Z.; Kobilarov, R. Stark broadening of lines from transition between states *n* = 3 to *n* = 2 in neutral helium. An experimental and computer-simulation study. *Astron. Astrophys.* **2014**, *561*, A135:1–A135:13.

12. Jovićević, S.; Ivković, M.; Zikic, R.; Konjević, N. On the Stark broadening of Ne I lines and static *vs.* ion impact approximation. *J. Phys. B Atom. Mol. Opt. Phys.* **2005**, *38*, 1249–1259.

13. Hitzschke, L.; Röpke, G.; Seifert, T.; Zimmermann, R. Green's function approach to the electron shift and broadening of spectral lines in non-ideal plasmas. *J. Phys. B Atom. Mol. Opt. Phys.* **1986**, *19*, 2443–2456.

14. Günter, S. *Optische Eigenschaften Dichter Plasmen.* Habilitation thesis, Rostock University: Rostock, Germany, 1995.

15. Sorge, S.; Wierling, A.; Röpke, G.; Theobald, W.; Sauerbrey, R.; Wilhein, T. Diagnostics of a laser-induced dense plasma by Hydrogen-like Carbon spectra. *J. Phys. B Atom. Mol. Opt. Phys.* **2000**, *33*, 2983–3000.

16. Omar, B.; Günter, S.; Wierling, A.; Röpke, G. Neutral helium spectral lines in dense plasmas. *Phys. Rev. E* **2006**, *73*, 056405:1–056405:13.

17. Günter, S.; Könies, A. Shifts and asymmetry parameters of hydrogen Balmer lines in dense plasmas. *Phys. Rev. E* **1997**, *55*, 907–911.

18. Calisti, A.; Mossé, C.; Ferri, S.; Talin, B.; Rosmej, F.; Bureyeva, L.A.; Lisitsa, V.S. Dynamic Stark broadening as the Dicke narrowing effect. *Phys. Rev. E* **2010**, *81*, 016406:1–016406:6.

19. Lorenzen, S.; Omar, B.; Zammit, M.C.; Fursa, D. V.; Bray I. Plasma pressure broadening for few-electron emitters including strong electron collisions within a quantum-statistical theory. *Phys. Rev. E* **2014** , *89*, 023106:1–023106:13.

20. Hitzschke, L.; Günter, S. The influence of many-particle effects beyond the Debye approximation on spectral line shapes in dense plasmas. *J. Quant. Spectrosc. Radiat. Transf.* **1996**, *56*, 423–441.

21. Beauchamp, A.; Wesemael, F.; Bergeron, P. Spectroscopicstudies of tudies DB white dwarfs: Improved Stark profiles for optical transitions of neutral helium. *Astrophys. J. Suppl. Ser.* **1997**, *108*, 559–573.

22. Bates D. R.; Damgaard, A. The calculation of the absolute strengths of spectral lines. *Phil. Trans. Roy. Soc. Lond. A* **1949**, *242*, 101–122.

23. Griem, H.R.; Baranger, M.; Kolb, A.C.; Oertel, G. Stark broadening of neutral helium lines in a plasma. *Phys. Rev.* **1962**, *125*, 177–195.

24. Halenka, J. Asymmetry of hydrogen lines in plasmas utilizing a statistical description of ion-quadruple interaction in Mozer-Baranger limit. *Z. Phys. D* **1990**, *16*, 1–8.

25. Kraeft, W.-D.; Kremp, D.; Ebeling, W.; Röpke, G. *Quantum Statistics of Charged Particle Systems*; Akademie-Verlag: Berlin, Germany, 1986.

26. Günter, S. Contributions of strong collisions in the theory of spectral lines. *Phys. Rev. E* **1993**, *48*, 500–505.

27. Baranger, M. General impact theory of pressure broadening. *Phys. Rev.* **1958**, *112*, 855–865.

28. Bray, I.; Stelbovics, A.T. Convergent close-coupling calculations of electron-hydrogen scattering. *Phys. Rev. A* **1992**, *46*, 6995–7011.

29. Fursa, D.V.; Dmitry V.; Bray, I. Calculation of electron-helium scattering. *Phys. Rev. A* **1995**, *52*, 1279–1297.

30. Bray, I.; Fursa, D.V.; Kadyrov, A.S.; Stelbovics, A.T.; Kheifets, A.S.; Mukhamedzhanov, A.M. Electron- and photon-impact atomic ionisation. *Phys. Rep.* **2012**, *520*, 135–174.

31. Debye, P.; Hückel, E. *Phys. Zeitschrift* **1923**, *24*, 185–206.

32. Kardar, M. *Statistical Physics of Particles;* Cambridge University Press, Cambridge, UK, 2007. pp. 268–273.

33. Zammit, M.C.; Fursa, D.V.; Bray, I. Convergent-close-coupling calculations for excitation and ionization processes of electron-hydrogen collisions in Debye plasmas. *Phys. Rev. A* **2010**, *82*, 052705:1–052705:7.

34. Zammit, M.C.; Fursa, D.V.; Bray, I.; Janev, R.K. Electron-helium scattering in Debye plasmas. *Phys. Rev. E* **2011**, *84*, 052705:1–052705:15.

35. Bray, I.; Fursa, D.V. Benchmark cross sections for electron-impact total single ionization of helium. *J. Phys. B Atom. Mol. Opt. Phys.* **2011**, *44* 6, 061001:1–061001:3.

36. Griaznov V.K. *Thermophysical Properties of the Working Gas of the Gas-Phase Nuclear Reactors*; Atomizdat: Moscow, Russia, 1980.

37. Kudrin, L.P. *Statistical Plasma Physics*; Atomizdat: Moscow, Russia, 1974.

38. Ecker, G.; Müller K.G. Plasmapolarisation und Trägerwechselwirkung. *Z. Phys.* **1958**, *153*, 317–330.

39. Iglesias, C.A.; Hooper, C.F. Quantum corrections to the low-frequency-component microfield distributions. *Phys. Rev. A* **1982**, *25*, 1049–1059.

40. Iglesias, C.A. Integral-equation method for electric microfield distribution. *Phys. Rev. A* **1983**, *27*, 2705–2709.

41. Ramazanov, T.S.; Dzhumagulova, K.N. Effective screened potentials of strongly coupled semiclassical plasma. *Phys. Plasmas* **2002**, *9*, 3758–3761.

42. Anderson, P.W. Pressure broadening in the microwave and infra-red regions. *Phys. Rev.* **1949**, *76*, 647–661.

43. Gigosos, M.A.; González, M.A.; Cardeñoso, V. Computersimulated Balmer-alpha, -beta and -gamma Stark line profiles for non-equilibrium plasmas diagnostics. *Spectrochim. Acta Part B* **2003**, *58*, 1489–1504.

44. Seidel, J.; Stamm, R. Effects of radiator motion on plasma-broadened hydrogen Lyman-β. *J. Quant. Spectrosc. Radiat. Transfer* **1982**, *27*, 499–503.

45. González, M.A.; Gigosos M.A. Analysis of Stark line profiles for non-equilibrium plasma diagnosis. *Plasma Sourc. Sci. Technol.* **2009**, *18(3)*, 034001:1–034001:5.

46. Bruggeman, P.; Schram, D.; González, M.A.; Rego, R.; Kong, M.G.; Leys, C. Characterization of a direct dc-excited discharge in water by optical emission spectroscopy. *Plasma Sourc. Sci. Technol.* **2009**, *18(2)*, 025017:1–025017:13.

47. Xiong, Q.; Nikiforov; A.Y.; González, M.A.; Leys, C.; Lu, X.P. Characterization of an atmospheric helium plasma jet by relative and absolute optical emission spectroscopy. *Plasma Sourc. Sci. Technol.* **2013**, *22(1)*, 015011:1–025017:13.

48. Dufour, E.; Calisti, A.; Talin, B.; Gigosos, A.M.; González, M.A.; Dufty, J.W. Charge correlation effects in electron broadening of ion emitters in hot and dense plasmas. *J. Quant. Spectros. Radiat. Transf.* **2003**, *81*, 125–132.

49. Dufour, E.; Calisti, A.; Talin, B., Gigosos, M.A.; González, M.A.; del Rio Gaztelurrutia, T.; Dufty, J.W. Charge-charge coupling effects on dipole emitter relaxation within a classical electron-ion plasma description. *Phys. Rev. E* **2005**, *71*, 066409:1–066409:9.

50. Konjević, N.; Ivković, M.; Jovićević, S. Spectroscopic diagnostics of laser-induced plasmas. *Spectrochim. Acta Part B: Atom. Spectrosc.* **2010**, *65*, 593–602.

51. Hooper, C.F., Jr. Low-frequency component electric microfield distributions in plasmas. *Phys. Rev.* **1968**, *165*, 215–222.

52. Potekhin, A.Y.; Chabrier, G.; Gilles, D. Electric microfield distributions in electron-ion plasmas. *Phys. Rev. E* **2002**, *65*, 036412:1–036412:12.

53. Omar, B.; Wierling, A.; Günter, S.; Röpke, G. Analysing brilliance spectra of a laser-induced carbon Plasma. *J. Phys. A: Math. Gen.* **2006**, *39*, 4731–4737.

54. Pérez, C; Santamarta, R.; de la Rosa, M.I.; Mar, S. Stark broadening of neutral helium lines and spectroscopic diagnostics of pulsed helium plasma. *Eur. Phys. J. D* **2003**, *27*, 73–75.

55. Kelleher, D.E. Stark broadening of visible neutral helium lines in a plasma. *J. Quantit. Spectrosc. Radiat. Transf.* **1981**, *25*, 191–220.

56. Gao, H.M.; Ma, S.L.; Xu, C.M.; Wu, L. Measurements of electron density and Stark width of neutral helium lines in a helium arc plasma. *Eur. Phys. J. D* **2008**, *47*, 191–196.

57. Kobilarov, R.; Konjević, N.; Popović, M.V. Influence of ion dynamics on the width and shift of isolated He I lines in plasmas. *Phys. Rev. A* **1989**, *40*, 3871–3879.

58. Morris, R.N.; Cooper, J. Stark Shifts of He I 3889, He I 4713, and He I 5016. *Can. J. Phys.* **1973**, *51*, 1746–1751.

59. Berg, H.F.; Ali, A.W.; Lincke, R.; Griem, H.R. Measurement of stark profiles of neutral and ionized helium and hydrogen lines from shock-heated plasmas in electromagnetic T tubes. *Phys. Rev.* **1962**, *125*, 199–206.

60. Milosavljević, V.; Djeniže, S. Ion contribution to the astrophysical important 388.86, 471.32 and 501.56 nm He I spectral lines broadening. *New Astron.* **2002**, *7*, 543–551.

61. Soltwisch, H.; Kusch, H.J. Experimental Stark profile Determination of some Plasma Broadened He I- and He II Lines. *Z. Naturforsch. A* **1979**, *34a*, 300–309.

62. Omar, B.; Wierling, A.; Reinholz, H.; Röpke, G. Diagnostic of laser induced Li II plasma. *Phys. Rev. Res. Int.* **2013**, *3*, 218–227.

63. Röpke, G.; Seifert, T.; Kilimann, K. A Green's function approach to the shift of spectral lines in dense. *Ann. Phys. (Leipzig)* **1981**, *38*, 381–460.

64. Kobzev, G.A.; Iakubov, I.T.; Popovich, M.M. *Transport and Optical Properties of Nonideal Plasmas*; Plenum Press: New York, NY, USA & London, UK, 1995.

65. Redmer, R. Physical properties of dense, low-temperature plasmas. *Phys. Rep.* **1997**, *282*, 35–157.

66. Günter, S. Stark shift and broadening of hydrogen spectral lines. *Contrib. Plasma Phys.* **1989**, *29*, 479–487.

67. Kadanoff L.P.; Baym, G. *Quantum Statistical Mechanics: Green's Function Methods in Equilibrium and Nonequilibrium Problems*; Addison-Wesley Publishing Co., Inc.: New York, NY, USA, 1989.

68. Zimmermann, R. *Many Particle Theory of Highly Excited Semiconductors*; BSB Teubner: Leipzig, Germany, 1987.

69. Zimmermann, R.; Kilimann, K.; Kraeft, W.-D.; Kremp D.; Röpke, G. Dynamical screening and self energy of excitons in the electron-hole plasma. *Phys. Stat. Sol. (B)* **1978**, *90*, 175–187.

70. Omar, B. Spectral line broadening in dense plasmas. *J. Atom. Mol. Opt. Phys.* **2011**, *2011*, 850807:1–850807:8.

71. Edmonds, A.R. *Angular Momentum in Quantum Mechanics*; Princeton University Press: Princeton, NJ, USA, 1960.

Reprinted from *Atoms*. Cite as: Lisitsa, V.S.; Kadomtsev, M.B.; Kotov, V.; Neverov, V.S.; Shurygin, V.A. Hydrogen Spectral Line Shape Formation in the SOL of Fusion Reactor Plasmas. *Atoms* **2014**, *2*, 195-206.

Review

Hydrogen Spectral Line Shape Formation in the SOL of Fusion Reactor Plasmas

Valery S. Lisitsa [1,2,]*, **Mikhail B. Kadomtsev** [1], **Vladislav Kotov** [3], **Vladislav S. Neverov** [1] and **Vladimir A. Shurygin** [1]

[1] National Research Center "Kurchatov Institute", P.O. Box 3402, Ploschad akademika Kurchatova 1, Moscow 123182, Russia; E-Mails: mkadomtsev@mail.ru (M.B.K.); vs-never@hotmail.com (V.S.N.); va-sh@yandex.ru (V.A.S.)

[2] Moscow Institute of Physics and Technology, Dolgoprudny, Moscow Region 141700, Russia

[3] Forschungszentrum Jülich GmbH, IEK-4-Plasma Physics, Jülich 52425, Germany; E-Mail: v.kotov@fz-juelich.de

* Author to whom correspondence should be addressed; E-Mail: vlisitsa@yandex.ru; Tel.: +7-499-196-73-34; Fax: +7-095-943-00-73.

Received: 28 March 2014; in revised form: 6 May 2014 / Accepted: 6 May 2014 / Published: 15 May 2014

Abstract: The problems related to the spectral line-shape formation in the scrape of layer (SOL) in fusion reactor plasma for typical observation chords are considered. The SOL plasma is characterized by the relatively low electron density (10^{12}–10^{13} cm^{-3}) and high temperature (from 10 eV up to 1 keV). The main effects responsible for the line-shape formation in the SOL are Doppler and Zeeman effects. The main problem is a correct modeling of the neutral atom velocity distribution function (VDF). The VDF is determined by a number of atomic processes, namely: molecular dissociation, ionization and charge exchange of neutral atoms on plasma ions, electron excitation accompanied by the charge exchange from atomic excited states, and atom reflection from the wall. All the processes take place step by step during atom motion from the wall to the plasma core. In practice, the largest contribution to the neutral atom radiation emission comes from a thin layer near the wall with typical size 10–20 cm, which is small as compared with the minor radius of modern devices including international test experimental reactor ITER (radius 2 m). The important problem is a strongly non-uniform distribution of plasma parameters (electron and ion densities and temperatures). The distributions vary for different observation chords and ITER operation regimes. In the present report, most attention is paid to the problem of the VDF calculations. The most correct method for solving the problem is an application of the Monte Carlo method for atom motion near the wall. However, the method is sometimes too complicated to be combined with other numerical codes for plasma modeling for various regimes of fusion reactor operation. Thus, it is important to develop simpler methods for neutral atom VDF in space coordinates and velocities. The efficiency of such methods has to be tested via a comparison with the Monte Carlo codes for particular plasma conditions. Here a new simplified method for description of neutral atoms penetration into plasma is suggested. The method is based on the ballistic motion of neutrals along the line-of-sight (LoS) in the forward–back approximation. As a result, two-dimensional distribution functions, dependent on the LoS coordinate and the velocity projection on the LoS, and responsible for the Doppler broadening of the line shape, are calculated. A comparison of the method with Monte Carlo calculations allows the evaluation of the accuracy of the ballistic model. The Balmer spectral line shapes are calculated for specific LoS typical for ITER diagnostics.

Keywords: tokamak; fusion reactor plasmas; Balmer spectral lines shapes; neutral atoms; velocity distribution function

1. Introduction

The penetration of neutral atoms into plasmas from the wall of thermonuclear devices is important for interpretation of the luminosity of the edge plasmas. This is essential for calculation of both the plasma radiation losses and the diagnostics of basic plasma components (*i.e.*, hydrogen isotopes) and impurity ions. The luminosity of the Balmer spectral lines of hydrogen isotopes is the major experimental method to measure the isotopic composition of plasmas, which is an important parameter for the International Test Experimental Reactor (ITER). One of the main problems here is the relationship between the luminosity of Balmer spectral lines and the neutral flux from the wall. Besides the neutral atom spatial distribution, another important parameter of plasma is the velocity distribution function (VDF) of neutral atoms. The latter determines the spectral line shapes of hydrogen isotopes, including the deuterium and tritium. The observed spectral line shapes of hydrogen isotopes are the integral characteristics which are determined by the integrals of the line shapes along the observation line, weighted with their local luminosities. Thus, we will consider calculations of the 2D distributions of neutral atoms in space coordinates and velocities as well as their intensities at certain observation lines in the edge plasma.

We performed the numerical calculation for certain observation chords as well as certain scenarios of discharge in ITER. We used the data from the Monte-Carlo simulation with the EIRENE code [1] stand-alone simulations of neutral deuterium VDF, applied on the SOL&divertor plasma background calculated by the SOLPS4.3 (B2-EIRENE) code [1–4]. These calculations, however, are time consuming. At the same time we need such calculations to be performed on a massive scale for various observation chords and scenarios in ITER. Therefore, the development of a simple and reliable computational model is of practical interest. Here we develop a kinetic model for calculation of neutral atom penetration into plasmas. This model is both accurate enough and fast for computation of characteristics of neutral atom flux from the wall. Unlike the complicated computational models, it enables one to retrace explicitly the impact of various physics processes upon formation of the spectral line shapes.

2. Description of Ballistic Model for Penetration of Neutral Atoms into Fusion Plasmas

The main sources of neutral atoms which come from the wall are the molecular dissociation and the recombination of ions on the wall. The slow atoms are formed by the dissociation of both the neutral molecules and the ionized molecular ions. The steady state neutral flux is the bases for consideration below which is available for steady operation in ITER. The effect of ITER strong magnetic field is ignored for neutral particle motion whereas for charged particles it results in limitation of charge particle motion across the magnetic field. The atoms which are formed in such a way produce the first-generation flux. These atoms become a source of atoms of the next generation due to their charge exchange on the plasma ions.

The present calculations are based on a ballistic model (BM) of penetration of neutrals into plasmas [5,6]. Since the neutral atoms are localized mainly in a thin layer in the edge plasma with the typical width ~10 cm, which is small as compared to the minor radius of plasma in modern tokamaks, as well as in ITER, we can use a 1D-model of neutral's motion (in the direction along the LoS perpendicular to the wall). In the frame of such an approximation the ballistic model for deuterium plasmas reduces to the following model:

The initial object of the theory is the ballistic flux of D_2 molecules which start from the wall and penetrate into the plasma with velocities determined by the wall temperature.

- Their collisions with electron result in the molecular dissociation and ionization. It is necessary to take into account the heating of molecular ions D_2^+ due to their elastic Coulomb collisions with plasma ions (the heating of neutral molecules is essentially smaller and is not taken into account); atoms created in dissociation processes move in two directions—into plasma and out the wall where they elastically reflected back;
- The other part of neutral atoms created due to ions recombination at the wall is also taken into account; so the total initial flux is formed by the dissociated and recombined atoms;
- The initial flux decreases due to ionization and charge exchange on plasma ions; the last process results in generation of secondary atomic flux with a temperature corresponding to the local one at the charge exchange point;
- The resulting atoms of the second generation are moving along ballistic trajectories together with those of the first generation. They undergo the ionization and the charge exchange that results in an attenuation of the flux and production of the next-generation atoms.

So the total neutral density is the sum of the all neutral atom generations. Calculations demonstrate a fast convergence due to a decrease of neutral fluxes at every step of fluxes generation.

The typical equations for connection of the next generation of neutrals (N_2) with the previous one (N_1) takes the form:

$$N_2(x, v_{D2} > 0) =$$

$$\frac{1}{v_{D2}} \left[\int_{-\infty}^{\infty} dv_{D1} \int_{0}^{x} dN_1(v_{D1}, y) P\left[v_{D2}, v_{D1,}y\right] \exp(-\phi(x,y)/v_{D2}) \right]$$

$$dN_1(v_{D1}, y) = N_e \left\langle \sigma_{cx} v_i \right\rangle N_1(v_{D1}, y) dy \tag{1}$$

$$\varphi(x, y) = \int_{y}^{x} dz N_e \left[\left\langle \sigma_{cx} v_i \right\rangle + \left\langle \sigma_{al} v_e \right\rangle\right]$$

Here x is the coordinate along line of sight, v_{D2} is a velocity of deuterium neutral atoms arising from atoms with velocity v_{D1}, $P(v_{D2}, v_{D1}, y)$ is velocity distribution function in a specific space point y with account for charge exchange, $dN_1(v_{D1}, y)$ is a source of neutral atom first generation, connected with the second one by the charge exchange processes with cross section σ_{cx} and ion thermal velocity v_i on ions with density equal to electron density N_e, $\varphi(x, y)$ is the propagation function describing the propagation of neutrals along ballistic trajectories from the point y to the point x with account of their charge exchange and electron ionization (with the rate $\left\langle \sigma_{al} v_e \right\rangle$) processes.

The distribution function $P(v_{D2}, v_{D1}, y)$ is obtained from a solution of kinetic equation averaged over components of velocities perpendicular to the line of sight (LoS). The problem is that charge exchange rate depends on all three components of colliding particles velocities whereas the ballistic model is based on one dimensional motion along LoS. Such averaging shows that the distribution function can be expressed in terms of velocity projections with small corrections of the distribution function as compared with a Maxwellian one.

We perform the successive calculation of neutral atom density along the ballistic trajectories of the atoms. Note that the atoms of the former generation are considered to be the sources for atoms of the next generation. Such a procedure leads to a successive reduction of neutral atom density on every step, so that only a few iterations are sufficient to obtain a good result. The results obtained with the ballistic model are compared with those simulated by the EIRENE code [1] stand-alone simulations of neutral deuterium VDF, applied on the SOL&divertor plasma background calculated by the SOLPS4.3 (B2-EIRENE) code [1–4] for certain observation chords. This enables us to draw a conclusion on the accuracy of the ballistic model.

To meet the requirements for the accuracy of the Balmer spectral lines diagnostic of the edge plasma parameters, it is important to take into account the population of the excited atomic states.

The electron excitation of atoms in the atomic ground state obviously does not change the ballistic velocity distribution function of neutrals. In contrast, the charge exchange from the excited atoms on the plasma ions results in the production of the excited neutrals with the local Maxwellian velocity distribution function.

Indeed, the cross section of the charge exchange of an atom in the excited atomic state n is proportional to n^4, so it increases sharply with increasing n. Therefore, the radiation emission of the Balmer lines goes via two major channels: the first one is due to population of the emitting atomic level by the charge exchange in the ground state of neutrals and the second one is due to the resonance charge-exchange in the excited states. The difference between these two channels, and the corresponding luminosities of Balmer spectral lines, is due to the difference between the velocity distribution function of the Maxwellian generation of the neutrals and that of the ballistic one. The distribution function of the initial ballistic generation is essentially nonlocal and determined mainly by the ballistic transport of neutrals from one point to another. The distribution function of the Maxwellian generation is a local one and determined by the Maxwellian distribution in a given spatial point that is caused by the high rate of the charge exchange in the excited atomic states (an instant transfer of excitation).

3. The Typical Spatial Distribution of Plasma Parameters along Observation Chords in ITER

General view of some diagnostic chord in ITER is shown in Figure 1 (*cf.* [4,7]). The coordinate x on Figure 2 and below are marked off the wall (the point $x = 0$ is the position of the wall).

Figure 1. General view of some diagnostic chords of observation of the scrape of layer (SOL) in International Test Experimental Reactor (ITER) with designations of high field side (HFS) and low field side (LFS) domains [4,7].

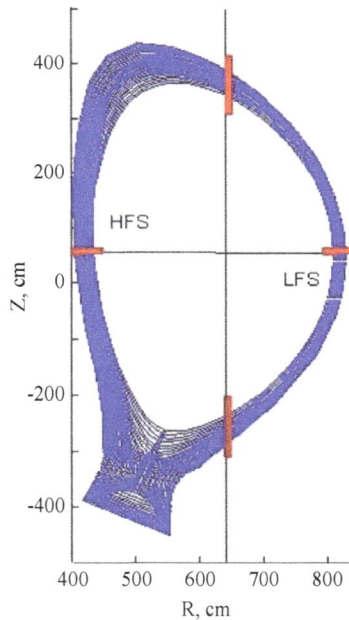

The distributions of temperature and density along the horizontal chord is given in Figure 2 on the basis of simulations with the codes [4,7].

4. The Numerical Modeling Results

We tested the ballistic model for the following plasma parameters: For systematical test of ballistic model there were taken the plasma parameters along the outer part of horizontal chord for the ITER SOL&divertor case #1514 (*cf.* [3]) with density and temperature profiles presented above in Figure 2.

We performed the following calculations in the frame of developed ballistic model:

(1) radial space distribution of molecules and molecular ions;
(2) radial space distributions of initial and secondary neutral atomic densities;
(3) radial space distribution of atomic velocity distribution function.

Figure 2. Electron and ion temperature (**left**) and electron (and ion) density (**right**) along the horizontal observation chord according to the data from simulations with the codes [4].

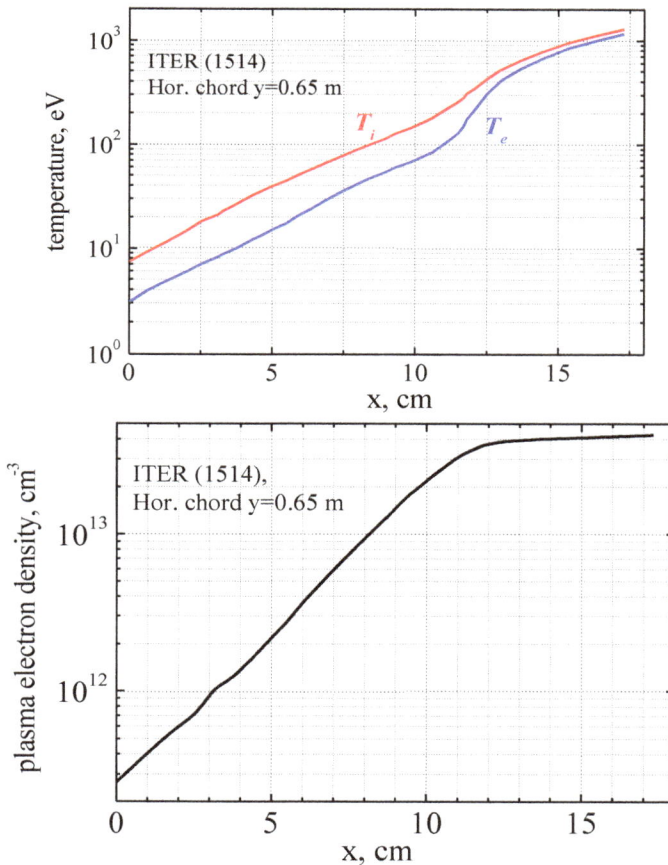

The initial source of fluxes in the ballistic model are molecules and molecular ions of deuterium. A comparison of these calculations with those performed with Monte-Carlo EIRENE code is shown in Figure 3. One can see a good agreement between the functions of molecular sources computed with foregoing two methods. This encourages us towards making further comparisons of the data computed with developed ballistic model and the EIRENE code numerical simulations.

Figure 3. Comparison of space profiles of molecules and molecular ions of deuterium for the outer part of horizontal chord. Solid line—ballistic model, histogram (stepped line)—EIRENE code simulations.

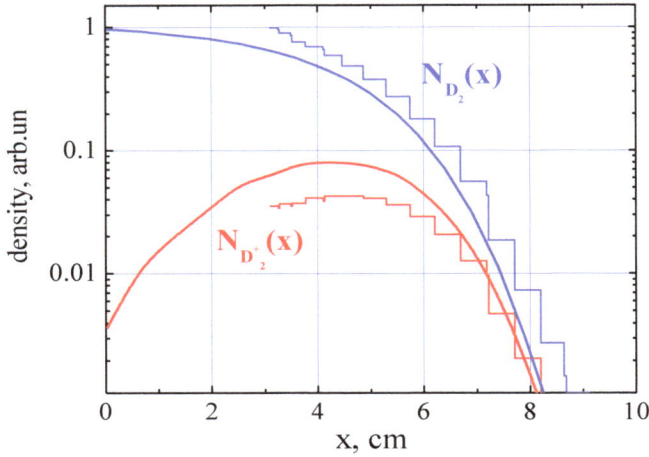

Figure 4 shows the impact of the initial neutral flux from the wall N_1 and the impacts of fluxes of various generations of neutrals in the ground state N_2, N_3, N_4. One can see a good agreement between the Monte-Carlo EIRENE code simulations data and those from the ballistic model.

Figure 4. Comparison of space profiles of molecules and molecular ions of deuterium for the outer part of horizontal chord. Solid line—ballistic model, histogram (stepped line)—EIRENE code simulations.

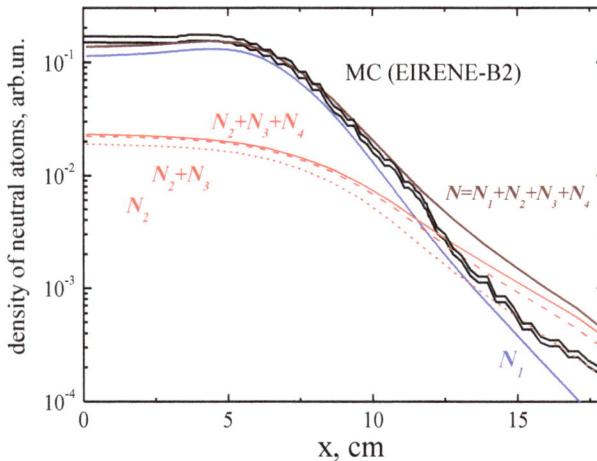

Figures 5 and 6 show the velocity distribution function of neutral deuterium atoms along the distance from the wall computed with ballistic model as compared with the EIRENE code simulations data. One can see a good agreement between two models both at relatively small and large distances.

Figure 5. Atomic velocity distribution function at different distances from the wall at the outer horizontal chord: thick solid lines—ballistic model, thin lines—Monte-Carlo simulations with the EIRENE code.

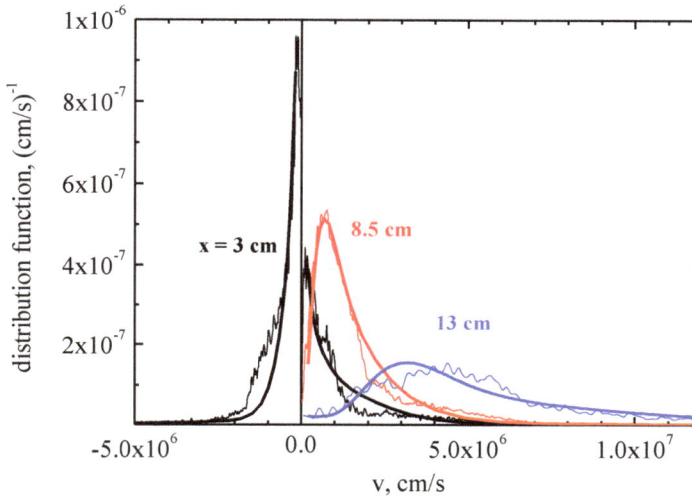

The velocity distribution function of neutrals in the edge plasma calculated in the frame of ballistic model is of importance for fast line shape calculations along lines of sights needed for spectral line diagnostics of neutrals in ITER.

Figure 6. Atomic velocity distribution function at large distances from the wall at the outer horizontal chord: thick solid lines-ballistic model, thin lines-MC modeling with the EIRENE code.

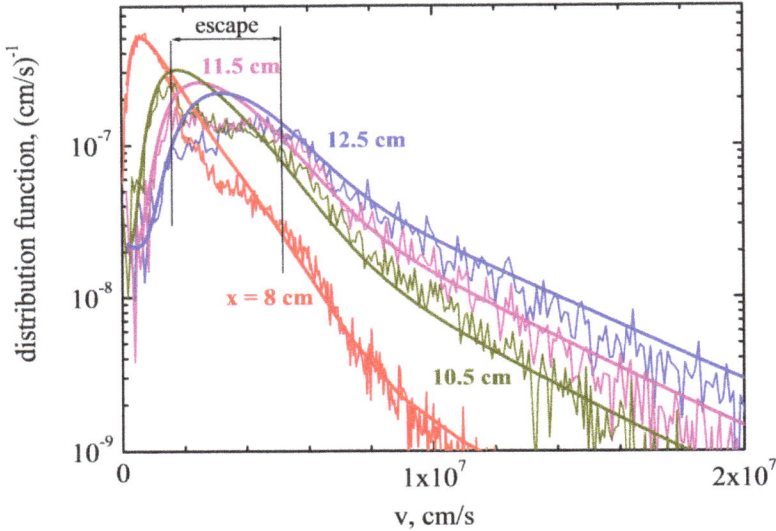

5. The Charge Exchange from the Excited Atomic States

The calculation scheme of the charge-exchange impact may be reduced as follows. The initial source of atoms population is the excitation of ballistic neutrals of the first generation by an electron impact. In order to take into account the population of atomic levels of these neutrals by the charge exchange we add the charge exchange contribution to the ionization rates that allows for electron transfer from a given excited atomic level to the atomic levels of a plasma ion. We can neglect the inverse process since the neutral density is sufficiently smaller than that of the ambient plasma ions. Then the population of the atomic levels of the first-generation atoms with account of the charge exchange is decreased as compared with those without the charge exchange.

$$N_{BAL}^{cx} = N_{BAL}\left(1 - K\right) \qquad (2)$$

where K takes into account a decrease of the population of atomic levels of ballistic (non-Maxwellian) atoms due to the charge exchange.

The local populations of atoms, which arise from the charge exchange of the ballistic atoms on ions are calculated in the frame of the radiative-collisional model. The source of atomic levels population is the charge exchange from the levels of ballistic atoms. Thus, the local population takes the form:

$$N_{LOC}^{cx}\left(n\right) = N_{BAL}\left(n\right)K \qquad (3)$$

The sum of populations (2) and (3) remains constant and equals to the population without charge exchange in the excited states because it is determined only by the processes of collision with electrons and radiative decay. The charge exchange process results only in a redistribution of populations between the ballistic (non-Maxwellian) atoms and the "local" (Maxwellian) ones.

In the calculations of radiation spectra it should be remembered that the ballistic emitting atoms and the plasma ones have different velocity distribution functions. The velocity distribution function of the ballistic atoms is determined by their motion from the wall, while that of the "local" atoms is determined by ion temperature in a given point.

The population redistribution coefficient $K(n, T_e, N_e)$ depends on the principal quantum number of the atomic level as well as on the temperature and density of the ambient plasma. The values of K for the atomic levels with principal quantum numbers $n = 3$ and $n = 4$ for typical plasma conditions in the SOL in ITER are presented in Tables 1 and 2.

Table 1. The population redistribution coefficient $K(n, T_e, N_e)$ for the hydrogen atomic level $n = 3$ for various plasma temperatures and densities.

T_e, eV\N_e, m^{-3}	10^{18}	3×10^{18}	10^{19}	10^{20}	10^{21}
20	0.04163	0.08648	0.1675	0.319	0.4185
50	0.0442	0.09318	0.1862	0.3746	0.491
100	0.04588	0.09774	0.1991	0.4185	0.5477
200	0.0473	0.1018	0.2106	0.4622	0.6036

Table 2. The population redistribution coefficient $K(n, T_e, N_e)$ for the hydrogen atomic level $n = 4$ for various plasma temperatures and densities.

T_e, eV\N_e, m^{-3}	10^{18}	3×10^{18}	10^{19}	10^{20}	10^{21}
20	0.1861	0.2824	0.3735	0.4972	0.5723
50	0.2046	0.3219	0.4322	0.5798	0.6602
100	0.2169	0.3503	0.476	0.6398	0.7221
200	0.2274	0.3764	0.518	0.6953	0.7771

One can see that the magnitude of redistribution coefficients is essential especially for the level $n = 4$ (Balmer-beta line).

6. Radiation Spectra Calculations

Radiation spectra calculations were performed on the basis of population kinetic models described above. The main mechanisms of the spectral line broadening is the Doppler effect. The presence of a strong magnetic field with the field strength of 5T, typical for ITER plasma, is taken into account. In the frame of above approximations the radiation atomic spectra are described by the formula:

$$J_n\left(\Delta\omega, B, T_e, T_i, N_e\right) = J_{BAL}^n\left(\Delta\omega, B, T_i\right)\left[1 - K\left(n, T_e, N_e\right)\right] + J_{LOC}^n\left(\Delta\omega, B, T_i\right) K\left(n, T_e, N_e\right) \qquad (4)$$

where the line shape $J^n_{BAL}(\Delta\omega, B, T_i)$ is calculated with the help of the ballistic atoms velocity distribution function obtained with the ballistic model and appeared to be close to the data of the Monte-Carlo simulations, and the line shape $J^n_{LOC}(\Delta\omega, B, T_i)$ is calculated for a Maxwellian velocity distribution function with the ionic temperature in a given spatial point.

The results for the Balmer-alpha and Balmer-beta line shapes are presented in Figures 7 and 8.

Figure 7. Spectral intensity of Balmer-alpha (**a**) and Balmer-beta (**b**) deuterium lines in SOL for the bottom part of vertical observation chord ($R = 65.5$ cm) for the ITER SOL&divertor case #1514. The red line shows contribution to the radiation from neutrals obtained from charge exchange of neutrals in the exited state $n = 3$ (for the Balmer-alpha line) and $n = 4$ (for the Balmer-beta line) on fast ions. Blue lines—from ballistic atoms. Black line is the total spectrum.

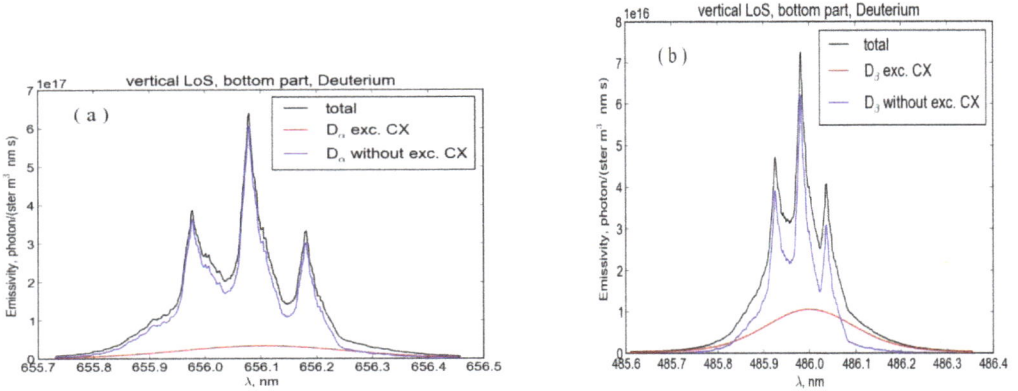

Figure 8. The same as in Figure 7 but for the top part of vertical observation chord.

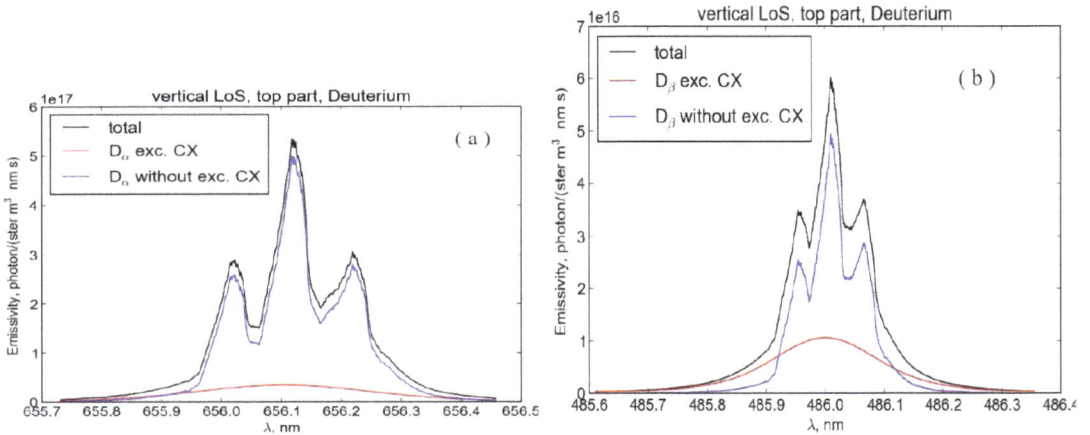

7. Conclusions

A new simplified method and numerical code for calculations of neutral atoms penetration into plasma is suggested. The method is based on the ballistic motion of neutrals along the line of sight. As a result, two dimensional distribution functions, dependent on the LoS coordinate and the velocity projection on the LoS, and responsible for the Doppler broadening of the line shape, are calculated.

The interrelation between MC simulation and fast ballistic code is as following: the MC code provides the distributions of main plasma components (electron and ions temperatures and densities, see Figure 2) whereas the ballistic code provides the fast calculations of much less neutral atomic density distribution together with atomic kinetics and line shapes along LoS.

The distributions on Figures 3–6 are quite satisfactory but still not refine. The reason for discrepancies between Monte-Carlo (MC) and ballistic model (BM) seems to be related to 3D simulation in MC and 1D simulation in BM. Such discrepancies exist in the radiation transport theory between 3D calculations and "forward-back" photons propagation approximation [8]. Nevertheless the 1D simulation is quite reasonable for determination of photon space distribution [8] as well as neutral component distribution. So the application of 1D BM seems to be successful on average taking into account the great advantage in calculation time. Some differences of the velocity distribution function is connected with charge exchange collision integral. Really there is a difference between particles moving forward (from the wall) and back (to the wall) after charge exchange in some space point; the first ones meet more hot plasma particles whereas the second ones—more cold. So the charge exchange rates are slightly different for these particles that can be a reason for discrepancy between both types of data on Figure 5 (for the curve $x = 3$ cm) near velocity points $\pm$ $(1–2) \times 10^6$ cm/c. The difference between data on Figure 6 is less important for line shape calculations because of strong decrease in atomic densities with increasing of distance from the wall.

The test calculations of the Balmer spectral line shapes are performed for plasma parameters distributions typical for ITER diagnostic tasks. A comparison of the method with the respective (Monte Carlo) EIRENE code [1] stand-alone simulations of neutral deuterium velocity distribution function, applied on the SOL&divertor plasma background calculated by the SOLPS4.3 (B2-EIRENE) code [1–4], is made to evaluate the accuracy of the ballistic model. The results of the model reproduce numerical data obtained by the EIRENE code with an average accuracy of 10%–20%. The calculation time is smaller by the orders of magnitude as compared with the Monte Carlo simulation one. This makes it possible to apply the code for calculations of background plasma radiation properties of SOL plasma in various regimes of ITER operation.

Acknowledgments

This work was partially supported by the grant No. 3328.2014.2 of President of Russian Federation for Leading Scientific Schools. The authors thank K.Yu. Vukolov and A.G. Alekseev for stimulating discussions and support, A.S. Kukushkin, for providing the ITER SOL&divertor background plasma data, and A.B. Kukushkin, for helpful discussions.

Author Contributions

Valery S. Lisitsa—development of models, analytical calculations of ballistic model, writing of manuscript.

Mikhail B. Kadomtsev—development of models, analytical calculations of charge exchange kinetics, writing of manuscript.

Vladislav Kotov—Monte Carlo calculations of neutral space and velocity distribution functions.

Vladislav S. Neverov—numerical calculations of hydrogen line shapes along SoL in ITER.

Vladimir A. Shurygin—numerical calculations of neutral space and velocity distribution functions in ballistic model.

Conflicts of Interest

The authors declare no conflict of interest.

References

1. Reiter, D.; Baelmans, M.; Boerner, P. The EIRENE and B2-EIRENE Codes. *Fusion Sci. Tech.* **2005**, *47*, 172–186.
2. Braams, B.J. Computational studies in tokamak equilibrium and transport. Ph.D. Thesis. Utrecht University, Netherlands, 1986.
3. Kukushkin, A.S.; Pacher, H.D.; Loarte, A.; Komarov, V.; Merola, M.; Pache, G.W.; Reiter, D. Analysis of performance of the optimized divertor in ITER. *Nucl. Fusion* **2009**, 49, 075008.
4. Kotov, V.; Reiter, D.; Kukushkin, A.S. Numerical Study of the ITER Divertor Plasma with B2-EIRENE Code Package. Available online: http://www.eirene.de/kotov_solps42_report.pdf, (accessed on 13 May 2014).
5. Kadomtsev, M.B.; Kotov, V.; Lisitsa, V.S.; Shurygin, V.A. Ballistic model for neutral hydrogen distribution in ITER edge plasma. In Proceedings of the 39th EPS Conference on Plasma Physics & 16th International Congress on Plasma Physics, Stockholm, Sweden, 2–6 July 2012; P4.093.
6. Kadomtsev, M.B.; Kotov, V.; Lisitsa, V.S.; Shurygin, V.A. Shurygin. Kinetics of hydrogen atom radiation emission of the SOL plasma in ITER. In Proceedings of the 40th EPS Conference on Plasma Physics, Espoo, Finland, 1–5 July 2013; P1.135.
7. Kukushkin, A.B.; Lisitsa, V.S.; Kadomtsev, M.B.; Levashova, M.G.; Neverov, V.S.; Shurygin, V.A.; Kotov, V.; Kukushkin, A.S.; Lisgo, S.; Alekseev, A.G.; *et al.* Theoretical issues of high resolution H-α spectroscopy measurements in ITER. In Proceedings of the 24th IAEA Fusion Energy Conference, San Diego, CA, USA, 8–13 October 2012.
8. Biberman, L.M.; Vorob'ev, V.S.; Yakubov, I.T. *Kinetics of Nonequilibrium Low Temperature Plasmas*; Consultants Bureau: New York, NY, USA, 1987.

Reprinted from *Atoms*. Cite as: Dalimier, E.; Oks, E.; Renner, O. Review of Langmuir-Wave-Caused Dips and Charge-Exchange-Caused Dips in Spectral Lines from Plasmas and their Applications. *Atoms* **2014**, *2*, 178-194.

Review

Review of Langmuir-Wave-Caused Dips and Charge-Exchange-Caused Dips in Spectral Lines from Plasmas and their Applications

Elisabeth Dalimier [1,2]**, Eugene Oks** [3,]*** and Oldrich Renner** [4]

[1] Sorbonne Universités Pierre et Marie Curie, 75252 Paris CEDEX 5, France
[2] Ecole Polytechnique, LULI, F-91128 Palaiseau CEDEX, France
[3] Physics Department, 206 Allison Lab, Auburn University, Auburn, AL 36849, USA
[4] Institute of Physics v.v.i., Academy of Sciences CR, 18221 Prague, Czech Republic

* Author to whom correspondence should be addressed; E-Mail: goks@physics.auburn.edu; Tel.: +1-334-844-4362; Fax: +1-334-844-4613.

Received: 22 January 2014; in revised form: 25 March 2014 / Accepted: 30 April 2014 / Published: 13 May 2014

Abstract: We review studies of two kinds of dips in spectral line profiles emitted by plasmas—dips that have been predicted theoretically and observed experimentally: Langmuir-wave-caused dips (L-dips) and charge-exchange-caused dips (X-dips). There is a principal difference with respect to positions of L-dips and X-dips relative to the unperturbed wavelength of a spectral line: positions of L-dips scale with the electron density N_e roughly as $N_e^{1/2}$, while positions of X-dips are almost independent of N_e (the dependence is much weaker than for L-dips). L-dips and X-dips phenomena are important, both fundamentally and practically. The fundamental importance is due to a rich physics behind each of these phenomena. L-dips are a multi-frequency resonance phenomenon caused by a single-frequency (monochromatic) electric field. X-dips are due to charge exchange at anticrossings of terms of a diatomic quasi-molecule, whose nuclei have different charges. As for important practical applications, they are as follows: observations of L-dips constitute a very accurate method to measure the electron density in plasmas—a method that does not require knowledge of the electron temperature. L-dips also allow measuring the amplitude of the electric field of Langmuir waves—the only spectroscopic method available for this purpose. Observations of X-dips provide an opportunity to determine rate coefficient of charge exchange between multi-charged ions. This is an important reference data, virtually inaccessible by other experimental methods. The rate coefficients of charge exchange are important for magnetic fusion in Tokamaks, for population inversion in the soft x-ray and VUV ranges, for ion storage devices, as well as for astrophysics (e.g., for the solar plasma and for determining the physical state of planetary nebulae).

Keywords: dips in spectral line profiles; diagnostic of Langmuir waves amplitude; diagnostic of electron density in plasmas; method for measuring charge exchange rates

PACS: 32.70.Jz; 52.35.Fp; 34.70.+e; 32.30.Rj; 33.15.-e

---

1. Langmuir-Wave-Caused Dips

Langmuir-wave-caused dips (L-dips) in profiles of spectral lines in plasmas were discovered experimentally and explained theoretically for dense plasmas, where one of the electric fields **F** experienced by hydrogenic radiators is quasi-static. This field can be the ion micro-field and (or) a low frequency electrostatic turbulence.

There is a rich physics behind each of these phenomena. L-dips result from a resonance between the Stark splitting $\omega_F = 3n\hbar F/(2Z_r m_e e)$ of hydrogenic energy levels, caused by a quasistatic field **F** in a plasma, and the frequency ω_L of the Langmuir waves, which practically coincides with the plasma electron frequency $\omega_p(N_e) = (4\pi e^2 N_e/m_e)^{1/2} \approx 5.641 \times 10^4 \, [N_e(\text{cm}^{-3})]^{1/2}$:

$$\omega_F = s\,\omega_p(N_e), \quad s = 1, 2, \ldots \tag{1}$$

Even for the most common case of $s = 1$, it is actually a multi-frequency resonance phenomenon despite the fact that the electric field of the Langmuir wave is considered to be single-frequency (monochromatic): $\mathbf{E}_0 \cos\omega_p t$. Its multi-quantum nature has been revealed in paper [1]: it is a resonance between many quasienergy harmonics of the combined system "radiator + oscillatory field", caused simultaneously by all harmonics of the total electric field $\mathbf{E}(t) = \mathbf{F} + \mathbf{E}_0 \cos\omega_p t$; where vectors **F** and $\mathbf{E}_0$ are not collinear.

The history of dips covers a long period from 1977 to 2013, during which they have been studied experimentally in different plasma sources, such as gas-liner pinch, laser-produced plasmas, and Z-pinch plasmas. These experiments, performed by various groups, required specific configurations and improved high-resolution X-ray spectrometers. The theory of L-dips provided a diagnostic tool for measuring the electric field amplitude E_0 of the Langmuir waves and an independent method for measuring the electron density N_e.

1.1. Theory of L-dips

We present here a summary of the theoretical results from [2–5]. The resonance condition (1) is controlled by the principal quantum number n of the upper level involved in the radiative transition, the nuclear charge of the radiating ion Z_r, the electron density N_e and the quasi-static micro-field F. As the quasi-static field F has a broad distribution ΔF over the ensemble of radiators, there would always be a fraction of radiators for which the resonance condition (1) is satisfied. These radiators are subjected simultaneously to the resonance value F_{res} of the quasi-static field (defined by Equation (1)) and to the Langmuir wave field $\mathbf{E}_0 \cos\omega_p t$. The L-dips are possible only as long as $E_0 < F_{res}$. This imposes an upper limit on the ratio of the energy density of the Langmuir waves to the thermal energy density: $E_0^2/(8\pi N_e T_e) < 4U_{ion}/(9n^2 T_e)$, where U_{ion} is the ionization potential of the radiating ion.

The resonance condition translates to specific locations of L-dips in spectral line profiles, depending on N_e; the distance of a L-dip from the unperturbed wavelength is given by:

$$\Delta\lambda_{dip} = aN_e^{1/2} + bN_e^{3/4} \tag{2}$$

where coefficients a and b are controlled by quantum numbers and by the charges of the radiating and perturbing ions.

The primary term in Equation (2) reflects the dipole interaction with the ion micro-field. The second, much smaller term, takes into account—via the quadrupole interaction—a spatial non-uniformity of the ion micro-field.

Figure 1. The theoretically-expected structure of the L-dip.

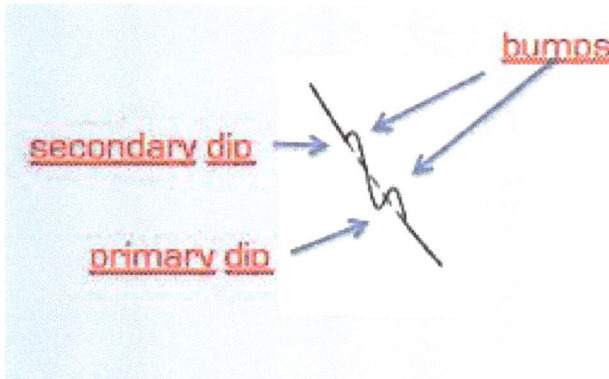

Each L-dip represents a structure consisting of the dip itself (the primary minimum of intensity) and two surrounding bumps (Figure 1). The bumps are due to a partial transfer of the intensity from the wavelength of the dip to adjacent wavelengths. The total structure can lead to a secondary dip (or a small shoulder).

The half-width of the L-dip, controlled by the amplitude E_0 of the electric field of the Langmuir wave, is given by:

$$\delta\lambda_{1/2} \approx (3/2)^{1/2}\lambda_0^2 n^2\, E_0/(8\pi m_e ecZ_r), \tag{4}$$

where λ_0 is the unperturbed wavelength of the spectral line. Thus, by measuring the experimental half-width of L-dips, the amplitude E_0 of the Langmuir wave can be determined.

205

1.2. Experimental Observations of L-dips

Gas-Liner Pinch Experiments in Kunze's Group (1980-1990). The first observations of L-dips were made in a gas-liner pinch [6]—a modified z-pinch with a special gas inlet system—which is injected concentrically along the pinch wall and accelerated into the center of the discharge chamber to form the plasma. The plasma is mostly the driver gas plasma. It is characterized by a density $(1–3) \times 10^{18}$ cm^{-3}, a relatively low temperature 10–13 eV, and a quasi-monochromatic field $E_0 \cos(\omega_p t + \phi)$ (Langmuir wave) of the same order as the average ion micro-field F*. With the injection of hydrogen, the Stark broadened profiles of Lyα were observed along the z axis at different times after the compression using a concave grating spectrometer [6].

The spectra (Figure 2) were taken 110 ns (top) and 200 ns (bottom) after the maximum compression. The electron densities were obtained by an independent diagnostic, *i.e.*, coherent Thomson scattering. The first-order dips (due to the one-quantum resonance $\omega_F = \omega_p$) and the second-order dips (due to the two-quantum resonance $\omega_F = 2\omega_p$) were observed and their positions were in agreement with the theory. The shift of the dips with increasing densities was also visible and consistent with Equation (2); the effect of the spatial non-uniformity of the ion micro-field on the L-dip positions was revealed experimentally for the first time. Moreover, the detailed bump-dip-bump structure in the profile of each Stark component was also revealed for the first time (Figure 3).

Figure 2. Comparison of the experimental and theoretical positions of the dips in the profiles of the hydrogen Lyα line in the gas-liner pinch. The theoretical positions are shown by vertical solid lines connected by dashed lines.

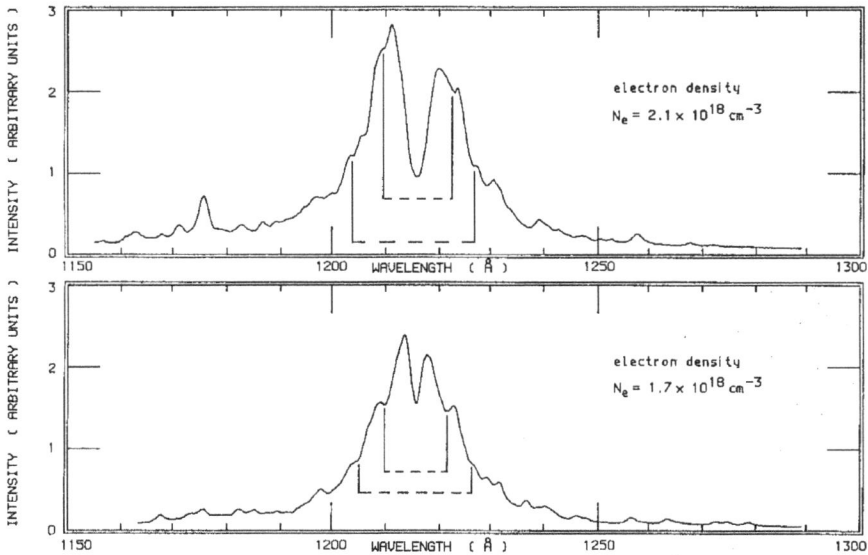

The gas-liner pinch experiments allowed two important diagnostics: the electron density from the location of the dips, using Equation (2), and the Langmuir wave amplitude from the half-width of the L-dips, using Equation (3). Figure 4 shows the electron densities calculated from the separation of the first-order dips *vs.* the electron density obtained by coherent Thomson scattering; the agreement is very good.

Laser Produced Plasma Experiments at LULI (Laboratoire pour l'Utilisation des Lasers Intenses, O. Renner et al. 2006). The experiment was performed at the nanosecond Nd:glass laser facility at LULI. A single laser beam with an intensity of 2×10^{14} W/cm^2 was focused onto a structured target; an Al strip sandwiched between magnesium substrate [7,8]. The Al plasma was well confined and the transverse emission was optimized. The plasma parameters $N_e = 5 \times 10^{22}$ cm^{-3} and $T_e = 300$ eV were estimated by hydrodynamic simulations. They confirm a resonant coupling between the ion micro-field F and the Langmuir field E. A Vertical-geometry Johann Spectrometer VJS with high spectral (4200) and spatial resolution was specially designed for the assessment of fine structures in the red wing of the Al XIII Lyγ line (Figure 5).

Figure 3. Zoom on the observed bump-dip-bump structure from the top part of Figure 2, the farthest dip in the blue wing.

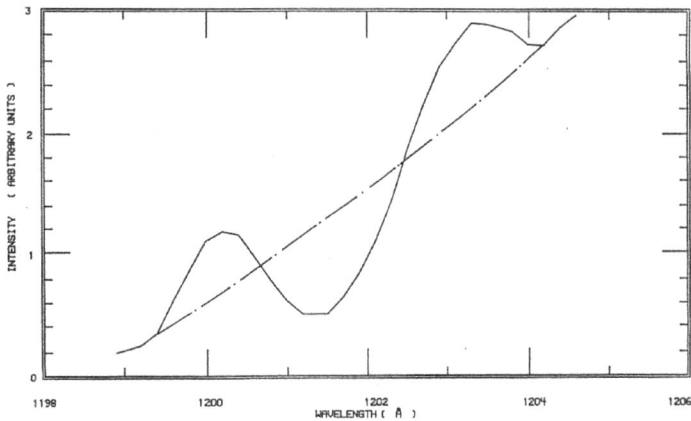

Figure 4. Comparison of the electron density deduced from coherent Thomson scattering and from the first-order dips, *i.e.*, the dips like the pairs of dips closest to the line center in Figure 2.

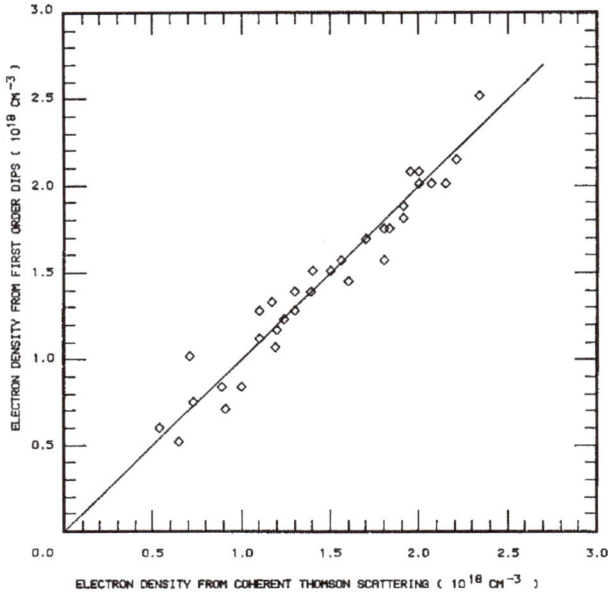

Figure 5. Experimental setup showing the structured target ($CH/Al_4C_3/CH$ or $Mg/Al/Mg$), the Vertical-geometry Johann Spectrometer (VJS), and a sample experimental record consisting of two identical (except for noise) sets of spatially resolved spectra, symmetrically located with respect to the central wavelength.

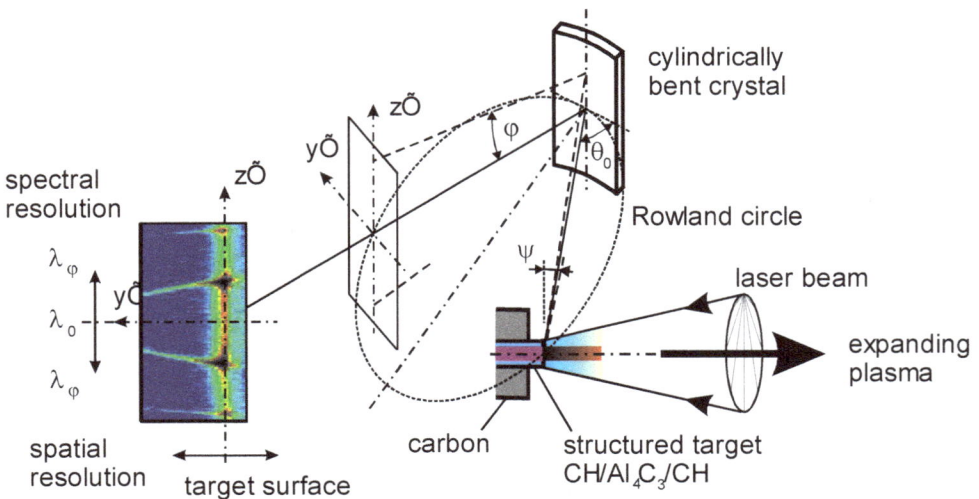

Theoretically expected L-dip positions (Figure 6) were confirmed by the experiment (Figure 7). The simultaneous production of a pair of symmetric spectra with this spectroscopic setup provides a reference point for the computational reconstruction of raw data, thus increasing the confidence in the identification of the dips.

Figure 6. Theoretically expected positions of L-dips in profiles of the Al XIII Lyγ line *vs.* the electron density.

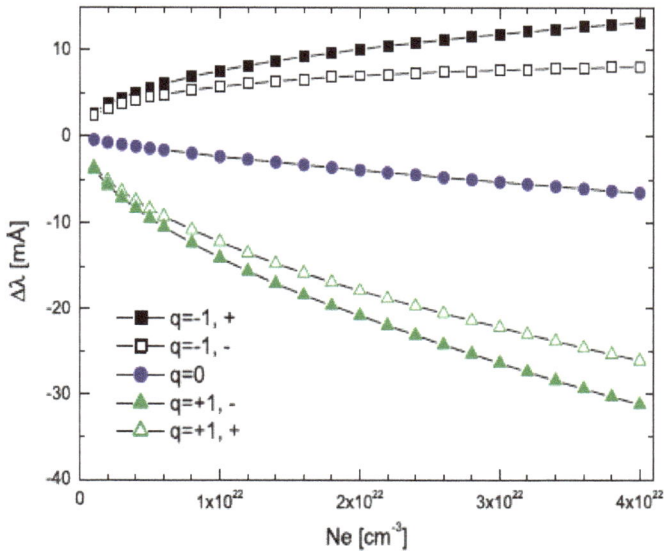

Figure 7. L-dips in the experimental profiles of the Al XIII Lyγ line from a plasma produced by the LULI laser. Only the red dips (corresponding to q = 1 in Figure 6) are visible, the blue dips (q = −1) are merged with the noise. The different spectra correspond to the emission from different distances from the target (*i.e.*, different densities).

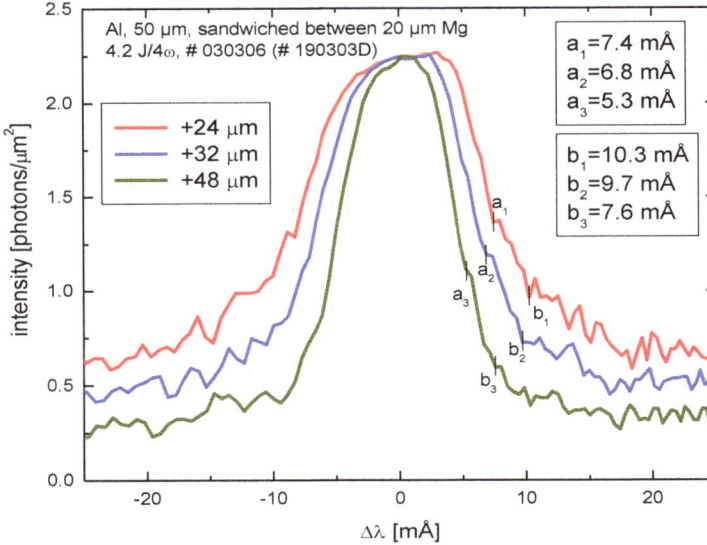

This experiment in a laser-produced plasma confirms the dependence of the dips positions on the density, thus allowing a density diagnostic. The densities can be compared to the ones obtained from line broadening simulations (IDEFIX code at LULI and PIM PAM POUM code at PIIM-Marseille).

The latest experimental application of the theory of L-dips in Z-pinch plasma at Chongqing University, China (Jian et al. 2013). At the Yang accelerator Z-pinch aluminum plasmas, characterized by a density $N_e = 5 \times 10^{21}$ cm^{-3}, the electron temperature $T_e = 500$ eV and the ion temperature $T_i = 10$ keV, were obtained by an imploding aluminum wire-array [9]. These conditions were favorable to a resonant coupling between the ion micro-field F and the Langmuir field E. The experimental Stark broadened profiles of the Al XIII Lyα and Al XIII Lyγ lines were recorded with a high spectral resolution (2500) on a "Uniform Dispersion mica Crystal Spectrograph" UDCS [9]. These lines, emitted from different regions, exhibit bump-dip-bump first-order resonance structures, as shown in Figure 8. They correspond to slightly different densities and to different radial temperature gradients. The relationship between the positions from the line center of red Langmuir dips and electron density was studied. The electron densities deduced from the fine spectral features (the red dips) were compared to those derived from measurements of the Plasma Polarization Shift (PPS) of the entire spectral line. There was a 5% difference in electron density obtained by the two methods, which was due to the uncertainty of determining the average T_e required for the PPS method (but not required by the method based on the L-dips).

Figure 8. The experimental red dips L$^+$ and L$^-$ in the profiles of the Al XIII Lyα and Al XIII Lyγ lines, observed at a Z-pinch facility in China, are compared to the theoretical predictions. The blue dips merge in the noise.

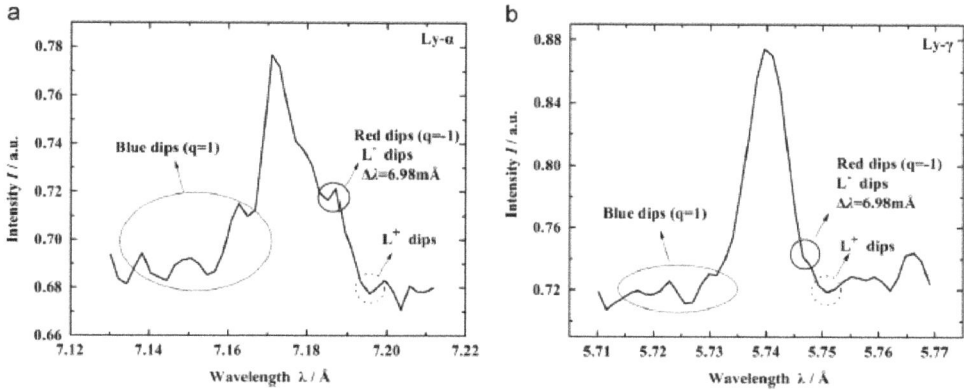

2. Charge-Exchange-Caused Dips

The charge-exchange-caused-dips (X-dips) in profiles of spectral lines from dense plasmas are due to the Charge Exchange (CE) atomic process inside the plasma—distinct from L-dips caused by a resonant coupling between the plasma micro-field and Langmuir waves. Up to now, the process and its spectroscopic manifestation was studied in H-like radiating ions of the nuclear charge Z, exchanging an electron with a perturbing fully-stripped ion of a different nuclear charge Z' ≠ Z. The history of X-dips covers a long period from 1995 to 2012. These dips represent signatures of the CE process in spectral line profiles emitted from dense plasmas. Observations of X-dips allow measuring rate coefficients of CE between multi-charged ions in dense plasmas.

2.1. Theory of X-dips

We present here a summary of theoretical results from [10–15]. The X-dips are caused by CE at quasi-crossings of the energy terms E(R) of the quasi-molecule Z-e-Z', made up of a H-like radiating ion Z and a perturbing fully stripped ion Z'. Here, R stands for the interionic distance. It is worth noting that there is a theorem by Neumann-Wigner [10] stating that terms "of the same symmetry" cannot cross, where "the same symmetry" means that the terms are characterized by the same projection M of the angular momentum on the inter-nuclear axis. However, this theorem is not valid for the Z–e–Z' quasi-molecule [11], because this systems has a higher than geometric symmetry and therefore possesses an additional conserved quantity. The extra conserved quantity is the projection on the inter-nuclear axis of the super-generalized Runge-Lenz vector [12], *i.e.*:

$$\mathbf{A} = \mathbf{p} \times \mathbf{M} - M^2/R\ \mathbf{e}_z - Zr/r - Z'\ (\mathbf{R} - \mathbf{r})/|\mathbf{R} - \mathbf{r}| + Z'\ \mathbf{e}_z\ , \qquad \mathbf{e}_z = \mathbf{R}/R, \qquad (4)$$

where **p** and **M** are linear and angular momenta vectors, respectively; **r** is the radius vector of the electron. At the quasi-crossing, the terms are characterized by the same energy E and by the same projection M of the angular momentum on the inter-nuclear axis, but differ by the projection A_z of vector **A** on the inter-nuclear axis. This extra conserved quantity A_z controls the selection rule, allowing CE process at the quasi-crossing.

We note that CE could be possible not only due to quasicrossings, but also due to a rotational (Coriolis) coupling of terms (see, e.g., books [16,17]). Physically, the latter possibility is caused by a rotation of the internuclear axis; a simple yet general enough analytical treatment of this being given, e.g., in paper [18]. However, only CE due to quasicrossings causes X-dips – the formation of X-dips is intimately related to quasicrossings. We mention also that CE due to a rotational coupling usually occurs at significantly smaller internuclear distances than CE due to quasicrossings. Therefore, even if the rotational coupling would be able to cause an X-dip (though in reality it cannot), such dip would be practically impossible to observe; it would be located in a very far wing of the corresponding spectral line, where the experimental line profile merges with the noise. In short, an X-dip is a spectroscopic manifestation of CE, but not vice versa; not every CE causes an X-dip, but only CE due to a quasicrossing.

The transition energies $\Delta E(R) = \hbar\omega$ for the radiating ion Z, corresponding to the Stark components due to the static field produced by the Z' ion (averaged over the ion distribution), yield the quasi-static profiles. These profiles are subjected to the dynamical broadening $\gamma(R)$ due to electrons and ions.

Two independent complementary mechanisms explain the formation of X-dips in spectral line profiles [15]: one through the behavior of $\Delta E(R)$ and another through the dynamical broadening $\gamma(R)$. A sharp change of the slope of the transition energy $\Delta E(R)$ at the quasi-crossing R_c leads to a dip at large distances [14] that can be predicted using the expansion of $\Delta E(R)$ in powers of $1/R$. The other mechanism is based on the fact that CE provides an extra channel for the decay of the excited state of the radiating ion Z—in addition to the decay at the rate $\gamma_{nonCE}(R)$ caused by the dynamical Stark broadening [13]. As a consequence, at R_c, the intensity of the profile is smaller (the dip formation) and in the vicinity ΔR it is larger (the bump formation) than it would be without CE.

Figure 9 presents a magnified sketch of the quasi-crossing structure and the role of the dynamical broadening.

Figure 9. A magnified sketch of the quasi-crossing structure. The transition energy, the bold line, occupies a bandwidth $\gamma(R)$. In the interval ΔR the transition has two branches.

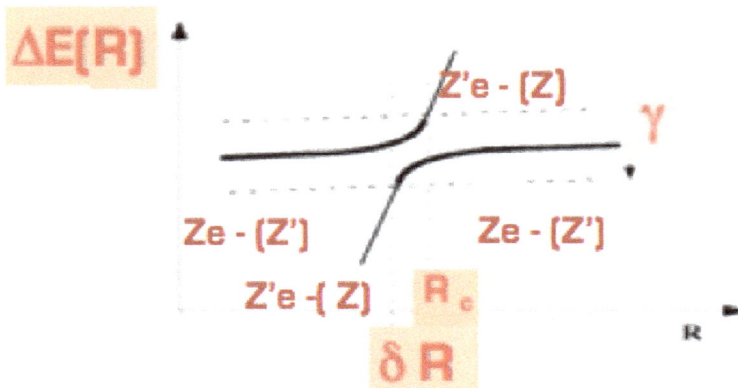

The transition energy Ze-Z' (the bold line) occupies a bandwidth $\gamma(R)$, which is controlled by the dynamical broadening $\gamma_{nonCE}(R)$ caused by electrons and ions, and by the charge exchange increase $\gamma_{CE}(R)$ $i.e.$:

$$\gamma(R) = \gamma_{CE}(R) + \gamma_{nonCE}(R). \tag{5}$$

It is important to emphasize that $\gamma_{CE}(R)$ rapidly decreases away from R_c, while $\gamma_{nonCE}(R)$ varies very slowly away from R_c. This explains that the dip itself is surrounded by two adjacent bumps.

The bump-to-dip ratio of intensities is a function of the ratio:

$$r = \gamma_{CE}(R_c)/\gamma_{nonCE}(R_c). \tag{6}$$

This function was calculated analytically in [15]. As r increases, the bumps move away from the center of the X-dip and their intensities decrease. The bump-to-dip ratio measurement r_{exp} allows deducing the rate coefficient of charge exchange [15] as follows:

$$< v \, \sigma_{CE}(v)> = \gamma_{nonCE}(R_c) \, r_{exp}/N_i, \tag{7}$$

where N_i is the ion density. The quantity $\gamma \equiv \gamma_{nonCE}(R_{cr})$ in Equation (7), representing the frequency of inelastic collisions with electrons and ions leading to virtual transitions from the upper state of the radiator to other states, can be calculated for given plasma parameters N_e, N_i, T_e, and T_i by using one of few contemporary theories (presented, e.g., in the book [19]).

The experimental determination of the rate coefficients of charge exchange from the dip structure is an important reference data that is virtually inaccessible by other experimental methods.

Figure 10. The X-dip structure in the blue side of the Hα line emitted by hydrogen atoms perturbed by fully stripped helium.

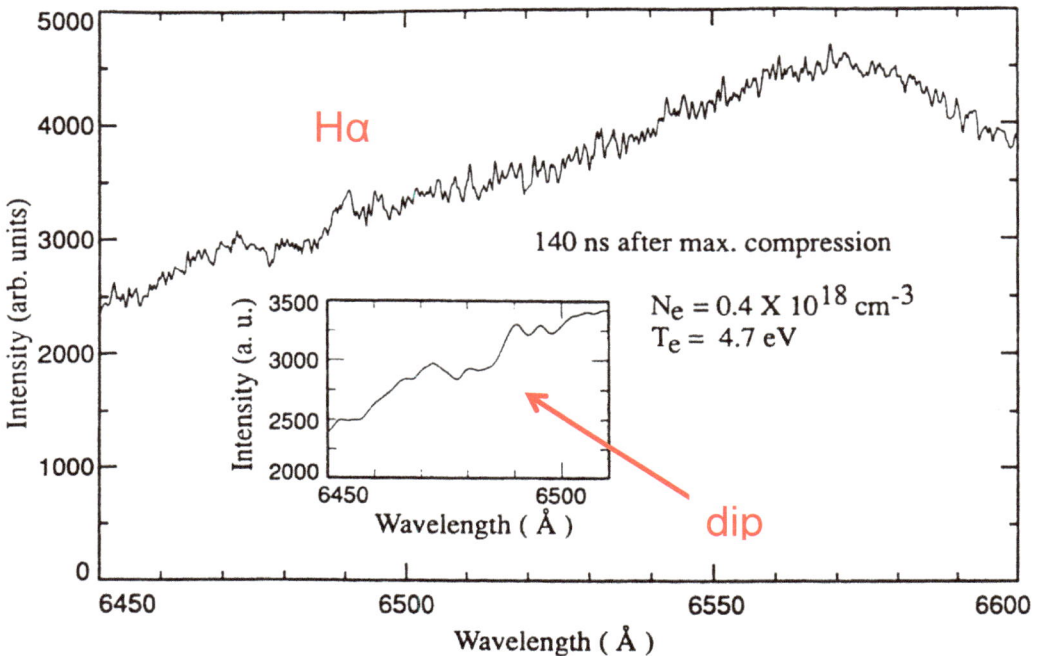

2.2. Experimental Observations of X-dips

The first observation in the gas-liner pinch (Kunze's group 1995). The first observation of X-dips was made in the gas-liner pinch [13] in the blue side of the Hα line emitted by hydrogen atoms H (Z = 1) perturbed by fully stripped helium He (Z' = 2) (Figure 10). The dips were observed for a relatively small range of electron densities around 10^{18} cm^{-3}. For lower densities, the quasi-crossing distance R_c was much lower than the mean inter-ionic distance R_i and for higher densities, the dynamical broadening $\gamma_{nonCE}(R_c)$ was smoothing the dip structure, both reasons being unfavorable for experimental observations. In these experiments the electron density N_e was measured by Thomson scattering and it was verified that the positions of the X-dips did not depend on N_e.

Figure 11. Observation of the X-dip structures X_1, X_2, X_3 in the red wing of the Al XIII Lyγ line perturbed by carbon in laser-produced plasma experiments at LULI. The L-dips are also visible, but less pronounced than the X-dips.

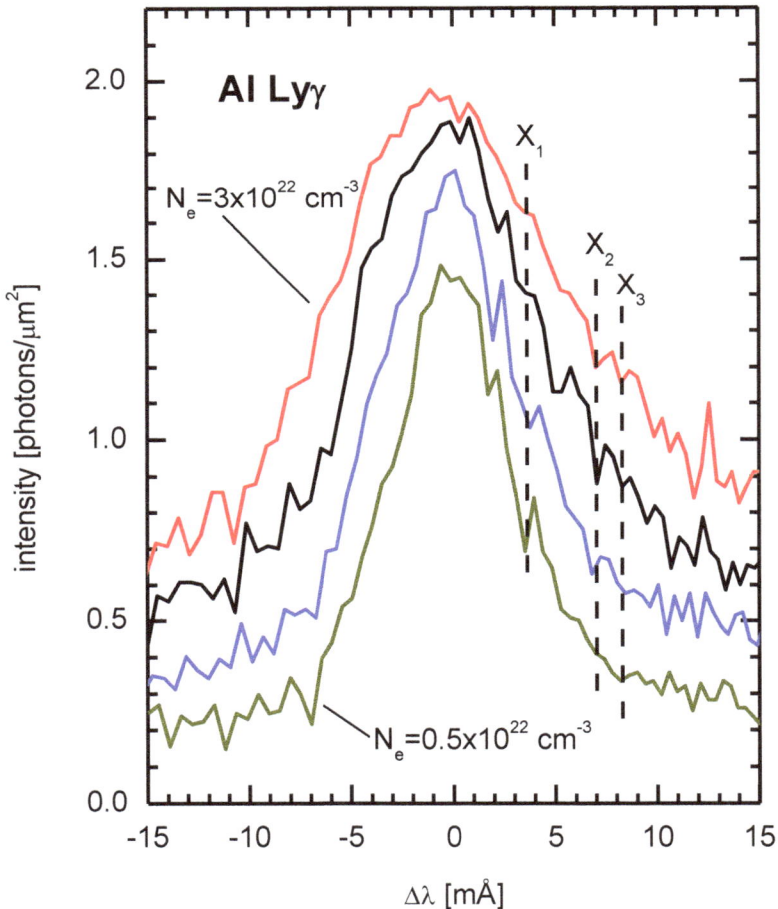

Figure 12. A magnified plot of the observed X_1-dip structure, used for deducing the rate coefficient of CE.

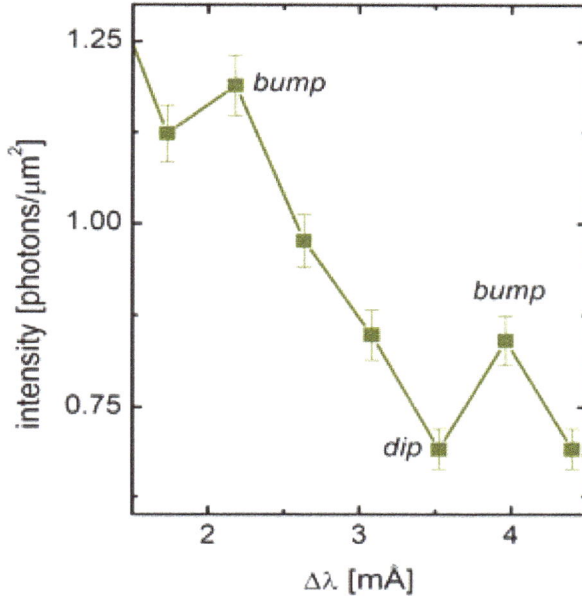

Laser Produced Plasma Experiments at LULI (Laboratoire pour l'Utilisation des Lasers Intenses) (E. Dalimier et al. 2001). Later, the X-dips were observed in laser-produced plasmas characterized by a high electron density $10^{22} - 3 \times 10^{22}$ cm^{-3} [15,20]. The setup was implemented at the laser facility LULI using the same nanosecond laser at 10^{14} W/cm^2 and the same high-resolution Vertical-geometry Johann Spectrometer (R = 8,000) as in the experiments devoted to the observation of L-dips [7,8]. The targets used for the observation of X-dips were aluminum carbide Al_4C_3 strips inserted in carbon substrate. The emission from the heterogeneous plasma made up of Al and C ions exhibited spectroscopic signatures of CE. X-dips were observed for the first time in the experimental profile of the Lyγ line of Al XIII (Z = 13) perturbed by fully stripped carbon CVI (Z' = 6), as shown in Figure 11. The positions of the dips did not vary significantly in the small density domain. For smaller densities the line is too narrow to allow the visibility of the exotic X-dip structures and at higher densities the dips are smoothed out.

From the experimental bump-to-dip ratio, the rate coefficient of charge exchange between a hydrogenic aluminum ion in the state n = 4 and a fully stripped carbon was found to be [15]:

$$< v\ \sigma_{CE}(v)> = (5.2 \pm 1.1)\ 10^{-6}\ cm^3/s \tag{8}$$

A magnified plot of the X_1-dip structure, used for deducing the rate coefficient of CE, is presented in Figure 12.

Laser-produced plasma experiments at PALS (O. Renner et al. 2012). The latest experimental study of X-dips was performed in 2012; in Plasma Wall Interaction (PWI) experiments performed at PALS [21]. A plasma jet of aluminum ions, produced by the nanosecond iodine laser of the intensity

3×10^{14} Wcm^{-2} incident on a foil, interacted with a massive carbon target. The high spectral and spatial resolution Vertical-geometry Johann Spectrometer was adjusted for the observation of X-dips in the experimental profile of the Lyγ line of Al XIII ($Z = 13$) (Figure 13). The dips X_1 and X_2, clearly visible in the red wing, are the signatures of CE phenomena accompanying the PWI. The electron densities involved in the experiment and simulated by codes PALE [22] and MULTIF [23] can reach 5×10^{22} cm^3, which is higher than those achieved at LULI during the study of the same line (Figure 11). It is important to note that a weak dependence of the positions of the dips on the density has been detected. Simulations can explain this dependence [21]. The dip, visible in the far-red wing, is an L-dip corresponding to a possible multi-quantum resonance, thus providing a spectroscopic diagnostic of the electron density.

Figure 13. Observation of the X-dip structures X_1 and X_2, in the red wing of the Al XIII Lyγ line in the Plasma-Wall Interaction experiments at PALS (aluminum plasma interacting with carbon wall).

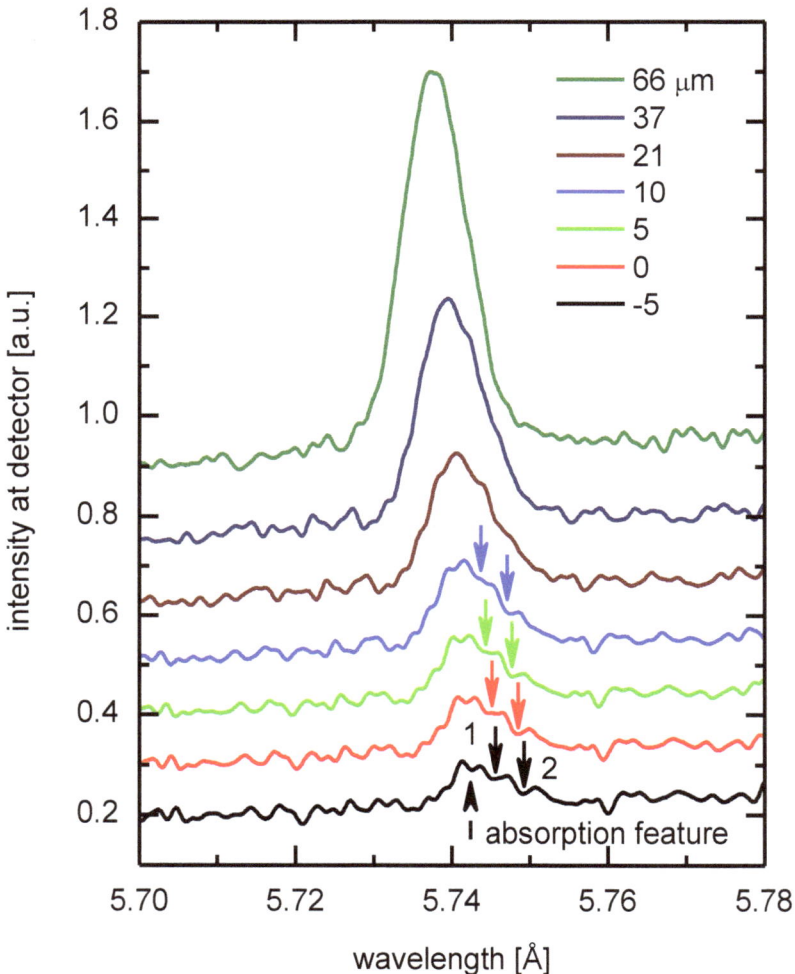

Table 1. Prospective He-like lines and reciprocal cases of H-like lines, and solid targets for observing X-dips in laser-produced plasmas.

Line	Perturber	Reciprocal Case	Target
O VII He-gamma 17.78 A	Li^{+3}	Li III L-alpha 135.0 A perturbed by O^{+7}	Lithium oxide (lithia) Li_2O (solid)
Si XIII He-beta 5.681 A	Be^{+4}		Be-Si binary alloy or beryllium-doped silicon
Mg XI He-gamma 7.473 A	B^{+5}	B V L-alpha 48.59 A perturbed by Mg^{+11}	Boracite $Mg_3B_7O_{13}Cl$ (mineral)
Si XIII He-gamma 5.405 A	C^{+6}	C VI L-alpha 33.74 A perturbed by Si^{+13}	Silicon carbide (carborundum) SiC (solid)
Ca XIX He-beta 2.705 A	C^{+6}		Calcium carbonate $CaCO_3$ (the common substance found in rocks and the main component of shells of marine organisms, snails, and egg shells)
S XV He-gamma 4.089 A	N^{+7}	N VII L-alpha 24.78 A perturbed by S^{+15}	Ammonium sulfate $(NH_4)_2SO_4$ (crystals/granules)
V XXII He-beta 2.027 A	N^{+7}		Vanadium nitride VN (solid)
Fe XXV He-beta 1.574 A	O^{+8}		Iron oxides FeO, Fe_3O_4, Fe_2O_3 (crystalline solid)
Ca XIX He-gamma 2. 57 A	F^{+9}	F IX L-alpha 14.99 A perturbed by Ca^{+19}	Calcium fluoride (fluorite) CaF_2 (mineral/solid)
Cu XXVIII He-beta 1.257 A	F^{+9}		Copper fluoride CuF_2 (crystalline solid)

2.3. The Latest Development on X-dips

Up to now, the X-dip phenomena were considered to be possible only in spectral lines of hydrogen-like ions perturbed by fully stripped ions. The reason is the existence of an additional conserved quantity for the corresponding two-center Coulomb system with one bound electron. Recently we showed that X-dip phenomena are possible in spectral lines of helium-like ions from laser-produced plasmas [24,25]. The reason is the existence of an approximate additional conserved quantity for the corresponding two-center Coulomb system with two bound electrons [26]. This offers a new opportunity to extend the range of fundamental data on charge exchange.

Table 1 presents 15 prospective He-like spectral lines and 10 corresponding solid targets for observing X-dips in laser-produced plasmas.

3. Conclusions

The conclusion of the review paper is focused on the importance of L-dips and X-dips phenomena.

First, there is rich physics in both phenomena, involving connections between different topics, *i.e.*, Atomic/Molecular Physics and Plasma Physics.

Second, high-resolution spectroscopy was crucial in all these studies and it resulted in new dense plasma diagnostic methods and in the production of new fundamental reference data, *i.e.*, the amplitude of the electric field of Langmuir waves from L-dips, the first spectroscopic signature allowing also an accurate measure of the electron density; the rate coefficient of CE between multi-charged ions from X-dips. The rate coefficient of CE is important for inertial fusion in laser-generated plasmas, for magnetic fusion in Tokamaks, for population inversion in soft X-ray and VUV ranges, for ion storage devices, and for astrophysics phenomena. The X-dips method presented here has shown its efficiency over a very large range of electron densities.

Acknowledgments

The participation of O. Renner in this work was supported by the Academy of Sciences of the Czech Republic, project M100101208.

Author Contributions

All authors contributed equally to this review.

Conflicts of Interest

The authors declare no conflict of interest.

References

1. Gavrilenko, V.P.; Oks, E. A new effect in the Stark spectroscopy of atomic hydrogen: Dynamic resonance. *Sov. Phys. JETP-USSR* **1981**, *53*, 1122–1127.
2. Zhuzhunashvili, A.I.; Oks, E. Technique of optical polarization measurements of the plasma Langmuir turbulence spectrum. *Sov. Phys. JETP-USSR* **1977**, *46*, 1122–1132.
3. Oks, E.; Rantsev-Kartinov, V.A. Spectroscopic observation and analysis of plasma turbulence in a Z-pinch. *Sov. Phys. JETP-USSR* **1980**, *52*, 50–58.
4. Gavrilenko, V.P.; Oks, E. Multiphoton resonance transitions between "dressed" atom sublevels separated by Rabi frequency. *Sov. Phys. Plasma Phys.* **1987**, *13*, 22–28.
5. Oks, E. *Plasma Spectroscopy: The Influence of Microwave and Laser Fields*; Springer Series on Atoms and Plasmas, Volume 9; Springer, New York, NY, USA, 1995.
6. Oks, E.; Böddeker, St.; Kunze, H.-J. Spectroscopy of atomic hydrogen in dense plasmas in the presence of dynamic fields: intra-Stark spectroscopy. *Phys. Rev. A* **1991**, *44*, 8338–8347.
7. Renner, O.; Dalimier, E.; Oks, E.; Krasniqi, F.; Dufour, E.; Schott, R.; Förster, E. Experimental evidence of Langmuir-wave-caused features in spectral lines of laser-produced plasmas. *J. Quant. Spectr. Rad. Transfer* **2006**, *99*, 439–450.
8. Krasniqi, F.; Renner, O.; Dalimier, E.; Dufour, E.; Schott, R.; Förster, E. Possibility of plasma density diagnostics using Langmuir-wave-caused dips observed in dense laser plasmas. *Eur. Phys. J. D* **2006**, *39*, 439–444.

9. Jian, L.; Shali, X.; Qingguo, Y.; Lifeng, L.; Yufen, W. Spatially resolved spectra from a new uniform dispersion crystal spectrometer for characterization of Z-pinch plasmas. *J. Quant. Spectr. Rad. Transfer* **2013**, *116*, 41–48.

10. von Neumann, J.; Wigner, E. Über das verhalten von Eigenwerten bei adiabatischen Prozessen. *Phys. Z.* **1929**, *30*, 467–470.

11. Gershtein, S.S.; Krivchenkov, V.D. Electron terms in the field of 2 different Coulomb centers. *Sov. Phys. JETP-USSR* **1961**, *13*, 1044.

12. Kryukov, N.; Oks, E. Super-generalized Runge-Lenz vector in the problem of two Coulomb or Newton centers. *Phys. Rev. A* **2012**, *85*, 054503.

13. Boeddeker, S.; Kunze, H.-J.; Oks, E. A Novel structure in the H-alpha line profile of hydrogen in a dense helium plasma. *Phys. Rev. Lett.* **1995**, *75*, 4740–4743.

14. Oks, E.; Leboucher-Dalimier, E. Spectroscopic signatures of avoided crossings caused by charge exchange in plasmas. *J. Phys. B: Atom. Mol. Opt. Phys.* **2000**, *33*, 3795–3806.

15. Dalimier, E.; Oks, E.; Renner, O.; Schott, R. Experimental determination of rate coefficients of charge exchange from X-dips in laser-produced plasmas. *J. Phys. B: Atom. Mol. Opt. Phys.* **2007**, *40*, 909–919.

16. Bransden, B.H.; McDowell, M.R.C. *Charge Exchange and the Theory of Ion-Atom Collisions*; Oxford University Press: Oxford, United Kingdom, 1992.

17. Smirnov, B.M., *Physics of Atoms and Ions*; Springer: Berlin, Germany 2003; Section 14.3.

18. Demkov, Yu.N.; Kunasz, C.V.; Ostrovskii, V.N. United-atom approximation in the problem of $\Sigma-\Pi$ transitions during close atomic collisions. *Phys. Rev. A*, **1978**, *18*, 2097–2106.

19. Oks, E., *Stark Broadening of Hydrogen and Hydrogenlike Spectral Lines in Plasmas: The Physical Insight*; Alpha Science International: Oxford, United Kingdom, 2006.

20. Leboucher-Dalimier, E.; Oks, E.; Dufour, E.; Sauvan, P.; Angelo, P.; Schott, R.; Poquerusse, A. Experimental discovery of charge-exchange-caused dips in spectral lines from laser-produced plasmas. *Phys. Rev. E*, **2001**, *64*, 065401(R).

21. Renner, O.; Dalimier, E.; Liska, R.; Oks, E.; Šmíd, M. Charge exchange signatures in X-ray line emission accompanying plasma-wall interaction. *J. Phys. Conf. Ser.* **2012**, *397*, 012017.

22. Liska, R.; Limpouch, J.; Kucharik, M.; Renner, O. Selected laser plasma simulations by ALE method. *J. Phys. Conf. Ser.* **2008**, *112*, 022009.

23. Chenais-Popovics, C.; Renaudin, P.; Rancu, O.; Guilleron, F.; Gauthier, J.C.; Larroche, O.; Peyrusse, O.; Dirksmöller, M.; Sondhauss, P.; Missalla, T.; *et al.* Kinetic to thermal energy transfer and interpenetration in the collision of laser-produced plasmas. *Phys. Plasmas* **1997**, *4*, 190–208.

24. Dalimier, E.; Oks, E. Dips in spectral lines of He-like ions caused by charge exchange in laser-produced plasmas. *Intern. Review of Atomic and Molec. Phys.* **2012**, *3*, 85–92.

25. Dalimier, E.; Oks, E. Analytical theory of charge-exchange-caused dips in spectral lines of He-like ions from laser-produced plasmas . *J. Phys. B: Atom. Mol. Opt. Phys.* **2014**, 105001.

26. Nikitin, S.I.; Ostrovsky, V.N. On the classification of doubly-excited states of the two-electron atom. *J. Phys. B: Atom. Mol. Opt. Phys.* **1976** *9*, 3141–3148.

MDPI AG
Klybeckstrasse 64
4057 Basel, Switzerland
Tel. +41 61 683 77 34
Fax +41 61 302 89 18
http://www.mdpi.com/

Atoms Editorial Office
E-mail: atoms@mdpi.com
http://www.mdpi.com/journal/atoms

www.ingramcontent.com/pod-product-compliance
Lightning Source LLC
Chambersburg PA
CBHW051921190326
41458CB00026B/6366